D1614247

WITHDRAWN

WITHDRAWN

Advanced Molecular Genetics

Edited by
A. Pühler and K. N. Timmis

With 98 Figures

Springer-Verlag
Berlin Heidelberg New York Tokyo 1984

Professor Alfred Pühler
Universität Bielefeld
Fakultät für Biologie
Postfach 8640
D-4800 Bielefeld
Fed. Rep. of Germany

Professor Kenneth N. Timmis
Department of Medical Biochemistry
University Medical Centre
9, Avenue de Champel
CH-1211 Geneva 4
Switzerland

ISBN 3-540-12740-2 Springer-Verlag Berlin Heidelberg New York Tokyo
ISBN 0-387-12740-2 Springer-Verlag New York Heidelberg Berlin Tokyo

Library of Congress Cataloging in Publication Data. Main entry under title: Advanced molecular genetics. 1. Molecular genetics. I. Pühler, A. II. Timmis, K. N. QH430.A38 1984 574.87'328 83-20340

This work is subject to copyright. All rights are reserved, whether the whole or part of the material is concerned, specifically those of translation, reprinting, re-use of illustrations, broadcasting, reproduction by photocopying machine or similar means, and storage in data banks. Under § 54 of the German Copyright Law, where copies are made for other than private use, a fee is payable to "Verwertungsgesellschaft Wort", Munich.

© by Springer-Verlag Berlin Heidelberg 1984
Printed in Germany

The use of registered names, trademarks, etc. in this publication does not imply, even in the absence of a specific statement, that such names are exempt from the relevant protective laws and regulations and therefore free for general use.

Offsetprinting and bookbinding: Brühlsche Universitätsdruckerei, Giessen
2131/3130-54321

Preface

The development of powerful new techniques and refinements of techniques in molecular genetics in recent years, and the surge in interest in biotechnology based on genetic methods, have heralded a new golden age in molecular genetics, and stimulated in diverse disciplines much interest in the technologies themselves and their potential uses in basic and applied biomedical sciences.

Although some excellent specialist laboratory manuals (especially the Cold Spring Harbor Laboratory manuals by J.H. Miller; R.W. Davies et al.; and T. Maniatis et al.) on certain chapters of molecular genetics exist, no general text that covers a broad spectrum of the subject has thus far been published. The purpose of this manual is to present most, though of necessity not all of the important methods of molecular genetics, in a series of simple experiments, many of which can be readily accomplished by the microbiologist, biochemist or biotechnologist that has had only limited exposure to genetics. The remainder of the experiments require either greater familiarity with the subject, or guidance by someone with such experience. The book should, therefore, not only enable individuals to acquire new procedures for ongoing projects, but also serve as a basis for the teaching of molecular genetic techniques in formal predoctoral and postdoctoral laboratory courses. The majority of protocols in this manual were developed for, and tried and tested in laboratory courses supported by and organised under the auspices of the European Molecular Biology Organisation. We hope they will prove to be as useful as their progenitors were for the students that participated in these courses.

It will be noted that plasmids are the experimental subjects in many of the protocols presented. Although the primary reason for this is the past and current research interests of the authors, the ease of manipulation of these simple genetic elements, and their importance as natural and artificial carriers of interesting genetic information, make them ideal for many molecular genetics experiments.

We are indebted to EMBO for its enthusiasm and support for the courses, to the course instructors who gave time and energy in designing experimental protocols that functioned well under class conditions and, most particularly, to the large number of our students, technicians and secretaries without whose infectious enthusiasm the courses and this manual would not have been possible.

A. PÜHLER and K.N. TIMMIS

Contents

Chapter 1 Basic Methods 1

1.1 Genetic Transfer in Prokaryotes: Transformation, Transduction, and Conjugation
M.P. O'Connell 2

1.2 Isolation of Plasmid DNA
U. Priefer 14

1.3 Characterization of Plasmid DNA by Agarose Gel Electrophoresis
U. Priefer 26

1.4 Characterization of Plasmid in Wild Strains of Bacteria
N. Datta and M.E. Nugent 38

1.5 Mobilization of Host Chromosomal Genes and Vector Plasmids in *E. coli* by the P-Type Plasmid R68.45
G. Rieß and A. Pühler 51

1.6 Characterization of Plasmid Relaxation Complexes
A. Nordheim and K.N. Timmis 64

1.7 Use of the BAL31 Exonuclease to Map Restriction Endonucleases Cleavage Sites in Circular Genomes
J. Frey and K.N. Timmis 74

1.8 IncP1 Typing of Plasmid DNA by Southern Hybridization
R. Simon, W. Arnold, and A. Pühler 80

1.9 Molecular Epidemiology by Colony Hybridization Using Cloned Genes
M.A. Montenegro, G.J. Boulnois, and K.N. Timmis 92

Chapter 2 Mutagenesis................................ 105

2.1 Isolation of Suppressible Mutations by Hydroxylamine Mutagenesis in Vitro
R. Eichenlaub.................................... 106

2.2 Transposition of Tn*1* to the Phage P1 Genome: Isolation of Restriction Deficient Mutants
A. Pühler, V. Krishnapillai, and H. Heilmann 111

2.3 In Vivo Genetic Engineering: Use of Transposable Elements in Plasmid Manipulation and Mutagenesis of Bacteria Other Than *E. coli*
R. Simon....................................... 125

2.4 Generation of Deletion Mutations in Vitro with the BAL31 Exonuclease
J. Frey, M. Bagdasarian, and K.N. Timmis 141

Chapter 3 Gene Cloning................................ 151

3.1 Gene Cloning: An Introduction
K.N. Timmis..................................... 152

3.2 Ligation of Cohesive-Ended and Flush-Ended DNA Fragments
G. Brady and K.N. Timmis 154

3.3 Cloning with Plasmid Vectors
K.N. Timmis and J. Frey............................ 160

3.4 Mini-Plasmid Formation: The Cloning of Replication Origins
K.N. Timmis and J. Frey............................ 171

3.5 Gene Cloning with Bacteriophage λ
A. Pühler....................................... 176

3.6 Cloning with Cosmids
U. Priefer, R. Simon, and A. Pühler.................... 190

Chapter 4 Gene Expression.............................. 203

4.1 Synthesis of Plasmid-Encoded Polypeptides in Maxicells
G.J. Boulnois and K.N. Timmis 204

4.2 Synthesis of Bacteriophage and Plasmid-Encoded Polypeptides in Minicells
J.N. Reeve 212

4.3 Determination of Coding Regions on Multicopy Plasmids: Analysis of the Chloramphenicol Acetyltransferase Gene of Plasmid pACYC184
W. Klipp and A. Pühler 224

4.4 Identification of Gene Products by Coupled Transcription-Translation of DNA Fragments in Cell-Free Extracts of *E. coli*
J.M. Pratt, G.J. Boulnois, V. Darby, and I.B. Holland 236

Chapter 5 DNA Sequencing
G. Volckaert, G. Winter, and C. Gaillard 249

Chapter 6 Electron Microscopy
H. Burkardt and R. Lurz 281

Chapter 7 Transcription 303

7.1 Mapping of RNA Polymerase Binding Sites by Electron Microscopy
R. Lurz and H. Burkardt 304

7.2 Determination of the Startpoints and Orientation of in Vitro Transcripts by Electron Microscopy of R-Loops
R. Eichenlaub and H. Wehlmann 309

Chapter 8 DNA Replication 315

8.1 Localization of Origins of Plasmid DNA Replication
R. Eichenlaub 316

8.2 Plasmid DNA Replication in Vitro
W.L. Staudenbauer 325

Subject Index 339

Chapter 1 Basic Methods

1.1 Genetic Transfer in Prokaryotes: Transformation, Transduction, and Conjugation

M.P. O'CONNELL[1]

Contents

1. General Introduction 2
2. Experiment 1: Transformation of *E. coli* with pBR325 DNA 3
    a) Introduction 3
    b) Objectives 4
    c) Procedure and Results 4
3. Experiment 2: Transfection of *E. coli* with DNA of Bacteriophage λ 6
    a) Introduction 6
    b) Objectives 6
    c) Procedure and Results 6
4. Experiment 3: Transduction of Plasmid RP4 Using Bacteriophage P1 7
    a) Introduction 7
    b) Objectives 8
    c) Procedure and Results 8
5. Experiment 4: Conjugal Transfer of Plasmid RP4 Between Strains of *E. coli* 9
    a) Introduction 9
    b) Objectives 10
    c) Procedure and Results 11
6. Materials 12
7. References 12

1. General Introduction

Gene transfer in bacteria occurs by three main processes: transformation and transfection, which involve the direct uptake of free DNA by bacteria; transduction, which involves bacteriophage mediated transmission of DNA from one bacterium to another; and conjugation, which involves the direct transmission of genetic material from one bacterium to another and requires cell:cell contacts. These mechanisms, in conjunction with recombination processes, are largely responsible for the rapid evolution of bacteria as a consequence of mutation, variation, and reassortment of genetic traits. The exploitation by geneticists of all three systems of gene transfer has been of crucial importance for the construction of new strains and the analysis and manipulation of

---

1 Lehrstuhl für Genetik, Fakultät für Biologie, Universität Bielefeld, D-4800 Bielefeld, Fed. Rep. of Germany

a wide range of biological processes. In this chapter, each system is introduced and a typical procedure described. The following publications provide useful background information: transformation, Smith et al. (1981); bacterial systems available for transformation and transfection, Suzuki and Szalay (1979); P1-mediated transduction of markers in *E. coli*, Willetts et al. (1969), and conjugal transfer of plasmids, Clark and Warren (1979).

2. Experiment 1: Transformation of *E. coli* with pBR325 DNA

a) Introduction

In 1928 Griffith introduced the term transformation when he observed that avirulent strains of the pneumococcus *Diplococcus pneumoniae* could be "transformed" to virulence after contact with dead virulent cells. Subsequently DNA was identified as the active principle in transformation (Avery et al. 1944, McCarty et al. 1946).

Bacteria that are able to take up and incorporate free DNA into their genomes are said to be competent. Some bacterial strains are highly competent during one or more phases of growth under normal laboratory conditions, whereas others require special treatments to be rendered competent. Still others appear to be completely refractory to transformation. The factors that determine competence are not yet fully understood, but a lack of transformability could presumably be due to a number of factors, e.g., cell surface barriers, such as capsules, and the production of intra- and extracellular nucleases.

Bacteria that are highly transformable include *Streptococcus pneumoniae*, *Haemophilus influenzae* (Alexander and Leidy 1951), and *Bacillus subtilis* (Spizizen 1958). Competence in such bacteria is a transitory physiological state during the growth cycle.

The process of transformation in Gram-positive bacteria, e.g., *Streptococcus, Bacillus,* can be divided into the following stages:

1. Development of competence due to secretion of a small protein called competence factor.
2. Binding of double stranded DNA molecules to bacteria.
3. Uptake of single stranded DNA molecules.
4. Coating of single stranded molecules with a specific protein that protects DNA from nucleases (eclipse complex).
5. Integration by recombination of the transformed DNA strand into the recipient chromosome.
6. Replication of integrated DNA segments and segregation of recipient and donor alleles in progeny bacteria to yield transformed clones with new phenotypes.

In Gram-negative bacteria, e.g., *Haemophilus*, no competence factor appears to be produced and the process of DNA binding and uptake differs from that of Gram-positives. Thus, only homologous DNA is specifically bound and DNA is taken up in the form of intact duplex molecules.

Unlike the species described above, *E. coli* cannot be transformed effectively without special treatments, i.e., competence must be induced by experimental procedures. This may be accomplished by removal of most of the cell wall to produce spheroplasts (Henner et al. 1973) or, alternatively, the cell envelope can be rendered permeable to DNA by treatment with Ca^{2+} ions and subjection to heat shock (Mandel and Higa 1970). The latter method, originally developed for the transfection of *E. coli* by λ DNA, in which the transformed cells do not survive, was subsequently shown to be effective for the transformation of bacteria with plasmid DNA (Cohen et al. 1972), in which transformed cells are required to survive the treatment. The mechanism by which calcium promotes competence is unclear. However, transformation of large plasmids (> 30 kb) is not very efficient and the uptake of chromosomal DNA is limited to strains carrying specific mutations in their recombination systems (*recB$^-$*, *recC$^-$*, *sbc$^-$*), which lack exonuclease V and are unable to degrade incoming linear DNA fragments (Oishi and Cosloy 1972, Cosloy and Oishi 1973).

Transformation of *E. coli* is of profound importance for the DNA cloning technology since competent cells can easily be transformed by a variety of small replicons used in the construction of recombinant DNA molecules. Furthermore, the examination of some cloned genes in different genetic backgrounds is an essential feature of their analysis.

b) Objective

Introduction of the cloning vector plasmid pBR325 into *E. coli* by transformation.

c) Procedure and Results

Plasmid pBR325 is a small DNA molecule, 5.4 kb in length, which encodes resistance to ampicillin (Ap), chloramphenicol (Cm), and tetracycline (Tc). Purified plasmid DNA is prepared by one of the procedures described in Chap. 1.2. The strain to be transformed is an *E. coli* K-12 mutant defective in restriction (HB101 = r_K^- m_K^- *recA$^-$*), which does not degrade DNA taken up and lacks the *E. coli* K specific modification. Transformed cells are obtained by selection for one of the pBR325 encoded resistances, e.g., Tc^R.

Day 1

Inoculate a colony of strain HB101 in 5 ml of L-broth (LB) and incubate overnight.

Day 2

Pre-cool tubes, centrifuge tubes, and solutions in ice before commencing the experiment.

Add 0.2 ml of the overnight culture to 10 ml fresh LB in a 50 ml Erlenmeyer flask and incubate with shaking at 37°C until the culture reaches an $O.D._{590}$ of 0.5.

Chill the culture by placing flask in ice for at least 10 min; centrifuge culture at 7000 rpm in Sorvall centrifuge at 4°C.

Resuspend pellet in 5 ml of cold 50 mM $CaCl_2$ and place on ice for 20 min.

Pellet cells at 4°C and resuspend in 1 ml of cold 50 mM $CaCl_2$. The cells should be competent and ready for transformation.

In three tubes, labeled A, B, and C, standing in an ice-water bath, mix 50 µl of DNA (1 µg dissolved in TEN buffer) with 0.2 ml of cold competent cells (A); 50 µl of DNA with 0.2 ml 50 mM $CaCl_2$ (B); and 50 µl of TEN buffer with 0.2 ml of competent cells (C). B and C are controls for the sterility of the reagents.

Place the transformation mix and the two controls on ice for 30 min.

Transfer the three tubes to a 42°C water bath for 2 min. This heat shock aids the passage of DNA into the cells.

Add 0.7 ml of LB to each tube and incubate at 37°C for 30 min.

Make serial 1:10 dilutions to 10^{-4} of the transformation mixtures in LB. Plate 0.1 ml of the undiluted culture and of dilutions 10^{-1} and 10^{-2} on plates containing tetracycline (15 µl/ml).

In addition, spread 0.1 ml of dilution 10^{-4} on LB plates to count viable cells.

Incubate plates overnight at 37°C.

Day 3

Score transformants for the presence of non-selected markers by transferring onto plates containing Ap and Cm.

Calculate the transformation frequency (number of transformants/µg DNA/viable cell); it should be about 10^{-3}.

Confirm the presence of pBR325 in selected transformants by producing minipreparations of plasmid DNA and analyzing these by electrophoresis on an agarose gel (see Chap. 1.3, 2).

Note: The preparation and testing of competent cells is time consuming; preparation and maintenance of large stocks preserved at $-20°C$ is therefore very convenient if repeated transformations of one or a few strains are carried out (Morrison 1979). To preserve competent cells add glycerol, final concentration 15%, to the $CaCl_2$ solution used for resuspending the competent cells and store cells in this solution at $-20°C$ for up to 3 months. Prior to use, slowly thaw cells at 0°C and perform transformation as described above.

Rapid Procedure for Preparation of Competent Cells

Grow cells in 50 ml LB to an $O.D._{590}$ of 0.5; cool on ice for 5 min and add $CaCl_2$ to 0.1 M. Hold cells on ice for 30–60 min, centrifuge, and resuspend in 2 ml of $CaCl_2$. Add 0.2 ml cells to DNA and continue as described above.

3. Experiment 2: Transfection of *E. coli* with DNA of Bacteriophage λ

a) Introduction

Transfection is the uptake of naked phage DNA by competent bacteria and is thought to occur by a mechanism similar to that of transformation. Transfected DNA is biologically active and capable of replication and of directing normal phage development. Transfection of *E. coli* is of considerable practical relevance as a result of the importance of phage λ in gene cloning. The efficiency of transfection is quite low, however, and in vitro packaging of recombinant λ phage DNA molecules in phage heads and their use for infection of host cells, is a useful alternative procedure for the introduction of DNA into host bacteria.

b) Objective

Transfection of *E. coli* K-12 C600 with λ DNA.

c) Procedure and Results

Day 1

Grow the recipient strain, *E. coli* C600, in 10 ml LB overnight at 37°C.

Day 2

Dilute the overnight culture 1:50 in 10 ml of fresh LB and incubate until an $O.D._{590}$ of 0.5 is reached.
Prepare competent cells as described in previous section.
Add 0.2 ml of cold competent cells to 0.1 ml of λ DNA (about 0.1 μg; prepared as described in Chap. 3.6) dissolved in SSC.
Prepare a DNA and a cell control equivalent to those used in transformation.
Leave on ice for 30 min.
Heat shock for 2 min at 42°C.
Add 0.1 ml of the suspension to 2.5 ml of LB soft agar and pour on LB base plates.
Incubate plates overnight at 37°C.

Day 3

Count the number of plaques and calculate the transfection frequency obtained, which should be about 2×10^5 plaques/μg DNA.

4. Experiment 3: Transduction of Plasmid RP4 Using Phage P1

a) Introduction

Transduction is the transfer of genetic information between bacteria mediated by a bacteriophage and was discovered by Lederberg et al. (1951) during attempts to demonstrate conjugation in *Salmonella typhimurium*. Their studies revealed that recombinants were produced not only when two mutant strains were mixed, but also when one mutant was mixed with a cell-free extract of the other. The transducing agent was identified as a temperate bacteriophage designated P22 (Zinder and Lederberg 1952).

Two types of transduction can be distinguished. Specialized transduction involves the integration of a specific section of chromosomal DNA, such as the *gal* operon, in the phage genome, as a result of imprecise excision of a λ prophage from its primary attachment site in the *E. coli* chromosome. This system does not have a general application in the construction of new strains, although the possibility of obtaining integration of λ prophage at secondary attachment sites, in *gal*-deletion mutants of *E. coli*, provides a limited application.

Generalized transduction is not based on any physical association between phage and bacterial genomes (Clowes 1958), but occurs as a result of the occasional encapsidation of host instead of phage DNA. The transducing particle consisting of a viral capsid and host DNA is released upon cell lysis and can subsequently inject its DNA into a recipient cell. Since the transducing particle does not contain a phage genome, the infected cell will not lyse. If the injected DNA shares homology with the recipient chromosome, it can be integrated by recombination to form a transductant. Generalized transduction has been particularly useful for the construction of genetic maps: the cotransduction of two markers demonstrates physical and genetic linkage and cotransduction frequencies indicate relative distances between markers. Three factor crosses by transduction enable fine structure genetic analyses to be carried out:

Transduction has application also in plasmid manipulation, e.g.:

1. Encapsidation can select deletion derivatives of plasmid genomes that are too large to be packaged per se.
2. Phage-mediated plasmid injection overcomes surface exclusion, the reduction in recipient ability of plasmid-carrying bacteria towards donors that donate DNA by means of conjugal transfer functions related to those of the plasmid in the recipient.
3. A single plasmid can be transferred from donor bacteria carrying several plasmids, as a result of only one plasmid molecule being packaged per capsid. This type of biological separation of plasmid mixtures is also readily accomplished in transformation and, to a lesser extent, by conjugation.

Phage P1, a temperate phage the prophage of which exists as an extrachromosomal element in lysogenic bacteria, has been used extensively to mediate generalized transduction in *E. coli* (Lennox 1955). P1 derivatives that are temperature sensitive for the maintenance of lysogeny (Rosner 1972) or resistant to an antibiotic (e.g., P1Cm, Kondo and Mitsuhashi 1964) have been obtained and these facilitate genetic manipulations. Furthermore, host range derivatives that propagate on members of the

Enterobacteriaceae closely related to *E. coli* have been obtained (Murooka et al. 1978) and are useful for developing and manipulating such strains.

b) Objective

Transduction of plasmid RP4.

c) Procedure and Results

The bacterial strain used is *E. coli* K-12 C600 F$^-$ *thr leu thi ton*A *lac*Y *sup*E. The experiment is performed in three steps:

1. Preparation of transducing lysates. P1 is propagated on *E. coli* K-12 C600 (RP4) and the phage particles harvested after cell lysis.
2. Assay of lysate. The number of plaque forming units (pfu) per ml of the lysate is calculated by titration of the phage preparation on an indicator strain.
3. Transduction. Recipient bacteria are infected with the lysate and the number of cells that have required RP4 are calculated.

Preparation of a Transducing Lysate

Day 1

Add 0.1 ml of an overnight culture of *E. coli* K-12 C600 (RP4) to 0.9 ml LB and grow for 3 h at 37°C.

Add 0.5 ml of this culture to 0.1 ml of a P1 lysate diluted to 10^7 pfu/ml in saline (increase to 10^8 pfu/ml if a Rec$^-$ strain is being used, since the efficiency of plating on such a strain is low).

Add 3 ml of soft (0.7%) agar containing 5 mM $CaCl_2$ and pour onto a thick LB agar plate. Incubate plate overnight at 37°C.

Day 2

Transfer the soft agar top layer to a tube, add 0.5 ml chloroform and mix on a vortex.
Spin in a bench centrifuge and transfer the supernatant to a new tube. Add 0.5 ml chloroform. The lysate is now ready for assay.

Assay of Lysate

Centrifuge an overnight culture of C600 in LB and resuspend in 1/2 volume of LB containing 10 mM $MgSO_4$ and 5 mM $CaCl_2$.

Make 10-fold serial dilutions of the phage lysate in saline to 10^{-8} and mix 0.1 ml of each dilution with 0.1 ml of indicator bacteria.

Add 2.5 ml of soft LB agar containing 5 mM $CaCl_2$ to each mix and pour onto thick LB agar base plates.

Incubate plates overnight at 37°C.

Day 3

Count plaques and calculate titre of the P1 suspension.

Transduction

Add 1 ml of an overnight culture of recipient bacteria (e.g., C600) to 9 ml LB and incubate at 37°C until the culture reaches 1×10^9 cells/ml.

Mix 0.5 ml of bacterial suspension with 0.5 ml of P1 lysate (diluted to 5×10^7 pfu/ml in LB) and 0.5 ml of 15 mM $CaCl_2$-30 mM $MgSO_4$ (multiplicity of infection, *MOI*, about 0.05). Set two control mixtures that are similar, but lack bacteria (1st control) or P1 lysate (2nd control), respectively.

Incubate mixtures for 20 min at 37°C.

Pellet the cells and wash in an equal volume of phage buffer.

Pellet the cells and resuspend in 1 ml of LB. Incubate at 37°C for 1 h.

Pellet the cells and resuspend in 1 ml of phage buffer. Make serial 10-fold dilutions in phage buffer and plate 0.1 ml of each dilution on LB agar plates containing tetracycline (15 μg/ml).

Incubate plates at 37°C.

Day 4

Count the Tc^R transductants and score for non-selected RP4-specific antibiotic resistance markers by replica plating onto LB agar plates containing (a) kanamycin (30 μg/ml) and (b) ampicillin (100 μg/ml).

Calculate the transduction frequency, which is the ratio:number of transductants/number of pfu used for infection. The frequency is normally in the range of 10^{-4} to 10^{-5} transductants/pfu.

5. Experiment 4: Conjugal Transfer of Plasmid RP4 Between Strains of *E. coli*

a) Introduction

Bacterial conjugation was discovered by Lederberg and Tatum (1946) when they observed that mutant strains of *E. coli* K-12 have the ability to complement each other by transferring genetic material during cellular contacts. The transfer is unidirectional, namely, from a fertile donor to a recipient. Fertility was subsequently attributed to the presence of a conjugal factor (Fertility factor, F-plasmid) by W. Hayes (1953a). This factor exists as an autonomous replicon in the cell, promoting its own transfer from one bacterium to another at high frequencies and transfer of chromosomal genes at low frequencies. Alternatively, the F-plasmid can integrate into the chromosome and form a high frequency of recombination (Hfr) donor (Cavalli-Sforza 1950, Hayes 1953b), which then transfers at the high frequencies observed for the autonomous plasmid, i.e., recipients (transcipients or transconjugants) that have incorporated (recombined) markers from an Hfr donor are observed at frequencies of about 10^{-1}

compared with those of about 10^{-6} when an F^+ donor is employed. Conjugation requires the establishment of productive cell:cell contacts between donor and recipient bacteria, i.e., the formation of mating aggregates, and the subsequent transfer of DNA from donor cells to recipients in the aggregates. Both events are directed by F-plasmid gene products. A filamentous cell surface organelle on donor cells, the pilus, appears to cause the formation of mating aggregates (see Tomoeda et al. 1975, for review), which in turn triggers the process of DNA transfer. This process involves the nicking of the plasmid at a specific site (oriT = origin of transfer replication: Willetts et al. 1975), in a specific strand, and the oriented transfer of the nicked strand from the donor to a recipient cell, in which it is recircularized. Double stranded plasmids are subsequently regenerated in both donor and transconjugant cells. In matings involving Hfr donor cells, in which the plasmid is integrated into the chromosome, the oriented transfer occurs of a single stranded DNA molecule composed of a segment of the F-plasmid, followed by the chromosome, then the remainder of the plasmid. Conjugal transfer of the entire genome of *E. coli* takes approximately 100 min; the spontaneous disruption of mating aggregates and consequent fragmentation of the chromosomal DNA being transferred results in the low frequency transfer of *oriT*-distal markers. Transferred markers are subsequently integrated into the recipient chromosome by reciprocal recombination.

With the discovery of self-transmissible resistance elements called R-factors or R-plasmids (Watanabe 1867, Helinski 1973), the study of conjugation became fused to the increasing interest in the emergence of multiple antibiotic resistance. The number of different R-plasmids isolated increased rapidly and variations in their properties became apparent. Some are now known to have a chromosome mobilizing ability similar to that of the F-plasmid. This property of one plasmid, R68.45, is reviewed in Chap. 1.5. Many R-plasmids have a narrow host range, transferring only intra- and interspecifically. Others are capable of intergeneric transfer. Initially, intergeneric transfer studies concentrated on the Enterobacteriaceae, but have since been extended to other genera, e.g., *Pseudomonas* and *Rhizobium*. However, while conjugation is the principal mechanism of gene transfer in Gram-negative bacteria, it appears to be much less important in Gram-positives.

The frequency at which conjugation occurs is an intrinsic plasmid property, but it is also influenced by the types of donor and recipient bacteria that participate in the mating and by the experimental conditions of the mating. For example, the transfer of plasmids belonging to compatibility groups N and P (e.g., RP4), which specify rigid pili, occurs best when donor and recipients are maintained in close proximity, i.e., are incubated together on an agar surface. Other plasmids, including F, also transfer when the mating is carried out in liquid medium.

b) Objectives

To demonstrate the transfer of R-plasmid RP4 between *E. coli* strains and to compare the transfer efficiencies obtained with surface and broth matings.

c) Procedure and Results

The donor used in this experiment is a derivative of *E. coli* K-12 strain C600 carrying the self-transmissible IncP plasmid RP4 that mediates resistance to ampicillin (Ap), kanamycin (Km), and tetracycline (Tc) (Datta et al. 1971). Mating this donor with a streptomycin resistant (Sm^R) plasmid-free strain gives rise to transconjugant cells carrying plasmid RP4, which can be selected on medium containing tetracycline plus streptomycin.

Day 1

Inoculate single colonies of parental strains into 5 ml LB and culture overnight at 37°C.
  Donor: *E. coli* K-12 C600 F^- *thr leu thi tonA lac*Y *sup*E (RP4)
  Recipient: *E. coli* K-12 C600 F^- *thr leu thi ton*A *lac*Y *sup*E Sm^R.

Day 2

The best transfer efficiencies are obtained from mating mixtures composed of actively growing (exponential phase) donor bacteria and stationary phase recipients.

Dilute donor culture 1/10 in LB and incubate at 37°C until the exponential phase of growth is reached.

Surface Mating

Mix 5×10^7 donor cells and 1×10^8 recipients in an Eppendorf tube, centrifuge, resuspend pellet in 50 µl of LB and transfer cells to a 0.45 µm Millipore filter on a LB agar plate. Incubate at 37°C for 1–2 h.
Resuspend cells by placing filters in a tube containing 0.5 ml of 0.85% saline and agitating the tube on a vortex.

Broth Mating

Mix 5×10^7 donor cells and 1×10^8 recipients in a glass tube and incubate for 1–2 h at 37°C without agitation.
Mix mating mixture vigorously on a vortex for 1 min to separate donors and recipients.

Subsequent Steps for Both Mating Mixtures

Make serial 1/10 dilutions to 10^{-8} of the mating mixtures.
Spread 0.1 ml of the dilutions 10^{-1} to 10^{-6} on LB agar containing streptomycin (150 µg/ml) and tetracycline (15 µg/ml).
Spread 0.1 ml of original overnight cultures of donor and recipient bacteria on the same medium to determine the frequency of spontaneous mutation to antibiotic resistance.
Spread 0.1 ml of the dilutions 10^{-6} and 10^{-8} on LB agar containing streptomycin (150 µg/ml) to estimate the numbers of recipient cells in the mating mixtures.
Incubate plates overnight at 37°C.

Day 3

Count the colonies that have grown on the selection plates and calculate the number of bacteria per ml in the mating mixtures exhibiting the phenotypes Sm^R (total recipients) and $Sm^R Tc^R$ (transconjugants). The transfer frequency is normally expressed as the number of transconjugants obtained per recipient (e.g., if 10^6 transconjugants were obtained from a total of 10^9 recipients, the transfer frequency was 10^{-3}). Occasionally, it is expressed per donor cell. For broth matings, RP4 usually transfers with a frequency of 10^{-3} to 10^{-4} per recipient, whereas in surface matings it transfers at much higher frequencies, approaching or even exceeding 100% as a result of transconjugant bacteria acting as donors.

6. Materials

LB medium	Tryptone	10 g
	Yeast extract	5 g
	NaCl	5 g
	Distilled H_2O	1 liter
L-agar	LB medium containing 1.5% agar	
L-soft agar	LB medium containing 0.7% agar	
TEN buffer	10 mM	Tris
	1 mM	EDTA
	50 mM	NaCl
	pH 7.5	
SSC	150 mM	NaCl
	15 mM	Na-citrate
Phage buffer	20 mM	Tris
	100 mM	NaCl
	10 mM	$MgSO_4$
	pH 7.5	

7. References

Alexander HE, Leidy G (1951) Determination of inherited traits of *H. influenzae* by deoxyribonucleic acid fractions isolated from typespecific cells. J Exp Med 93:345–359

Avery OT, MacLeod CM, McCarty M (1944) Studies on the chemical nature of the substance inducing transformation of pneumococcal types. I. Induction of transformation by a deoxyribonucleic acid fraction isolated from pneumococcus type III. J Exp Med 79:137

Cavalli-Sforza LL (1950) La sessulita nei batteri. Boll Ist Sieroter Milan 29:281

Clark AJ, Warren GJ (1979) Conjugal transmission of plasmids. Annu Rev Genet 13:99–125

Clowes RC (1958) The nature of the vector involved in bacterial transduction. In: Almqvist P, Wiksell T (eds) Abstr 8th Int Congr Microbiol, Stockholm, p 53

Cohen SN, Chang ACY, Hsu L (1972) Nonchromosomal antibiotic resistance in bacteria: genetic transformation of *Escherichia coli* by R-factor DNA. Proc Natl Acad Sci USA 69/8:2110–2114

Cosloy SD, Oishi M (1973) Genetic transformation in *Escherichia coli* K12. Proc Natl Acad Sci USA 70/1:84–87

Datta N, Hedges RW, Shaw EJ, Sykes RB, Richmond MH (1971) Properties of an R factor from *Pseudomonas aeruginosa*. J Bacteriol 108:1244

Griffith F (1928) Significance of pneumococcal types. J Hyg Camb 27:113

Hayes W (1953a) Observations on a transmissible agent determining sexual differentiation in Bact. coli. J Gen Microbiol 8:72

Hayes W (1953b) The mechanism of genetic recombination in *Escherichia coli*. Cold Spring Harbor Symp Quant Biol 18:75

Helinski DR (1973) Plasmid determined resistance to antibiotics: molecular properties of R factors. Annu Rev Microbiol 27:437–470

Henner WD, Kleber I, Benzinger R (1973) Transfection of *Escherichia coli* spheroplasts. J Virol 12:741–747

Kondo E, Mitsuhashi S (1964) Drug resistance of enteric bacteria. VI. Active transducing bacteriophage P1Cm produced by the combination of R factor with bacteriophage P1. J Bacteriol 88:1266–1276

Lederberg J, Tatum EL (1946) Novel genotypes in mixed cultures of biochemical mutants of bacteria. Cold Spring Harbor Symp Quant Biol 11:113

Lederberg J, Lederberg EM, Zinder ND, Lively ER (1951) Recombination analysis of bacterial heredity. Cold Spring Harbor Symp Quant Biol 16:413

Lennox ES (1955) Transduction of linked genetic characters of the host by bacteriophage P1. Virology 1:190–206

McCarty M, Taylor HE, Avery OT (1946) Biochemical studies of environmental factors essential in transformation of pneumococcal types. Cold Spring Harbor Symp Quant Biol 11:177

Mandel M, Higa A (1970) Calcium-dependent bacteriophage DNA infection. J Mol Biol 53:159–162

Morrison DA (1979) Transformation and preservation of competent bacterial cells by freezing. Methods Enzymol 68:326–331

Murooka Y, Higashiura T, Harada T (1978) Genetic mapping of tyramine oxidase and acrylsulfatase genes and their regulation in intergeneric hybrids of enteric bacteria. J Bacteriol 136:714–722

Oishi M, Cosloy SD (1972) The genetic and biochemical basis of the transformability of *Escherichia coli* K12. Biochem Biophys Res Commun 49/6:1568

Rosner JL (1972) Formation, induction and curing of bacteriophage P1 lysogens. Virology 48:679–689

Smith HO, Danner DB, Deich RA (1981) Genetic transformation. Annu Rev Biochem 50:41–68

Spizizen J (1958) Transformation of a biochemically deficient strain of *Bacillus subtilis* by deoxyribonucleate. Proc Natl Acad Sci USA 44:1072–1078

Suzuki M, Szalay AA (1979) Bacterial transformation using temperature-sensitive mutants deficient in peptidoglycan synthesis. Methods Enzymol 68:331–342

Tomoeda M, Inuzuka M, Date T (1975) Bacterial sex pili. Prog Biophys Mol Biol 30:25–36

Watanabe W (1967) Infectious drug resistance. Sci Am 217/6:19–27

Willetts NS, Clark AJ, Low B (1969) Genetic location of certain mutations conferring recombination deficiency in *E. coli*. J Bacteriol 97:244–249

Willetts NS, Maule J, McIntire S (1975) The genetic location of traO, finP and tra-4 on the *E. coli* K12 sex factor F. Genet Res 26:255–263

Zinder ND, Lederberg J (1952) Genetic exchange in *Salmonella*. J Bacteriol 64:679

1.2 Isolation of Plasmid DNA

U. PRIEFER [1]

Contents

1. General Introduction 14
2. Experiment 1: Large Scale Isolation of Plasmid DNA via CsCl-EtBr Density Gradient
   Centrifugation 16
      a) Introduction 16
      b) Objectives 18
      c) Procedure and Results 18
3. Experiment 2: Small Scale Techniques for Rapid Isolation of Plasmid DNA 21
      a) Introduction 21
      b) Objectives 22
      c) Procedure and Results 22
4. Materials 24
5. References 24

1. General Introduction

In molecular biology, plasmids serve as excellent model systems for the study of structure and function of DNA at the molecular level. In particular, they facilitate the investigation of DNA replication and its control, since their replication seems to be regulated similarly to that of the host bacterial chromosome. But unlike the host chromosome, they are usually small and can easily be isolated as intact molecules. Furthermore, they are particularly useful in DNA cloning, e.g., a variety of the commonly used cloning vectors are plasmids. Many small plasmids (e.g., ColE1) are maintained at a high copy number in the cell (multicopy plasmids). In addition, in contrast to the host chromosome, some plasmids can replicate in the absence of protein synthesis. This ability allows plasmid amplification up to a level of several thousand copies per cell, by growing plasmid-containing cells in the presence of chloramphenicol (Clewell 1972). These properties have obvious advantages: they not only permit a relatively easy isolation of plasmid DNA in large amounts, but also allow the synthesis of large quantities of plasmid-coded products (or that of DNA fragments cloned in them).

---

1 Lehrstuhl für Genetik, Fakultät für Biologie, Universität Bielefeld, D-4800 Bielefeld 1, Fed. Rep. of Germany

The first studies on plasmid DNA used the technique described by Marmur (1961) and Mamur et al. (1961), but this method led to DNA breakage and so only linear DNA fragments with a molecular weight of about 10×10^6 could be isolated. Later, in 1967, Hickson et al. succeeded in extracting intact circular plasmid molecules. A large number of isolation methods are now available, only some of which are described in this chapter. However, there are some basic features, which are outlined below.

Cell Growth: Bacteria are usually grown in rich medium (e.g., PA or LB) with vigorous agitation. It is recommended that the cultivation should always be started with a single colony, which has been proven to contain the plasmid. If the strain is likely to loose the plasmid readily, then the culture should be grown under selective conditions to assure its maintenance.

In the case of plasmids with a relaxed mode of replication (e.g., ColE1 and its derivatives), the yield of plasmid DNA can be increased by the addition of chloramphenicol (150 $\mu g/ml$) or spectinomycin (300 $\mu g/ml$) to the growth medium. These inhibitors of protein synthesis are usually added when the cell concentration has reached about 5×10^8 per ml and the culture is further incubated for 12–15 h. The addition of these antibiotics causes a relative enrichment of plasmid DNA per host chromosome, since chromosomal replication stops, while that of "relaxed" plasmids continues.

Cell Lysis is carried out in two steps:

1. Plasmid containing cells are exposed to the enzyme lysozyme that effects the integrity of the outer cell wall, leading to the formation of spheroplasts. This treatment is performed in the presence of sucrose, which prevents the spheroplasts from lysis.

2. The lysis is then completed by adding a detergent. In the case of ionic detergents (e.g., SDS, Sarcosyl), this treatment results in the release of total cellular DNA. The mixture of chromosomal and plasmid DNA is subsequently subjected to shearing forces so that the chromosomal DNA will be fragmented, whereas plasmid DNA (because of its smaller size and supercoiled nature) is left intact. If a non-ionic detergent (e.g., Triton X100, Brij 58) is used, the bulk of the chromosomal DNA remains attached to cellular components and will sediment with the residual cell debris during low speed centrifugation.

Separation of Plasmid CCC-DNA: Double stranded DNA molecules may exist in three different conformations (Vinograd et al. 1965, Burton and Sinsheimer 1965, Fiers and Sinsheimer 1962) (Fig. 1):

1. In a linear double stranded form
2. In an open circular (OC) form, which has one of the DNA strands intact while the other is nicked
3. As a covalently closed circle (CCC) with both strands covalently bonded and supertwisted (supercoiled DNA)

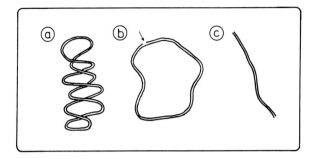

Fig. 1a–c. Conformations of DNA. a *covalently closed circular* DNA (CCC-DNA) with supertwist; b *open circular* (OC) DNA obtained from form *a* by introducing a nick *(arrow)* in one strand of the DNA; c linear DNA generated by breakage of both strands of the duplex molecule

The supercoiled CCC-configuration shows a physical behavior which differs from that of OC- and linear DNA. For example, conformation of DNA affects sedimentation behavior, molecular shearing, binding capacity of ligands, and melting (Clowes 1972). These differences can be used to separate plasmid CCC-DNA from OC- and linear DNA molecules present in crude and cleared cell extracts. Several techniques for this separation are available. A very reliable method for isolating plasmid CCC-DNA in highly purified form is by banding total or pre-purified DNA (see later) in a cesium-chloride (CsCl) gradient in the presence of ethidium bromide (EtBr). The rationale for this technique is outlined in Sect. 2. Alternatively, plasmid CCC-DNA can be separated by the more rapid method of alkaline extraction. This technique is described in Sect. 3. Most of the experimental procedures outlined are modified in some aspects from the original. The reader is referred to the original references which are given for further information.

2. Experiment 1: Large Scale Isolation of Plasmid DNA via CsCl-EtBr Density Gradient Centrifugation

a) Introduction

When a concentrated solution of CsCl is centrifuged at high speed (about $100,000 \times g$), the solution forms a steep and stable density gradient increasing with distance from the rotational axis. If DNA molecules are dissolved in a CsCl solution of suitable density, the DNA is forced to migrate through the density gradient established during centrifugation until it reaches equilibrium (i.e., the position at which the density of the surrounding liquid is equal to its own). Because the buoyant density of DNA is a function of its base composition, DNA molecules with different base ratios will be separated from one another with low density molecules near the top and DNA of higher density nearer the bottom of the gradient (Fig. 2). This method was first used by Marmur and colleagues in 1961 and somewhat later by Falkow et al. (1964). Both transferred a F'lac^+ plasmid from *E. coli* to a host with a different DNA base composition (*Serratia* and *Proteus,* respectively) and were able to identify the plasmid DNA as a "satellite" band in a density gradient.

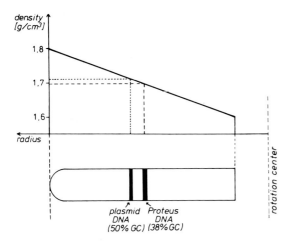

Fig. 2. Separation of DNA species with different base ratios (GC content) by CsCl density gradient centrifugation. DNA is isolated from *Proteus* (38% GC) harboring a plasmid with a GC content of 50% and ultracentrifuged in a CsCl solution of 1.7 g/cm³. The gradient which is established reaches 1.6 g/cm³ at the top to 1.8 g/cm³ at the bottom. The DNAs band at their equilibrium density, which is 1.698 for *Proteus* DNA and 1.710 for the plasmid

The use of ethidium bromide in CsCl density gradients is based on its differential effects on the different configurations of plasmid DNA (Radloff et al. 1967, Bauer and Vinograd 1968). EtBr is known to "stretch" the DNA helix by intercalation between adjacent base pairs, thus lowering the density. However, linear and OC-DNA intercalate much more EtBr than CCC-DNA. Intercalation into CCC-DNA is restricted by the capacity of the two supertwisted strands to unwind. Thus, in the presence of EtBr, the extension of CCC-DNA and, therefore, the decrease in its density is less (0.085 g/cm³) than that of OC- and linear DNA (0.125 g/cm³). In a CsCl-EtBr gradient, CCC-DNA therefore bands in a region of higher density and is readily separated from the other two forms even if there is no difference in base ratio (Fig. 3). In addition, EtBr makes it possible to visualize the DNA bands by irradiation with UV at 254 nm, since EtBr bound to DNA shows a change in its UV spectrum and a significant increase in fluorescence. UV light is absorbed by the DNA, transferred to the

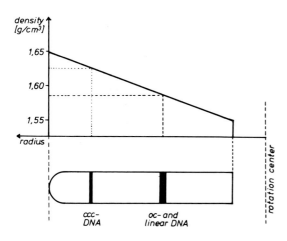

Fig. 3. Separation of DNA species with the same base ratios by CsCl-EtBr density gradient centrifugation. EtBr lowers the buoyant density of DNA by intercalation. However, the decrease is less for CCC-DNA (0.085) than for OC- and linear DNA (0.125). So, when DNA of a GC content of 50% (1.710 g/cm³) is ultracentrifuged in a CsCl gradient (1.65 g/cm³ to 1.55 g/cm³) in the presence of EtBr, the CCC form reaches equilibrium at a density of 1.625 g/cm³, while OC- and linear forms sediment at 1.585 g/cm³

intercalated dye, and re-emitted as fluorescence. Note that short wavelength UV causes damage of the DNA and destroys its function. Moreover, DNA-EtBr complexes have to be protected from visible light because EtBr effects a light dependent cleavage of DNA (Deniss and Morgan 1976).

Cell lysis may be performed by different techniques. One method is to extract total cellular DNA by ionic detergents (SDS, Sarcosyl) and to subject this lysate to shearing forces; the chromosomal DNA will be fragmented, whereas plasmid DNA is left intact because of its smaller size and CCC-configuration. This procedure is very useful particularly for the isolation of large plasmids (up to 200×10^6), although there is some loss of plasmid DNA, which is damaged during the shearing step.

An alternative method is the preparation of cleared lysates by treatment with non-ionic detergents (Brij 58, Triton X100) prior to the selective precipitation of chromosomal DNA complexes by "clearing spins". This low speed centrifugation, usually carried out at about 35,000 × g for 30 min, results in a supernatant fluid, called "cleared lysate", which contains pre-purified plasmid DNA. High molecular weight plasmids will necessarily co-precipitate to some extent so that these procedures preferentially extract small multicopy plasmids; for larger plasmids this treatment is somewhat ineffective due to increasing plasmid entrapment in the pellet.

b) Objectives

Isolation of highly purified plasmid DNA in concentrated form, especially for preparative purpose.

c) Procedure and Results

Production of Lysates

Extraction of Total Cellular DNA

Two protocols are outlined. The second method differs from the simple Sarcosyl lysis mainly, in that there is a detergent wash/osmotic shock treatment prior to lysis, which facilitates isolation of plasmid DNA from lysozyme refractory bacteria (e.g., fast-growing Rhizobia).

Note: In both cases, all samples and solutions should be kept cold trough all stages.

Sarcosyl Lysis (Bazaral and Helinski 1968)

Bacteria are grown in 10 ml of LB to saturation.
Harvest bacterial cells by centrifugation (10,000 rpm, 10 min).
Wash twice in equal volumes of cold TES buffer.
Resuspend pellet in 0.25 ml 20% sucrose in TES.
Add 0.25 ml of lysozyme (2 mg/ml)/RNase (1 mg/ml) in TES buffer; incubate at 37°C for 30 min; invert from time to time.

Add 0.1 ml Sarcosyl (10% in 0.25 M EDTA), mix gently by inversions; lysate should become clear and viscous.
Shear lysate 10–20 times with a fine-needled syringe until lysate looses viscosity.
Adjust volume to 5.0 ml and subject to CsCl-EtBr density gradient centrifugation.

Sarcosyl Lysis of Bacteria Sensitized to Lysozyme (Schwinghamer 1980)

Pellet 12 ml bacterial culture, grown to late log phase.
Detergent-wash (this step predisposes the cell surface to lysozyme penetration): Resuspend cells in cold TES buffer; add Sarcosyl (or SDS) to give a final concentration of 0.1%; mix on a vortex 15–30 s; centrifuge as above; remove supernatant very thoroughly.
Osmotic-shock (achieved by quick dilution from sucrose into water or 10 mM EDTA): Resuspend cells in 0.4 ml TES; add 0.35 ml sucrose mix (1.6 M sucrose, 0.55 M Tris, 0.10 M EDTA); hold at 5°C for 10–20 min; add 0.15 ml lysozyme (5 mg/ml in Tris, 0.05 M, pH 8,0); mix; add 3.6 ml cold EDTA (0.01 M); incubate 5–20 min at 5°C.
Add 2.5 ml of 2.5% Sarcosyl (or SDS), mix slowly to clear.
Shear lysate on a vortex for 15 s.
Subject to CsCl-EtBr density gradient centrifugation.

Extraction of Cleared Lysates

Again two protocols are described which, in principle, differ only in the non-ionic detergent used for cell lysis.

Lysis with Brij 58 (Clewell and Helinski 1969)

Inoculate bacteria into 250 ml LB and incubate overnight.
Pellet cells by centrifugation (8000 rpm, 10 min) and wash in TE buffer.

Note: Subsequent steps should be conducted at 4°C!

Resuspend in 3 ml 25% sucrose in 0.05 M Tris, pH 8.0.
Add 0.5 ml lysozyme (5 mg/ml in 0.25 M Tris, pH 8.0), incubate for 5 min.
Add 4 ml Brij 58 lysis mixture:

Brij 58	8.0 g
Na-deoxycholate (DOC)	3.2 g
1 M Tris pH 8.0	40 ml
0.5 M EDTA pH 8.0	100 ml
H_2O to	1000 ml

Solution should become viscous and opalescent.
Clear the lysate by centrifugation (30,000 × g, 30 min, 4°C).
Take 8 ml of "cleared lysate" for CsCl-EtBr density gradients.

Lysis with Triton X100 (Cannon et al. 1974)

Grow bacteria in 100 ml rich medium (e.g., PA, LB) overnight.
Harvest cells by low speed centrifugation.
Wash twice in cold TE buffer.

Note: The following step are normally carried out at 4°C!

Resuspend pellet in 1.5 ml 25% sucrose in 0.05 M Tris pH 8.0.
Add 0.5 ml lysozyme (10 mg/ml)/RNase (5 mg/ml) in TE buffer, incubate for 5 min.
Add 0.5 ml EDTA 0.5 M, pH 8.0, incubate for 5 min.
Add 2.5 ml Triton X100 lysis mixture (0.1% Triton X100 in 0.06 M EDTA, 0.05 M Tris), mix gently by inversions.
Clear the lysate by centrifugation (30,000 × g, 30 min, 4°C).
Adjust volume with TE buffer to 5.0 ml for CsCl-EtBr gradients.

CsCl-EtBr Density Gradient Centrifugation

The lysates extracted by one of the methods described in the preceding section are subjected subsequently to ultracentrifugation in CsCl-EtBr density gradients. Adjust lysate to a final volume of 5.0 ml and add 5.0 g of CsCl (or 8 ml of lysate + 8.0 g of CsCl). Add EtBr (20 mg/ml in H_2O) to a final concentration of 1 mg/ml.

Centrifugation is performed usually in a fixed angle rotor at 100,000 × g (e.g., 34,000 rpm in a Ti50 Beckman rotor). DNA will reach its equilibrium density in about 48 h, but longer sedimentation times will yield sharper bands and higher purity of DNA. DNA bands are visualized by exposure to long wavelength UV due to the fluorescence of intercalated EtBr. The lower band containing CCC-DNA is collected by puncturing the side of the centrifuge tube with a syringe.

Further Purification and Enrichment

Prior to use, it is necessary to remove EtBr and CsCl from DNA solution.

EtBr is removed from DNA in CsCl by extraction with isopropanol or butanol saturated with aqueous 5 M NaCl, 10 mM Tris, and 1 mM EDTA (pH 8.5). The extraction is repeated until no more color is visible.

CsCl is extracted by dialysis against TE buffer (pH 7.5). Dialysis tubing is prepared as follows:
Boil tubing in 0.1 M Na_2CO_3, 0.01 M EDTA.
Wash with distilled water.
Repeat boiling until buffer solution is no longer colored.
Rinse exhaustively with H_2O.
Store at 4°C.
Before use, wash inside of tubing with distilled water. Seal DNA solution in sterile tubing, change buffer at least twice.

It is also advisable to perform a protein extraction by *phenol treatment:* add 1 vol. phenol saturated with 100 mM Tris, pH 8.0, mix thoroughly on a vortex, centrifuge to separate the two phases, and transfer aqueous phase to new tubes. Alternatively, a mixture composed of 50% phenol and 50% $CHCl_3$ can be used.

DNA may be concentrated by *ethanol-precipitation*. Treat pure DNA (with CsCl already removed) as follows:

Add 1/10 vol. 3 M Na-acetate pH 5.2 and 1/100 vol. tRNA (1 mg/ml) (to facilitate precipitation if the DNA concentration is very low).
Till up with at least 2 vol. of refrigerated ethanol ($-20°C$).
Precipitate either in $-70°C$ dry ice methanol bath for 5 min or at $-20°C$ for 1 h or longer.
Sediment precipitation (e.g., spin down in Eppendorf centrifuge for 10 min at 10,000 rpm if Eppendorf tubes are used), pour off EtOH very carefully.
Wash pellet twice with 70% EtOH.
Drain and dry pellet; all liquid should either drain off or evaporate.
Dissolve DNA pellet in desired volume of buffer.

It is also possible to precipitate DNA in CsCl solution. For this purpose, dilute solution with 2 vol. H_2O and add EtOH. Subsequent steps are as above.

3. Experiment 2: Small-Scale Techniques for Rapid Isolation of Plasmid DNA

a) Introduction

For many purposes, there is no need to prepare highly purified plasmid DNA via CsCl-EtBr ultracentrifugation — a method which is relatively lengthy and costly. Less purified DNA is often sufficient. In particular, recombinant DNA research depends on the availability of techniques which allow rapid analysis of a large number of clones. Hence, methods have been developed which permit rapid DNA isolation from only a few cells resulting in a small amount of plasmid DNA (mini-lysates, mini-preps). These techniques are simple enough to allow screening and also yield plasmid DNA in a form sufficiently pure to be used for restriction enzyme analysis or for transformation experiments. For example, cleared lysates may be treated with phenol and subsequently precipitated with ethanol to yield a plasmid DNA preparation suitable for transformation and sensitive to restriction enzymes (preparation of mini-cleared-lysates). However, this simple protocol is applicable only to multicopy plasmids with a molecular weight of about 20×10^6.

Many of the mini-preparations developed achieve the extraction of plasmid CCC-DNA by alkali treatment. This alkaline extraction procedure is based on the differential behavior of CCC-, OC-, and linear DNA under alkaline conditions. Alkali treatment leads to breakage of the hydrogen bonds between the two strands of the DNA double helix, but there is a narrow range of pH (about 12.0–12.5), within which only complementary strands of linear and open circular duplexes are separated, while the strands of supercoiled CCC-DNA remain bonded. So exposing crude or cleared cell extracts to a pH value of 12–12.5 results in selective denaturation of plasmid

OC- and linear chromosomal fragments. Upon neutralization the denatured DNA renatures and aggregates to form an insoluble network that can be removed by centrifugation. Plasmid CCC-DNA is then precipitated from the supernatant by ethanol-treatment and can subsequently be used for analytical purposes. This alkaline extraction can also be performed in large scale isolation procedures as a pre-purification step before CsCl-EtBr density gradient centrifugation.

Prior to digestion with restriction endonucleases, it is advisable to remove contaminating RNA (which can inhibit DNA metabolizing enzymes) by dissolving the DNA pellet in buffer containing RNase (10 μg/ml) or by adding RNase directly to the incubation buffer. The digested RNA need not be removed.

Four mini-scale isolation methods are described. As mentioned above, the first method (mini-cleared-lysate) is only suitable for multicopy plasmids. The alkaline extraction procedures also extract single copy plasmids of higher molecular weight (up to 40×10^6 or more), although it has been reported that even F' plasmids ($60-120 \times 10^6$) may be isolated (Birnboim and Doly 1979). Of course, these protocols can be scaled up to isolate DNA by ultracentrifugation.

b) Objectives

Rapid isolation of small amounts of plasmid DNA, suitable for analytical purposes (such as restriction enzyme analysis) and for transformation experiments.

c) Procedure and Results

Preparation of mini-cleared lysates:

Start from 5 ml of an overnight culture.
Centrifuge at 8000 rpm, 10 min, 5°C.
Resuspend pellet in 100 μl 20% sucrose in 40 mM Tris, pH 8.0, transfer solution to Eppendorf tubes, leave on ice for 5 min.
Add 20 μl lysozyme (10 mg/ml), 10 min, 0°C.
Add 10 μl CDTA 0.5 M, pH 8.0, cool for 10 min.
Add 100 μl lytic mix (50 mM Tris, pH 8.0, 10 mM EDTA, 1% Triton X100).
Centrifuge at 30,000 $\times$ g, 30 min, 4°C.
Transfer supernatant to a new tube and add 100 μl H_2O and 100 μl phenol saturated with 100 mM Tris, pH 8.0, centrifuge 3 min in Eppendorf centrifuge.
Remove aqueous phase and mix with 100 μl NaCl (1 M) and 100 μl phenol, centrifuge as above.
Remove aqueous phase, add 1 ml of cold ethanol and precipitate at $-70°C$ for 5 min.
Spin down in Eppendorf centrifuge, pour off EtOH, and wash pellet twice with 80% EtOH.
Dry pellet, resuspend in 50 μl buffer.

Alkaline Extraction Procedure (Birnboim and Doly 1979)

Bacteria are grown in 2.5 ml of LB overnight.
Take 0.5 ml of culture, spin down for 15 s (all centrifugation steps are carried out in Eppendorf tubes).
Remove supernatant carefully, resuspend pellet in 100 µl solution I (see below), incubate 30 min at 0°C.
Add 200 µl of solution II, mix gently on a vortex (suspension becomes clear and slightly viscous), keep on ice for 5 min.
Add 150 µl of solution III, mix gently by inversions, keep on ice for 60 min.
Centrifuge 5 min.
Transfer the almost clear supernatant to a second tube, add 1 ml of cold EtOH, hold at $-20°C$ for 30 min.
Centrifuge 2 min.
Dry pellet and dissolve DNA in 100 µl of 0.1 M Na-acetate/0.05 M Tris, pH 8.0.
Reprecipitate with 2 vol. of cold EtOH, keep at $-20°C$ for 10 min.
Centrifuge.
Drain off liquid and dissolve DNA pellet in 40 µl water or buffer.

Solution I: 50 mM glucose, 10 mM CDTA, 25 mM Tris (pH 8.0), lysozyme (2 mg/ml)
Solution II: 0.2 N NaOH, 1% SDS
Solution III: 3 M Na-acetate, pH 4.8

Simplified Alkaline Extraction Procedure (Ish-Horowicz and Burke 1981)

Centrifuge 1 ml of an overnight bacterial culture.
Resuspend in 100 µl 50 mM glucose, 25 mM Tris, pH 8.0, 10 mM EDTA; incubate 5 min at room temperature.
Add 200 µl 0.2 N NaOH, 1% SDS, mix gently, put on ice for 5 min.
Add 150 µl pre-cooled 5 M K-acetate pH 4.8, mix gently, leave on ice for 5 min.
Remove precipitate by centrifugation for 1 min.
Add 2 vol. EtOH to the supernatant, hold 2 min at room temperature, precipitate DNA by centrifugation.
Wash pellet with 70% EtOH.
Drain and dry pellet, resuspend in 50 µl buffer.

Note: To achieve higher sensitivity to restriction enzymes, one can insert a phenol extraction step before precipitating DNA with EtOH.

Lysis by Boiling (based on Holmes and Quigley 1981, modified by R. Simon)

Transfer 1.5 ml of an overnight culture into an Eppendorf tube.
Harvest cells by centrifugation in an Eppendorf centrifuge for 30 s.
Wash cells with 200 µl of TE buffer.
Suspend cells in 25 µl S1 buffer (see below) + lysozyme (2 mg/ml), mix on a vortex, incubate for 2–3 min.
Add 25 µl S2 buffer (see below), mix on a vortex.
Incubate in a boiling water bath for 40–50 s.
Chill quickly in ice water.

Add 250 µl S3 buffer (see below), mix carefully by inverting several times.
Centrifuge for 10–12 min.
The pellet should be slightly gelatinous, transfer liquid supernatant into a new Eppendorf tube.
Add 50 µl phenol/chloroform mixture, mix on a vortex, and centrifuge for 3–5 min.
Transfer aqueous phase into a new Eppendorf tube.
Add 500 µl isopropanol, mix thouroughly, and precipitate DNA at $-20°C$ for 10–20 min.
Pellet precipitated DNA by centrifugation for 5 min.
Wash DNA pellet twice in 70% EtOH.
Drain and dry pellet, resuspend in 50–100 µl TE buffer.

S1: 8% sucrose, 50 mM CDTA, 50 mM Tris pH 8.0
S2: same as S1, additionally 10% Triton X100
S3: 500 mM NaCl, 10 mM Tris pH 8.0

4. Materials

TES buffer for Sarcosyl lysis:	0.050 M Tris 0.005 M EDTA 0.050 M NaCl	pH 8.0
TES buffer (Schwinghamer lysis):	0.050 M Tris 0.010 M EDTA 0.050 M NaCl	pH 8.0
TS buffer (Schwinghamer lysis):	0.050 M Tris 0.050 M NaCl	pH 8.0
TE buffer for cleared lysates:	0.010 M Tris 0.001 M EDTA	pH 8.0

5. References

Bauer W, Vinograd J (1968) The interaction of closed circular DNA with intercalative dyes. I. The super-helix density of SV40 DNA in the presence and absence of dye. J Mol Biol 33: 141–171

Bazaral M, Helinski DR (1968) Circular DNA forms of colicinogenic factors E1, E2 and E3 from *Escherichia coli*. J Mol Biol 36:185–194

Birnboim HC, Doly J (1979) A rapid alkaline extraction procedure for screening recombinant plasmid DNA. Nucleic Acids Res 7/6:1513–1523

Burton A, Sinsheimer RL (1965) The process of infection with bacteriophage ØX174. VII. Ultracentrifugal analysis of the replicative form. J Mol Biol 14:327–347

Cannon FC, Dixon RA, Postgate JR, Primrose SB (1974) Chromosomal integration of *Klebsiella* nitrogen fixation genes in *Escherichia coli*. J Gen Microbiol 80:227–239

Clewell DB (1972) Nature of ColE1 plasmid replication in *Escherichia coli* in the presence of chloramphenicol. J Bacteriol 110:667–676

Clewell DB, Helinski DR (1969) Supercoiled circular DNA-protein complex in *Escherichia coli*: purification and induced conversion to an open circular DNA form. Proc Natl Acad Sci USA 62:1159–1166

Clowes RC (1972) Molecular structure of bacterial plasmids. Bacteriol Rev 36/3:361–405

Deniss IS, Morgan AP (1976) Studies on the mechanism of DNA cleavage by ethidium. Nucleic Acids Res 3:315–232

Falkow S, Wohlhieter JA, Citarella RV, Baron LS (1964) Transfer of episomic elements to *Proteus*. I. Transfer of F-linked chromosomal determinants. J Bacteriol 87:209–219

Fiers W, Sinsheimer RL (1962) The structure of DNA of bacteriophage ØX174. III. Ultracentrifugal evidence for a ring structure. J Mol Biol 5:424–434

Hickson FT, Roth TR, Helinski DR (1967) Circular DNA forms of a bacterial sex factor. Proc Natl Acad Sci USA 58:1731–1738

Holmes DS, Quigley M (1981) A rapid boiling method for the preparation of bacterial plasmids. Anal Biochem 114:193–197

Ish-Horowicz D, Burke JF (1981) Rapid and efficient cosmid cloning. Nucleic Acids Res 9/13:2989–2998

Marmur J (1961) A procedure for the isolation of deoxyribonucleic acid from micro-organisms. J Mol Biol 3:208–218

Marmur J, Rownd R, Falkow S, Baron LS, Schildkraut C, Doty P (1961) The nature of intergeneric episomal infection. Proc Natl Acad Sci USA 47:972–979

Radloff R, Bauer W, Vinograd J (1967) A dye-buoyant-density method for the detection and isolation of closed circular duplex DNA: the closed circular DNA in HeLa cells. Proc Natl Acad Sci USA 57:1514–1520

Schwinghamer EA (1980) A method for improved lysis of some Gram-negative bacteria. FEMS (Fed Eur Microbiol Soc) Microbiol Lett 7:157–167

Vinograd J, Lebowitz J, Radloff R, Watson R, Laipis P (1965) The twisted circular form of polyoma viral DNA. Proc Natl Acad Sci USA 53:1104–1111

1.3 Characterization of Plasmid DNA by Agarose Gel Electrophoresis

U. PRIEFER [1]

Contents

1. General Introduction 26
2. Experiment 1: Rapid Sizing of Intact Plasmid Molecules on Agarose Gels 28
   a) Introduction 28
   b) Objectives 29
   c) Procedure and Results 30
3. Experiment 2: Molecular Weight Determination of Restriction Fragments 33
   a) Introduction 33
   b) Objectives 35
   c) Procedure and Results 35
4. Materials 36
5. References 36

1. General Introduction

Since the introduction of gel electrophoresis as a means of studying nucleic acids some 10 years ago, this method has become a powerful and versatile tool in the investigation and characterization of DNA molecules. It is rapid, precise, and inexpensive and requires only small amounts of DNA. Gel electrophoresis allows the rapid separation and high resolution of DNA species on the basis of molecular size and conformation. It is therefore an extremly useful method for the estimation of molecular weights of both intact plasmid molecules and linear DNA fragments produced by restriction endonucleolytic cleavage, provided one has standards of known length. The rationale for separating DNA species by electrophoresis is discussed in detail by Gould and Matthews (1976), but is directed more toward electrophoresis of RNA. Moreover Southern (1979) reviews gel electrophoresis of DNA restriction fragments extensively. With respect to the sizes of the DNA molecules to be separated, either polyacrylamide or agarose is used for gel electrophoresis. Only agarose gels will be discussed in this chapter since the resolution capacity of this method is sufficient in most cases.

---

1 Lehrstuhl für Genetik, Fakultät für Biologie, Universität Bielefeld, D-4800 Bielefeld 1, Fed. Rep of Germany

Agarose is a polysaccharide extracted from various red algae. It is a polymer of repeating disaccharide units composed of β-D-galactopyranose and 3.6-anhydro-L-galactose joined in a 1-3-β-glycosidic linkage. Interchain hydrogen bonds are presumed to form the cross-links which lead to polymerization. When DNA is subjected to agarose gel electrophoresis, it is forced to migrate through the interstices of this network toward the anode (due to its negatively charged phosphate residues) with a migration velocity depending on the following:

Molecular weight of DNA,
Conformation of DNA (CCC, OC, linear),
Pore sizes of the gel (determined by the agarose concentration),
Voltage gradient employed to the gel,
Electrophoretic buffer.

All these parameters interfere with one another, but there are some general guidelines. For example:

High voltage gradients are employed for separating small DNA fragments.
Better resolution of high molecular weight DNAs is achieved at low voltage gradients.
Small pore sizes (between 0.7 and 1% agarose) are used for small DNA fragments (up to 10×10^6).
Large DNA molecules ($15-40 \times 10^6$) are usually submitted to electrophoresis in low concentration gels (0.3–0.5% agarose).
Restriction enzyme digested DNA is usually run in Tris-acetate gels.
For separation of plasmid DNA in partially purified whole-cell lysates, either Tris-acetate or Tris-borate buffer can be used (largely dependent on the plasmid size).

Nevertheless, it is necessary to adapt electrophoretic conditions to the specific circumstances of the particular experiment.

Agarose gels can be prepared at concentrations ranging from 0.3–2% (w/v). Either vertical or horizontal slab gels can be used. Horizontal gels are more convenient to handle and are more stable, especially at agarose concentrations lower than 0.8%, although it has been reported that vertical gels give sharper bands. Gels are prepared by suspending and boiling the calculated amount of agarose powder in an electrophoretic buffer until the solution is completely homogenous and clear. The molten agarose is cooled to 50–60°C before pouring the gel. After the gel is completely set, samples are loaded. To increase the density, they should contain glycerol or sucrose (to a final concentration of 5–10%). Addition of Ficoll (1–2%) avoids the formation of U-shaped bands. A tracking dye may be added as a visible marker; usually bromophenol blue is used at a final concentration of 0.025%. However, DNA fragments can have the same electrophoretic mobility as the marker dye (depending on agarose concentration and voltage gradient: the lower the agarose concentration and the higher the voltage gradient, the larger the DNA fragment that will co-migrate with dye), thus, this DNA band may be missed, since the dye absorbs fluorescence from DNA-bound EtBr and obscures the DNA. DNA bands can be visualized by staining with EtBr (the rationale of EtBr-staining is outlined in Chap. 2). The sensitivity of this staining technique depends on the amount of DNA, i.e. the smaller the DNA fragments, the weaker the fluorescence. Gels can be stained either by including EtBr in the gel and electrophoretic buffer (Note: EtBr influences the mobility of DNA to a small degree) or after the run by soaking it in EtBr solution (0.5 µg/ml) for 30–60 min. To diminish the background of fluorescence, it is advisable to destain the gel with redistilled water. DNA bands are then detected by irradiation with UV at 254 nm, either from the side or on a transilluminator (Note: irradiation with short wavelength UV causes damage to the DNA and destroys its function). The illuminated gel is photographed, usually with a polaroid camera, equipped with a red filter to exclude UV light.

As mentioned above, the mobility of DNA in gels is a function of its molecular weight and — within a limited range — inversely proportional to its size: the larger the size, the slower the rate of migration. Thus, agarose gel electrophoresis allows the determination of relative and approximate molecular weight on the basis of electrophoretic mobility in relation to the mobility of DNA standards of known size. Mobilities are usually calculated from the photograph. Greatest accuracy is achieved by tracing the photograph in a microdensitometer. Alternatively, measurement can be taken from a photographic enlargement. The distance from the sample to the top of each band is measured (migration distance) and the relative mobility is calculated (migration distance divided by the length of the gel). A standard curve is constructed by plotting the log of molecular weight of standards against their relative mobilities and drawing a line through these points. In the small size range, linear relationship is obtained. However, migration of very large DNA molecules becomes more or less independent of size, so that the relationship is no longer linear and a curve is observed at the top of the scale. Molecular weight of the samples are determined by placing the measured mobilities on the standard curve and interpolating the size.

2. Experiment 1: Rapid Sizing of Intact Plasmid Molecules on Agarose Gel

a) Introduction

To determine by classical means the number and size of plasmid species isolated preparatively, it is necessary to either examine the plasmid DNA under the electron microscope or sediment the DNA through a sucrose gradient. Neither method is well-suited to routine work. Since gel electrophoresis has found application in the analysis of DNA, it is possible to use this method for molecular weight determinations of purified DNA. Aaji and Borst (1972) found that the migration rates of purified bacteriophage CCC-DNAs ranging from 3.4×10^6 to 10×10^6 daltons were related inversely to the logarithm of their molecular weights. In addition, they observed that in every case open circular molecules migrated considerably slower than covalently closed circles, whereas the mobility of linear DNA largely depended on the conditions employed (the higher the gel concentration, the faster the linear DNA in proportion to closed circles). However, there is a basic difficulty in deducing molecular weight from the electrophoretic mobility of purified CCC-DNA. Since CCC-DNA molecules are subjected to considerable stress during extraction and purification, which converts them to the OC- and linear forms, plasmid DNA preparations contain differential amounts of all three conformations. Consequently, three bands can occur after electrophoresis of *one* plasmid species. This complex banding pattern may result in the failure to identify the CCC-form, especially in the case of plasmids which easily form polymers or in DNA preparations extracted from strains bearing more than one plasmid. A relatively simple method of distinguishing CCC-, OC-, and linear DNA on gels is reported by Hintermann et al. (1981).

Molecular weight determination of CCC-plasmid DNA by agarose gel electrophoresis is not restricted to purified plasmid DNA, but is also suitable for the separation of closed circular plasmid DNA from the chromosome in partially purified whole-cell lysates (Meyers et al. 1976). Currently, a variety of procedures are available which are based on the electrophoresis of total cell extracts resulting in the resolution of circular plasmid DNA and which allow rapid detection and sizing of plasmid DNA in a large number of clones. Three of these methods are described here. The first two methods exhibit considerable disadvantages compared with the third method. The method described by Hansen and Olsen (1978) is a relatively gentle procedure and allows visualization of plamids with a molecular weight up to 300×10^6, but it requires several steps and is very time-consuming. Although the first type (SDS-lysis) is very simple and rapid, there are some important limitations. Since this method involves extensive shearing to fragment host chromosomal DNA, large plasmids will also be damaged. Therefore, this method is not suitable for the detection of plasmids greater than 50×10^6 or for small low copy number plasmids. There is a second essential drawback in this method (and to a lesser degree in the second one): the large amount of contaminating host chromosomal fragments, migrating as a broad smear, may obscure plasmid DNA banding in the same region. The procedure described by Eckhardt (1978) is extremely rapid and requires a minimum of manipulations since lysis is achieved directly in the gel well. Thus, shearing of DNA is largely avoided so that very large plasmid molecules will stay intact and form a sharp CCC-DNA band and the bulk of the unbroken host chromosome remains in the gel well. Almost no chromosome (less than 0.5%) will enter the gel as a band of linear DNA.

All three procedures can be applied not only to *E. coli*, but also – with slight modifications – to other Gram-negative bacteria.

b) Objectives

The methods described are suitable for screening a large number of bacteria for the presence of plasmids. The migration velocity of CCC-plasmid DNA in agarose gels is inversely proportional to its size so that this relation can be used for molecular weight determinations.

c) Procedure and Results

SDS Lysis Procedure

Resuspend a large single colony in 100 µl 20% sucrose in TES.
Add 100 µl lysozyme (5 mg/ml), keep at room temperature for 5 min.
Add 100 µl EDTA 0.25 M, 5 min room temperature.
Add 100 µl SDS 2.5%.
Mix with 100 µl saturated phenol.
Shear the lysate with a fine-needled syringe.
Spin down in Eppendorf centrifuge.
Take 20–50 µl of the supernatant for agarose gel electrophoresis. Use Tris-acetate gels for plasmids with a molecular weight less than 20 kb and Tris-borate gels for larger plasmids (Fig. 1).

Enriched Crude Plasmid Preparation (Hansen and Olsen 1978; modified)

Pellet about 4 ml of late log phase culture by centrifugation.
Resuspend cells in 150 µl sucrose (25% in 0.05 M Tris, pH 8.0).

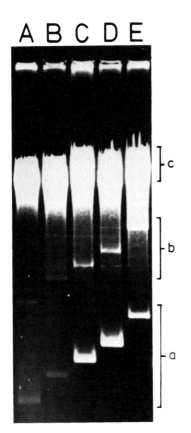

Fig. 1. Agarose gel electrophoresis of SDS lysates (20 µl loaded) isolated from *E. coli* strains harboring plasmids of different sizes. *a* CCC forms of the monomeric plasmids which migrate, in this size range, in front of the chromosomal DNA *(c)*. The bands *(b)* probably correspond to the CCC forms of dimeric molecules. Plasmids and plasmid sizes are: *A* pBR327, 3.3 kb (Covarrubias et al. 1981), *B* pACYC177, 3.7 kb (Chang and Cohen 1978), *C* pBR322, 4.3 kb (Bolivar et al. 1977), *D* pBR328, 4.9 kb (Covarrubias et al. 1981), and *E* pBR325, 5.4 kb (Bolivar 1978)

Add 100 μl lysozyme (5 mg/ml in 0.25 M Tris, pH 8.0), mix gently by repeated inversions, leave on ice for 5 min.

Add 200 μl EDTA (0.25 M, pH 8.0), mix by inversions, keep on ice for 5 min.

Add 250 μl SDS (20% in TE, 70°C). Heat shock the lysate for 5 min in a 70°C waterbath.

Note: To simplify this procedure, the following step of alkali denaturation and neutralization can be omitted.

Add 3 M NaOH to give a pH value of 12.1–12.3, invert repeatedly for 2–3 min.

Lower pH to 8.5–9.0 by addition of 2 M Tris, pH 7.0.

Add 250 μl of cold 5 M NaCl, mix thoroughly, leave on ice for 4 h or longer.

Spin down precipitate, transfer supernatant (measure the volume!) to new tubes.

Add 1/3 vol. of PEG 6000 (42% in 0.01 M Na-phosphate buffer, pH 7.0), mix carefully, leave on ice for 4 h or longer.

Centrifuge at 5000 rpm for 10 min at 4°C.

Drain and dry pellet.

Resuspend in 50–100 μl 10% sucrose + bromophenol blue.

Load 10–40 μl onto the gel (0.7–0.9%, Tris-borate) (Fig. 2).

This protocol can be adapted for preparative DNA isolation. For this purpose, increase input volume of cells to about 200 ml.

Fig. 2. Agarose gel electrophoresis of crude plasmid preparations extracted by the Hansen-Olsen method (made by R. Simon). The strong band migrating in front of the plasmids corresponds to linear chromosomal DNA. Lanes A RP4 (38×10^6), B mRP4 = deletion derivative of RP4 (20×10^6), C mRP4-ColE1 (24.6×10^6), and D RP4::Mu (64×10^6). Electrophoresis conditions: vertical slab gel, 0.9% agarose, Tris-borate buffer, 4 h at 150 V

Gentle Lysis in the Gel Well (Eckhardt 1978)

The original protocol involves the following steps:

Put 15 µl of lysozyme mixture (lysozyme 7500 U/ml, RNase 0.3 U/ml), 0.05% bromophenol blue, 20% Ficoll 40,000 in electrophoretic buffer) in each gel well.

Add bacteria; use only $10^7 - 10^8$ cells for lysis: either 1–2 single colonies scraped from plates or alternatively a suitable volume of liquid culture, centrifuged and resuspended in 10 µl of electrophoretic buffer with 20% Ficoll 40,000; do not overload the gel, since this diminishes the yield of plasmid DNA and increases the amount of chromosomal smear; incubate 2–3 min.

Overlay with 30 µl SDS mixture (0.2% SDS, 10% Ficoll 40,000 in electrophoretic buffer); mix by moving a toothpick from side to side (not more than twice).

Overlay with 100 µl of 0.2% SDS, 5% Ficoll in electrophoretic buffer.

Seal the wells with hot agarose and fill electrophoretic chambers with buffer.

Running conditions are: 2 mA for 60 min; 40 mA for 60–150 min.

This method has been successfully simplified with the following modifications:

Resuspend 10^7 cells in 20 µl of 7% Ficoll, 20% sucrose, 1 U/ml RNase, 1 mg/ml lysozyme in Tris-borate buffer.

Put cell suspension after mixing immediately into the well already filled with electrophoretic buffer.

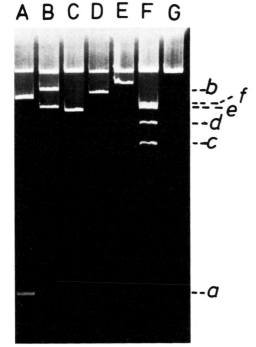

Fig. 3. Agarose gel electrophoresis of *Rhizobium meliloti* wild-type strains using the modified Eckhardt procedure (according to R. Simon). Prior to lysis, cells were washed with TE buffer incorporating 0.1% Sarcosyl. Plasmid sizes are (as determined by electron microscopy): Lane *A*, plasmid *a*, 4.8×10^6, Lane *D*, plasmid *b*, 260×10^6 (Spitzbarth et al. 1979), Lane *F*, plasmid *c*, 36×10^6, *d*, 54×10^6, *e*, 90×10^6, *f*, 104×10^6 (Horn 1981). The large plasmids present in all slots are supposed to have a molecular weight of about $600 \pm 200 \times 10^6$. Electrophoresis conditions: 0.8% agarose, Tris-borate buffer, 30 min at 5 mA, 2.5 h at 100 mA

Overlay carefully with about 50 μl of 1% SDS, 5% sucrose bromophenol blue in Tris-borate buffer.
It is not necessary to seal the wells with agarose.
Run 30 min at 5 mA, 2–3 h at 100 mA (Fig. 3).

3. Experiment 2: Molecular Weight Determination of Restriction Fragments

a) Introduction

The fragmentation of DNA molecules by restriction enzymes and the subsequent separation and characterization of the digestion products by electrophoresis in agarose gels has developed into an important tool in molecular biology. Restriction endonucleases are part of the prokaryotic restriction-modification system that modifies and, thus, marks host cell DNA by specific methylation (methylases) and recognizes and cleaves non-specifically methylated DNA (endonucleases) thus, protecting the host cell against invasion by foreign DNA (Luria and Human 1952, Bertani and Weigle 1953, Arber and Linn 1969).

All restriction enzymes recognize specific DNA sequences. However, some of them cleave at specific sites within the recognition sequence, wheras others cleave DNA randomly. Those that are nonspecific in their cleavage are designated class I enzymes (Arber 1974, Boyer 1971); those that cleave site-specifically are called class II enzymes (Boyer 1971). Class I enzymes, which generate heterogenous DNA products, cannot be used for restriction enzyme analysis, but cleavage of DNA with class II enzymes, resulting in a characteristic pattern of restriction fragments, is becoming increasingly useful for studying the structure and function of DNA. Some type II enzymes cleave both strands at a single site, thus generating fragments with fully complementary terminals ("flush" or "blunt" ends); others cleave the two strands at different sites within the recognition sequence ("staggered" nicks), thereby generating fragments with one protruding single strand terminal ("sticky" or "cohesive" ends).

The site specificity of cleavage and the availability of a wide range of endonucleases having different recognition sites make class II restriction enzymes important and versatile tools for investigating DNA structure. However, most of the restriction enzymes are still commercially expensive. Therefore, it is economical to isolate the more commonly used enzymes in the lab. Protocols for the purification of a variety of restriction enzymes are described in *Methods in Enzymology* (1980, Vol. 65, Part I).

Digestions are performed under conditions specific for the enzyme (Table 1). The reaction is stopped with an excess of EDTA (which complexes Mg-ions) or by heating the mixture. If fragments with longer sticky ends are present in the digestion mixture (e.g., lambda-DNA), heating and subsequent rapid cooling on ice is essential to separate hybridized sticky ends. Physical separation of the different size classes of restriction products is achieved by electrophoresis in agarose gels. The amount of DNA loaded onto the gel and the electrophoresis conditions depend largely on the number and size of the cleavage products. The gels are stained and photographed as described and

Table 1. Restriction endonucleases. Some of the most common restriction enzymes, their sequence specificity, and cleavage conditions are listed. The recognition sequences are written from 5' → 3' (only one strand is given); the cleavage site is indicated by a prime (')

Termini	Enzyme	Sequence	Incubation	
			Buffer [a]	Temperature °C
Flush ends	SmaI	CCC'GGG	Low	25
	HpaI	GTT'AAC	Low	37
Sticky ends	EcoRI	G'AAATTC	High	37
	HindIII	A'AGCTT	Med	37
	PstI	CTGCA'G	Med	37
	BamHI	G'GATCC	Med	37
	SalI	G'TCGAC	High	37
	KpnI	GGTAC'C	Low	37
	BclI	T'GATCA	Med	60
	XhoI	C'TCGAG	High	37
	BglII	A'GATCT	Low	37

[a] Recipes are given in the section on materials

the length of the fragments produced are calculated using molecular standards. In general, λ-DNA rectriction fragments generated by digestion with EcoRI, HindIII, or EcoRI/HindIII double digestion are used as standards (Table 2). The sizes of these fragments have been measured in the electron microscope to an accuracy of about 1% and range between 0.6 and 49 kb, including the full-sized phage DNA (Phillipsen et al. 1978, Szybalski and Szybalski 1979).

Table 2. λ-DNA restriction fragments. λ-DNA restriction fragments produced by cleavage with EcoRI, HindIII, and by EcoRI/HindIII double digestion. The sizes are given in bp. The full-size phage genome is determined to be 49 kb (Phillipsen et al. 1978, Szybalski and Szybalski 1979)

EcoRI	HindIII	EcoRI/HindIII
1 : 21.800	1 : 23.700	1 : 21.800
2 : 7.540	2 : 9.460	2 : 5.240
3 : 5.930	3 : 6.660	3 : 5.050
4 : 5.540	4 : 4.260	4 : 4.210
5 : 4.800	5 : 2.250	5 : 3.380
6 : 3.380	6 : 1.960	6 : 1.960
	7 : 590	7 : 1.910
	8 : 100	8 : 1.620
		9 : 1.320
		10 : 930
		11 : 880
		12 : 590
		13 : 100

b) Objectives

DNA from plasmid pBR325 is cleaved with several restriction enzymes. Digestion with *Eco*RI, *Hin*dIII, *Bam*HI, and *Pst*I, which each have single cleavage sites on pBR325, yield linear DNA molecules; double digestions with these enzymes result in the production of two fragments each. The cleavage products are subjected to electrophoresis and the fragment sizes are determined using *Eco*RI/*Hin*dIII restricted λ-DNA.

c) Procedure and Results

General protocol for restriction endonuclease cleavage:

1 μg of DNA and 1 U enzyme are mixed in an appropriate incubation buffer to give a final volume of 20–25 μl.

Incubate at recommended temperature for 60 min.

Stop reaction by heating the sample to 70°C for 5 min (in the case of λ *Eco*RI/*Hin*dIII digestion, transfer the heated sample subsequently onto ice!).

Before loading onto the gel, add 1/10 vol. of a solution containing 50% glycerol and 0.25% bromophenol blue.

Electrophoresis is carried out on Tris-acetate gels.

The gel is stained and photographed as described above (Fig. 4).

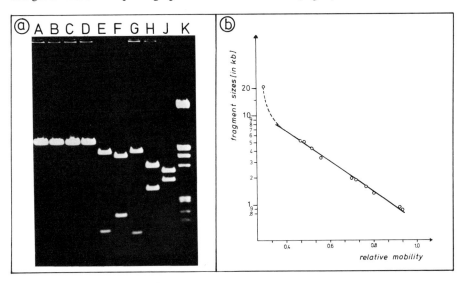

Fig. 4a, b. Cleavage of pBR325 DNA. a pBR 325 was digested with *Eco*RI *(A)*, *Pst*I *(B)*, *Hin*dIII *(C)*, *Bam*HI *(D)* to give the linearized molecule. Simultaneous cleavage with *Eco*RI/*Hin*dIII *(E)*, *Eco*RI/*Bam*HI *(F)*, *Eco*RI/*Pst*I *(G)*, *Pst*I/*Hin*dIII *(H)*, and *Pst*I/*Bam*HI *(J)* results in 2 fragments each. Lane *K* shows the restriction pattern of λ-DNA generated by *Eco*RI/*Hin*dIII double digestion. These fragments are used for standard curve construction (b). The sizes for pBR225 restriction products obtained by interpolation are: linear form about 5.4–5.5 kb, *E* 4.4 kb, 1.1 kb, *F* 4.2 kb, 1.3 kb, *G* 4.4 kb, 1.1 kb, *H* 3.3 kb, 2.2 kb, *J* 3.2 kb, 2.3 kb

4. Materials

TES buffer:		0.005 M EDTA	
		0.050 M Tris	
		0.050 M NaCl	pH 8.0
TE buffer:		0.050 M Tris	
		0.020 M EDTA	pH 8.0
Incubation buffer: (Davis et al. 1980)	Low:	10 mM Tris	pH 7.4
		10 mM MgSO$_4$	
		1 mM DTT	
	Med:	50 mM NaCl	
		10 mM Tris	pH 7.4
		10 mM MgSO$_4$	
		1 mM DTT	
	High:	100 mM NaCl	
		50 mM Tris	pH 7.4
		10 mM MgSO$_4$	
Enzyme dilution buffer:	10 m	10 mM Tris	pH 7.5
		250 mM NaCl	
		1 mM DTT	
		50% glycerol	
Electrophoretic buffer:			
Tris-acetate:		40 mM Tris	
		10 mM Na-acetate	
		1 mM EDTA	pH 7.8
Tris-borate:		90 mM Tris-base	
		90 mM Boric acid	
		2.5 mM EDTA	pH 8.3

5. References

Aaij C, Borst P (1972) The gel electrophoresis of DNA. Biochim Biophys Acta 269:192–200
Arber W, Linn S (1969) DNA modification and restriction. Annu Rev Biochem 38:467–500
Arber W (1974) DNA modification and restriction. Prog Nucleic Acid Res Mol Biol 14:1–37
Bertani G, Weigle JJ (1953) Host controlled variation in bacterial viruses. J Bacteriol 65:113
Bolivar F (1978) Construction and characterization of new cloning vehicles. III. Derivatives of plasmid pBR322 carrying unique EcoRI sites for selection of EcoRI generated recombinant molecules. Gene 4:121–136
Bolivar F, Rodriguez RL, Green PJ, Betlach MC, Heynecker HL, Boyer HW, Crosa JH, Falkow S (1977) Construction and characterization of new cloning vehicles. II. A multiple cloning system. Gene 2:95–113
Boyer HW (1971) DNA restriction and modification mechanisms in bacteria. Annu Rev Microbiol 25:153–176

Chang ACY, Cohen SN (1978) Construction and characterization of amplifiable multicopy DNA cloning vehicles derived from the P15A cryptic miniplasmid. J Bacteriol 134:1141–1156

Covarrubias L, Cervantes L, Covarrubias A, Soberón, Vichido I, Blanco A, Kupersztoch-Portnay YM, Bolivar F (1981) Construction and characterization of new cloning vehicles. V. Mobilization and coding properties of pBR322 and several deletion derivatives including pBR327 and pBR328. Gene 13:25–35

Davis RW, Botstein D, Roth JR (1980) Acvanced bacterial genetics. Cold Spring Harbor Lab, Cold Spring Harbor New York

Eckhardt T (1978) A rapid method for the identification of plasmid desoxyribonucleic acid in bacteria. Plasmid 1:584–588

Gould H, Matthews HR (1976) Separation methods for nucleic acids and oligonucleotides. In: Work TS, Work E (eds) Laboratory techniques in biochemistry and molecular biology, vol 4. North-Holland/American Elsevier, New York

Hansen JB, Olsen RH (1978) Isolation of large bacterial plasmids and characterization of the P2 incompatibility group plasmids pMG1 and pMG5. J Bacteriol 135/1:227–238

Hintermann G, Fischer HM, Crameri R, Hütter R (1981) Simple procedure for distinguishing ccc, oc, and l forms of plasmid DNA by agarose gel electrophoresis. Plasmid 5:371–373

Horn D (1981) Identifizierung und Charakterisierung von Plasmiden aus Wildtyp-Stämmen von *Rhizobium meliloti*. Diploma Thesis, Univ Erlangen

Luria SE, Human ML (1952) A non-hereditary host-induced variation of bacterial viruses. J Bacteriol 64:557

Meyers JA, Sanchez D, Elwell LP, Falkow S (1976) Simple agarose gel electrophoretic method for the identification and characterization of plasmid deoxyribonucleic acid. J Bacteriol 127: 1529–1537

Phillipsen P, Kramer RA, Davis RW (1978) Cloning of the yeast ribosomal DNA repeat unit in SstI and HindIII Lambda vectors using genetic and physical size selections. J Mol Biol 123: 371–386

Szybalski EH, Szybalski W (1979) A comprehensive molecular map of bacteriophage lambda. Gene 7:217–270

Southern E (1979) Gel electrophoresis of restriction fragments. Methods Enzymol 68:152–176

Spitzbarth M, Pühler A, Heumann W (1979) Characterization of plasmids isolated from *Rhizobium meliloti* Arch Microbiol 121:1–7

1.4 Characterization of Plasmid in Wild Strains of Bacteria

N. DATTA[1] and M.E. NUGENT[2]

Contents

1. General Introduction 38
2. Experiment 1: Detection of Plasmids in Wild Strain of *E. coli* Using the Method of Eckhardt (1978) 39
   a) Introduction 39
   b) Objective 39
   c) Procedure 39
   d) Materials 40
3. Experiment 2: Conjugative Transfer of Plasmids to *E. coli* K12 40
   a) Introduction 40
   b) Objective 40
   c) Procedure 41
   d) Materials 44
4. Experiment 3: Transfer of Plasmids to *E. coli* K12 by Transformation 45
   a) Introduction 45
   b) Objective 46
   c) Procedure 46
5. Experiment 4: Comparisons of Plasmids in Wild Strains and in the *E. coli* K12 Recipients 47
6. Experiment 5: Classification of Unknown Plasmids According to Incompatibility (Inc) Group 47
   a) Introduction 47
   b) Objective 48
   c) Procedure 48
   d) Materials 49
7. References 50

1. General Introduction

The experiments described below characterize antibiotic resistance plasmids that will replicate in *E. coli* K12. For other plasmids and plasmids of other bacterial species, methods must be adapted. No general instructions are possible for the characterization of any plasmid in any bacterial genus, although the availability of a plasmid-free

---

1 Dept. of Bacteriology, Royal Postgraduate Medical School, London, W.12 OHS, England
2 Dept. of Microbiology and Process Research, G.D. Searle and Co Ltd., High Wycombe, England

recipient strain which is able to propagate the plasmids under investigation greatly simplifies the analysis. Selection for recipient strains which have received plasmids that encode antibiotic resistance or the utilization of a nutrient is simple but for further characterization more elaborate selection/detection procedures may be required. For example, it may be necessary to eliminate the donor strain (with a phage or drug) and test the whole recipient culture, rather than purified colonies, for the character under test. Smith and Halls (1968) first described transfer of enterotoxin plasmids using ligated segments of intestine in experimental animals to demonstrate their presence in recipient bacteria. A generally applicable strategy for a plasmid carrying no easily selected marker is to convert it into an R plasmid by insertion of a transposon encoding an antibiotic resistance.

2. Experiment 1: Detection of Plasmids in a Wild Strain of *E. coli* Using the Method of Eckhardt (1978)

a) Introduction

The single colony lysate method described by Eckhardt (1978) provides a quick method of screening bacteria for the presence and approximate sizes of covalently closed circular (CCC) plasmid DNA (this volume). Eckhardt used this method to look for plasmids in *Bacillus subtilis, Escherichia coli, Pseudomonas aeruginosa, Pseudomonas pseudocaligenes* and *Neisseria gonorrhoea*. We have used the same method for *Acinetobacter calcoaceticus, Citrobacter koseri, Klebsiella pneumoniae, Serratia marcescens* and a modified version (with lysostaphin, Schwartz-Mann, 500 µg/ml, instead of lysozyme in the lysis mixture) for plasmids in *Staphylococcus aureus*.

The method of Eckhardt is useful for many purposes because of its sensitivity, simplicity, speed and the variety of bacterial species it can be applied to. In the present experiment we shall work only with strains of *E. coli*.

b) Objective

To determine the number and sizes of plasmids present in a wild strain of *E. coli*.

c) Procedure

E. coli 116 is a wild strain of *E. coli* that was isolated from infected urine and is resistant to tetracycline (Tc), chloramphenicol (Cm), streptomycin (Sm), sulphonamides (Su) and trimethoprim (Tp). Lysed colonies of this and *E. coli* K12 strains carrying single plasmids of known molecular weight are placed in separate wells of a 6 mm wide vertical 0.7% agarose slab gel containing 6.5 × 15 mm wells.
Run the gel for 4 h at a constant voltage of 40 mA.
Stain and photograph the gel.
Measure the migration distances of all plasmids.

Construct a calibration curve of the mobilities of the known plasmids versus their log molecular weights and calculate the molecular weights of the various plasmid species present in *E. coli* 116.

d) Materials

Nutrient agar plate cultures of *E. coli* 116 and *E. coli* K12 strains carrying the following standard plasmids:

Plasmid	Mol. wt. (10^6)	Reference
pBR322	2.6	Jacob et al. (1977)
R300B	5.7	Jacob et al. (1977)
AP201	9.6	Jacob et al. (1977)
RP4	36	Jacob et al. (1977)
R702	46	Jacob et al. (1977)
R1	62	Jacob et al. (1977)
RAl-1	85	Jacob et al. (1977)

3. Experiment 2: Conjugative Transfer of Plasmids to *E. coli* K12

a) Introduction

A medium or technique must be devised to allow separation of K12 clones that have acquired a plasmid from a large population of wild-type donors and plasmid-free *E. coli* K12. This generally requires the recipient strain to differ from the wild-type donor by at least two characters, in addition to the putative plasmid character. One distinguishing character allows counter selection of the donor and another, unselected one distinguishes mutants of the donor from recipient clones that have received plasmid characters from the donor. In the following experiment, resistances to Tc, Cm, Su, Sm and Tp are the characters to be tested for transfer. The rifampicin-resistance *(rpo)* of the recipient permits the use of rifampicin for donor counter selection and the *lac*, *his* and *trp* markers distinguish the K12 recipient from *rpo* mutants of the wild-type *E. coli* 116.

b) Objective

To transfer antibiotic resistance plasmids from *E. coli* 116 to *E. coli* K12 by conjugation.

c) Procedure

Broth Mating

Day 1

Dilute overnight culture of recipient strain *E. coli* K12 J62-2 *(pro, his, trp, lac, rpo)* with equal volume of warmed broth.
Mix broth culture of donor strain *E. coli* 116 (0.2 ml) with diluted recipient (1.8 ml).
Incubate at 37°C during working day without shaking.
Plate 0.1 ml vol. of mixture on the selection plate: MacConkey + rifampicin 50 µg/ml + Tc or Cm or Sm, MacConkey + 3% lysed horse blood + rifampicin + Tp and minimal sats agar + glucose + pro, his, trp + rif + Su.
Spread the inoculum evenly over half the plates and streak out over the other half.
Incubate overnight at 37°C.
Keep the mixed broths at room temperature.

Controls

Inoculate each type of selection plate with 0.1 ml pure broth culture of (a) donor and (b) recipient, separately.

Day 2

If colonies have grown, streak out one from each selection plate to the same medium to obtain single colonies. This purification step is necessary, since the selection inhibits but does not necessarily kill the large numbers of donors and recipients in the inoculum. Colonies are likely to be *E. coli* K12 R^+ but could be *E. coli* 116 rif-resistant mutants and must therefore be definitively identified. If they are lac^+ they could be *E. coli* 116 or K12 clones that have acquired lac^+R^+ plasmids (see Day 3).

If no colonies develop on the selection plates, inoculate more plates with the mixture that has been kept at room temperature; a plasmid that is temperature-sensitive for a conjugation function may be present.

Day 3

Suspend a well-separated colony from each selection plate in about 5 ml broth.
Spread about 0.1 ml of this suspension evenly on a plate of sensitivity test agar.
Place discs containing relevant antibiotics on the plate to reveal the sensitivity/resistance pattern of the clone.
Inoculate on nutrient agar and on minimal agar with and without pro + his + trp to confirm that the clone is J62-2.
Incubate broths and plates.
To check the activity of the discs, spread an area of a sensitivity test plate with a suspension of *E. coli* K12 R^- and an adjoining area with a suspension of the resistant *E. coli* 116 and place discs on the demarcation line.
After incubation, the difference in diameter of the zone of inhibition round the discs can be measured (Fig. 1).

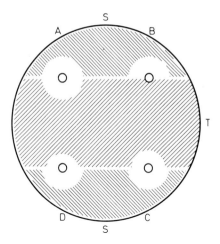

Fig. 1. Disc diffusion sensitivity test. *S* sensitive control culture; *T* test culture. Discs contain antibiotics *A*, *B*, *C* and *D*. The test culture is sensitive to *A*, resistant to *B*, *C* and *D*

Day 4

How many resistance patterns are exhibited by the transconjugant clones? (Table 1). Keep only one example of each.

Table 1. Patterns of resistance of transconjugants from *E. coli* 116 [a]

Selection	Patterns			
Sm	SmTcCmSuTp	SmSuTp		
Tc	SmTcCmSuTp	TcCmTp		
Cm	SmTcCmSuTp	TcDmSuTp	TcCmTp	
Su	SuTcCmSuTp	SmSuTp	SmSu	
Tp	SmTcCmSuTp	TcCmTp	SmSuTp	Tp

[a] See footnote to Table 2

Surface Mating

The method described above provides conditions for the transfer of some, but not all, plasmids. Bradley et al. (1980) and others have shown that some plasmids transfer more efficiently in surface matings of the type described below.

Day 1

Grow or adjust broth cultures of donor and recipient strains to a density of approx. 2×10^9 cells/ml (OD 1.0).

Mix 0.3 ml of each and spread at once on a well-dried nonselective nutrient agar plate.

Incubate for 1 h, then resuspend the bacteria and plate on selection plates as described for broth mating.

Day 2 and onwards

The same as for broth mating.

If J62-2 R^+ transconjugants are obtained with different resistance patterns, test each one for conjugative proficiency (tra^+) by mating with another strain of *E. coli* K12, J53, as follows:

Mix broth cultures of J62-2 R^+ and J53 R^- (diluted 1:2 in warmed broth as previously).
Incubate during the working day.
Plate on minimal agar with glucose, pro met and appropriate drug for selection of R plasmid.
Incubate plates.
Development of colonies indicates that J62-2 R^+ culture is tra^+.

Note: These methods are not intended to measure the frequency of plasmid transfer. The exact time of incubation of mating mixtures and the relative number of donors and recipients are therefore not important. However, the density of cultures is important; if too low, no transfer will be detectable. If the donor strain is rif-resistant, another counter selection must be employed; this could be phage- or colicin-resistance (Anderson and Threlfall 1974) or mutational drug resistance other than rif. Resistance to nalidixic acid, streptomycin or azide has also been used.

Mobilization of Nonconjugative Plasmids

Nonconjugative plasmids in wild strains of bacteria can often be transferred to *E. coli* K12 through conjugation brought about by a tra^+ (mobilizing) plasmid. There is no means of knowing which mobilizing plasmid is most likely to work. The choice will be for a plasmid that is readily transferable and that posesses convenient markers. Suitable examples are RP4, R64, and F-primes (Jacob et al. 1977). Two methods have been used: (a) the triparental cross (Anderson and Threlfall 1974), in which 3 strains are mixed; the wild strain under test, a K12 carrying the prospective mobilizer and a different K12 strain, the final recipient. The latter must have properties that allow its separation from the other two and its identification; (b) the prospective mobilizing plasmid is transferred to the wild strain as a preliminary to the test for mobilization. Both methods are described.

Note: With *E. coli* 116, there is no need to mobilize resistance genes using an introduced plasmid, as all resistances are transferable by conjugation. The method is given as if transfer had been unsuccessful.

Day 1

Mix broth cultures of *E. coli* 116, J53 (RP4), and J62-2, 0.5 ml of each, with 0.5 ml warm broth.
Incubate during working day.
Plate 0.1 ml volumes on selection plates for isolation of (a) *E. coli* 116 (RP4), minimal salts agar + glu + Km, and (b) J62-2 with R plasmids of *E. coli* 116, MacConkey agar + rif + Tp or Cm.

Day 2

If colonies have grown on the MacConkey plates, purify them and test as in experiment 3; discard the minimal plates.

If there is no growth on MacConkey but growth on either of the minimal plates, pick one colony and purify by streaking on the appropriate selective plate.

Day 3

Prepare broth cultures of such transcipients.

Day 4

Use in conjugation experiment with J62-2 as described in experiment 3.

Finally, show whether the J62-2 transconjugants carry (a) a nonconjugative plasmid from the wild strains; (b) a nonconjugative plasmid and RP4; (c) a recombinant plasmid composed of RP4 and a plasmid or transposon from the wild *E. coli*. This is done by further transfer tests combined with single colony gel electrophoresis.

d) Materials

Broth and Surface Mating

Day 1

Broth cultures of donor and recipient strains, grown aerated overnight. Nutrient agar plates for surface mating.

Selection Plates

MacConkey agar + rifampicin 50 μg/ml + Tc   10 μg/ml
MacConkey agar + rifampicin 50 μg/ml + Cm   15 μg/ml
MacConkey agar + rifampicin 50 μg/ml + Sm   15 μg/ml
MacConkey agar + rifampicin 50 μg/ml + Tp   10 μg/ml + 4% lysed horse blood
Minimal salt agar + glu + rif 50 + Su 100 μg/ml

MacConkey is used because it has the advantage of allowing recognition of the K12 strain by its lac⁻ character. Because it contains bile salts it inhibits the growth of many nonenteric bacteria. It can be used with all drugs except sulphonamides which are inactive in media containing thymine and thymidine. If trimethoprim is used in a nutrient medium such as MacConkey, lysed horse blood (4% by volume) is incorporated after the medium is cooled below 56°C. This makes the pH indicator more difficult, but not impossible, to see (Jobanputra and Datta 1974). Sulphonamide is active in minimal salts medium and some other special media e.g., Mueller-Hinton agar. Any other medium on which the K12 will grow can be used, made selective by the addition of rifampicin plus one other antibiotic.

Concentrations of drugs used for selection are the lowest required to inhibit effectively growth of plasmid-free K12 and may vary with the medium chosen. The following (in μg drug/ml medium) are usually suitable:

Streptomycin	(Sm) 15	Sulphonamide	(Su)	100	Ampicillin	(Ap)	25
Neomycin	(Nm) 15	Trimethoprim	(Tp)	10	Carbenicillin	(Cb)	100
Kanamycin	(Km) 15	Chloramphenicol (Cm)		15	Cephaloridine	(Cp)	50
Gentamicin	(Gm) 3	Tetracycline	(Tc)	10	Mercuric chloride	(Hg)	25

Day 2

More of the same selection plates, if no growth obtained on plates of previous day. Plates of the same medium for purifying transconjugants.

Day 3

Minimal salts agar with and without pro his trp. Antibiotic-containing discs for sensitivity tests, available commercially. Use concentrations per disc similar to those given above per ml medium. Sensitivity test plates: use commercially available media e.g., Mueller-Hinton (Difco), DST, direct sensitivity test (Oxoid), Welcotest (Welcome). Such special media are necessary to show sulphonamide or trimethoprim sensitivity and recommended for all sensitivity testing, thus allowing the use of one uniform medium.

Mobilization of Nonconjugative Plasmids

Day 1

Same as for direct transfer (p. 44) and an overnight, aerated broth culture of *E. coli* K12 J53 (RP4).
J53 is lac$^+$ *pro met*. RP4 (or RP1 or RK2) is an IncP plasmid that encodes reistance to ApTcKm.
Selection plates: minimal salts agar + glu + Km (no amino acids), MacConkey agar + rif + Cm.
MacConkey agar + rif + Tp + lysed horse blood (concentrations as above).

Test for Conjugative Ability of Transconjugants Obtained in Experiment 2

E. coli K12 J53 lac$^+$ *pro met*, overnight, aerated broth culture.
Minimal salts agar + glu, pro, met + drugs appropriate for selection of R transfer.

4. Experiment 3: Transfer of Plasmids to *E. coli* K12 by Transformation

a) Introduction

It is sometimes necessary or convenient to transfer resistance or other phenotypic markers by transformation. This is the case when transfer by direct conjugation or mobilization is not successful or to separate plasmids that have been transferred together by conjugation. The methods of transformation vary according to the bacterial genus; a review is given in this volume. In this section, only transformation experiments involving restrictionless *E. coli* strains, such as *E. coli* C and *E. coli* K12 r⁻ m⁺ are considered. Highest transformation frequencies are obtained using freshly isolated plasmid DNA that has been purified through caesium chloride-ethidium bromide (CsCl-EtBr) gradients and that contains a low proportion of nicked or broken molecules. However, DNA freshly isolated by other methods, such as that of Meyers et al. (1976), may be used if very pure DNA is not available (for a review of different isolation methods see this volume). The efficiency of plasmid transformations decreases with increasing size of plasmid DNA and for large plasmids ($> 100 \times 10^6$) may be too low for success. However, transformation is an ideal method for the transfer of small nonmobilizable plasmids.

Table 2. Patterns of resistance of transformants from *E. coli* 116 [a]

Selection	Patterns
Sm	SmSu
Tc	TcCmTp
Cm	TcCmTp
Su	SmSu, SuTp
Tp	SuTp, Tp

[a] *Note:* *E. coli* 116 carries a conjugative plasmid of 70×10^6, determining TcCmTp resistance. It readily undergoes deletion giving a conjugative plasmid of 40×10^6 with Tp-resistance only. It also contains small, nonconjugative plasmids determining SmSu-resistance, 4×10^6, and SuTp-resistance, 10×10^6. All transconjugants and transformants are found to carry one or more of these plasmids

b) Objective

To transfer antibiotic resistance plasmids from *E. coli* 116 to *E. coli* K12 by transformation.

c) Procedure

Days 1–4

Prepare plasmid DNA from *E. coli* 116 by centrifugation of a cleared lysate in a CsCl-EtBr gradient.

Day 5

Prepare competent cells of *E. coli* C; keep on ice.
Add 0.1 ml of *E. coli* 116 plasmid DNA in a TNE buffer (10 µg/ml) to 0.5 ml of competent cells in a *glass* tube.
Agitate the tube in a 37°C water bath for 60 s and then return to ice for 1 h.

Add 2.7 ml of nutrient broth and incubate in 37°C water bath for 1 h.
Plate cells on selective media (0.3 ml per plate), i.e., nutrient agar + Sm, Tc or Cm, minimal salts (supplemented with appropriate nutrients + Su or Tp). Use 2 plates for each selection.
Do controls for (1) sterility of DNA; (2) growth of nontransformed competent cells on selective media; (3) level of competence of cells (transform with a standard plasmid DNA preparation e.g., pBR322).

Day 6

Examine plates.
Ditch-test transformant colonies and replate them on nutrient agar.

5. Experiment 4: Comparisons of Plasmids in Wild Strains and in the *E. coli* K12 Recipients

Having completed the conjugation and transformation experiments, it should now be possible to determine which of the plasmids seen in single colony lysates of *E. coli* 116 encodes the various antibiotic resistances. Carry out Eckhardt gel analysis of the transformants and transconjugants obtained (one example of each resistance pattern) in parallel with an *E. coli* 116 lysate and compare the plasmid patterns to relate individual plasmids to the specific phenotypic traits investigated.

6. Experiment 5: Classification of Unknown Plasmids According to Incompatibility (Inc) Group

a) Introduction

Pairs of closely related plasmids are unable to coexist stably in a growing cell line; they are incompatible. Incompatibility between two plasmids, e.g., one unknown and one known, is therefore an indication of their relatedness and is currently the basis of plasmid classification.

Incompatibility is tested by attempting to construct clones carrying the unknown plasmid with each one of a series of plasmids of known Inc group (Jacob et al. 1977).

Plasmids of *E. coli* 116 can be tested with known plasmids (e.g., see Jacob et al. 1977) carrying Ap or Km resistance, since the wild strain has neither of these characters. Plasmids with very many resistances cannot be tested against a range of standard plasmids. However, variants that lack one or more resistance genes can often be found as spontaneous mutants by replica plating or after transduction by a phage whose head is too small to package the whole plasmid (Shipley and Olsen 1975).

Many standard plasmids are available from the Plasmid Reference Center, Department of Medical Microbiology, Stanford University, California.

b) Objective

To determine the incompatibility group of one or more of the plasmids from *E. coli* 116.

c) Procedure

Day 1

Mix a broth culture of J62-2 R^+ (carrying a single plasmid from *E. coli* 116, obtained in Experiment 2) with each of a series of broth cultures of J53 strains carrying known plasmids of different Inc groups.

Incubate during the working day.

Plate 0.1 ml of each mixture on minimal salts agar plates supplemented to select for plasmid transfer.

If the plasmid from *E. coli* 116 has been shown to be tra^+, select for plasmid transfer in both directions, i.e., the 116 plasmid into the J53 R^+ strain and the standard plasmid into the J62-2 R^+ strain.

If the 116 plasmid is tra^-, select for transfer of the standard plasmid only.

Day 2

Purify 10 transconjugant clones from each selection by streaking on MacConkey + appropriate drug, or minimal salts + appropriate nutrient + Su, to get single colonies.

Day 3

Streak purified transconjugants across double ditch plates to show markers of each pair of plasmids (Fig. 2).

Note: If a heavy inoculum is used on Su or Hg ditches, even sensitive strains will grow. The inoculum is not critical on the other drugs. Include R^+ and R^- controls.

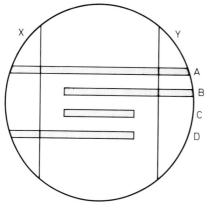

Fig. 2. Ditch plate sensitivity test. Ditches have been cut out and replaced with agar containing antibiotics X and Y, respectively. C sensitive control culture; A clone resistant to X and Y; B clone sensitive to X, resistant to Y; D clone resistant to X, sensitive to Y

Day 4

Read results. To which Inc group can plasmids be assigned?

Note: Where both plasmids appear to be present, they should be tested for stability and for separate replication.

d) Materials

Day 1

Overnight broth culture of:

	Inc gp	Res pattern
J53 (R1-16)	IncFII	Km
J53 (RP4)	IncP	ApKmTc
J53 (R447b)	IncN	ApKm
J53 (R7K)	IncW	Ap

Selection plates:

Minimal salts agar + glu + pro his trp + Km
Minimal salts agar + glu + pro his trp + Ap
Minimal salts agar + pro met + Tc, Cm, Tp, Sm or Su

(Number depending on results in Experiment 2).

Note: These plasmids represent only four Inc groups. About 20 groups are known.

Day 2

MacConkey plates incorporating Km, Ap, Tc, Cm, Sm or Tp + lysed blood, and minimal salts agar + pro, met, Su to purify transconjugant clones.

Day 3

Double ditch plates to test for resistance to the relevant drugs. Cut a slice from either side of prepared plates and refill the space with agar containing appropriate concentration of drug. The concentration can be higher than in selection plates:

Ap	100 µg/ml	Km	25	Su	1000
Sm	25	Nm	25	Cb	200
Tc	25	Tp	20		
Cm	25	Hg	50		

MacConkey or nutrient agar can be used for everything except Su and Tp. For these, utilize whatever medium is in general use for sensitivity tests (e.g., Mueller-Hinton) and fill ditches with same medium + drug.

7. References

Anderson ES, Threlfall EJ (1974) The characterisation of plasmids in the enterobacteria. J Hyg 72:471

Bradley DE, Taylor DE, Cohen DR (1980) Identification of surface mating systems among drug resistance plasmids in *Escherichia coli* K12. J Bacteriol 143:1466

Eckhardt T (1978) A rapid method of identification of plasmid DNA in bacteria. Plasmid 1:584

Jacob AE, Shapiro JA, Yamamoto L, Smith DI, Cohen SN, Berg D (1977) Plasmids studied in *Escherichia coli* and other enteric bacteria. In: Bukhari AI, Shapiro JA, Adjya SL (eds) DNA insertion elements, plasmids and episomes. Cold Spring Harbor Laboratory, USA

Jobanputra RS, Datta N (1974) Trimethoprim resistance factors in enterobacteria from clinical specimens. J Med Microbiol 7:169

Meyers JA, Sanchez D, Elwell LP, Falkow S (1976) Simple agarose gel electrophoretic method for the identification and characterization of plasmid deoxyribonucleic acid. J Bacteriol 127:1529

Shipley PL, Olsen RH (1975) Isolation of a nontransmissible resistance plasmid by transductional shortening of R factor RP1. J Bacteriol 123:30

Smith HW, Halls S (1968) The transmissible nature of the genetic factor in *Escherichia coli* that controls enterotoxin production. J Gen Microbriol 52:319

1.5 Mobilization of Host Chromosomal Genes and Vector Plasmids in *E. coli* by the P-Type Plasmid R68.45

G. RIESS and A. PÜHLER[1]

Contents

1. General Introduction 51
2. Experiment 1: Mobilization of Host Chromosomal Genes in *E. coli* 54
   a) Introduction 54
   b) Objectives 54
   c) Procedure 55
   d) Results 56
3. Experiment 2: Mobilization of *E. coli* Vector Plasmids 56
   a) Introduction 56
   b) Objectives 57
   c) Procedure 58
   d) Results 60
4. Materials 62
5. References 63

1. General Introduction

Host chromosomal gene transfer by conjugation was first discovered in *E. coli*. The fertility plasmid F was found to be responsible for this conjugation process. Using Hfr-strains, where the F plasmid is integrated into the chromosome, crosses could be carried out that helped to map the *E. coli* chromosome. Due to lack of suitable conjugation systems, many bacterial strains of scientific or commercial importance could not be analyzed. More recently, a plasmid was isolated by Haas and Holloway (1976), which is able to mobilize the bacterial chromosome in *Pseudomonas aeruginosa*. This plasmid called R68.45 is a derivative of R68, which confers resistance to ampicillin, kanamycin, and tetracycline. It is self-transmissable and belongs to the incompatibility group P. The importance of R68.45 lies in its ability to promote host chromosome transfer (Cma) in a very extensive range of Gram-negative bacteria following conjugation (Holloway 1979), unlike the F plasmid. It should be mentioned that on the basis of restriction analysis and heteroduplex studies, plasmids R68, RP1, RP4, and RK2 are identical (Burkhardt et al. 1979). All these plasmids show a very similar behavior.

---

[1] Lehrstuhl für Genetik, Fakultät für Biologie, Universität Bielefeld, D-4800 Bielefeld 1, Fed. Rep. of Germany

The following characteristics make R68.45 a useful tool in bacterial genetics:

1. Plasmid R68.45 belongs to the incompatibility group P with wide host range, including all Gram-negative bacteria tested so far (Datta and Hedges 1972).
2. Mobilization of the bacterial chromosome with R68.45 occurs from many origins (Haas and Holloway 1976).
3. Similar to the F plasmid, derivatives of R68.45 can be isolated carrying chromosomal segments. Such R68.45' (prime)-plasmids can be used to study gene expression in heterologous systems as was shown by Johnston et al. (1978).
4. R68.45 can also be used to mobilize *E. coli* vector plasmids at high frequency (Willetts et al. 1981, Rieß et al. 1980).

The molecular analysis of plasmid R68.45 revealed that a duplication of a DNA region already present on R68 is the only difference between the two plasmids. This region is located close to the kanamycin resistance gene. It starts 690 bp from the single *Hin*dIII restriction site in the kanamycin resistance gene and extends for a length of 2120 bp. The DNA region – duplicated in R68.45 – is characterized by a sequence of five restriction sites, namely *Sma*I, *Pst*I, *Pst*I, *Kpn*I, *Hpa*I (Rieß et al. 1980).

Figure 1 shows an overall restriction map of plasmid R68, while the *Hin*dIII/*Sal*I restriction fragments 3 of R68 and R68.45 are shown in detail in Fig. 2.

Recently, Willetts et al. (1981) showed that the DNA region of R68 characterized by the sequence of five restriction sites is an insertion element. This insertion element is called IS*21*. It has been postulated that the Cma property of R68.45 is due to the

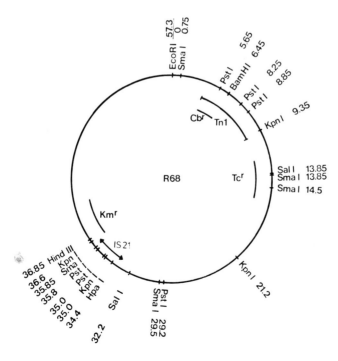

Fig. 1. Restriction map of plasmid R68 for the enzymes *Eco*RI, *Hin*dIII, *Bam*HI, *Pst*I, *Sma*I, *Kpn*I, *Hpa*I, and *Sal*I. The contour length of the plasmid was determined by electron microscopy to be 19.1 μm ≙ 57.3 kb (Burkardt et al. 1979). The distances of the individual restriction sites to the single *Eco*RI site are given in kb. The *arrowed line* indicates the location of insertion element IS*21* (Willetts et al. 1981)

tandem duplication of IS21. The location of IS21 on the restriction map of R68 is indicated in Fig. 1.

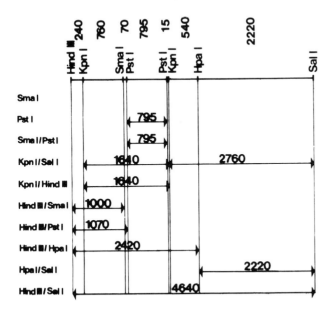

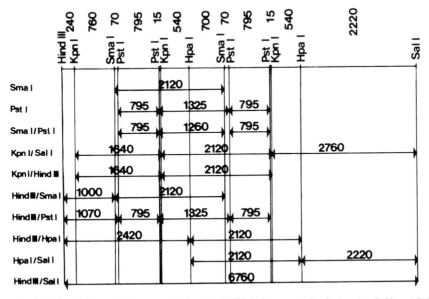

Fig. 2. A detailed restriction map of the HindIII/SalI fragments 3 of plasmids R68 and R68.45. Fragment lengths are given in bp. A sequence of five restriction sites, SmaI, PstI, PstI, KpnI, and HpaI already present in R68 *(upper part)* is duplicated in R68.45 *(lower part)*

2. Experiment 1: Mobilization of Host Chromosomal Genes in *E. coli*

a) Introduction

Plasmid R68.45 is well-known for its chromosome mobilizing ability (Cma) in various Gram-negative bacteria including *Pseudomonas aeruginosa* (Watson and Holloway 1978), *Escherichia coli* (Rieß et al. 1980), *Rhizobium meliloti* (Kondorosi et al. 1977), *Rhizobium leguminosarum* (Beringer and Hopwood 1976, Beringer et al. 1978), *Rhodopseudomonas sphaeroides* (Sistrom 1977), and *Agrobacterium tumefaciens* (Hamada et al. 1979).

In this experiment, the mobilization of the chromosomal *leu* gene in *E. coli* will be demonstrated. The crosses to be carried out are summarized in Table 1. First, two different donor strains are used, $recA^+$ (CSH51) and *rec*A (CSH52). Thus, the influence of the *E. coli* recombinant system on the mobilizing activity of R68.45 can be tested. In addition, donors are included that do not contain plasmid R68 or plasmid R68.45. In this way, it will become clear that the tandem duplication of IS*21* in R68.45 is responsible for mobilizing the chromosomal markers. A leucine auxotrophic strain is used as recipient. It is resistant to nalidixic acid, which can be used for counter selection. In this experiment, the frequency of plasmid transfer and of chromosome mobilization per donor cell shall be determined.

Table 1. Crosses, strains, and plasmids

Donor	Recipient			
1) CSH51	$recA^+$	CSH65	*leu, nal*A	
2) CSH51 (R68)	$recA^+$	CSH65	*leu, nal*A	
3) CSH51 (R68.45)	$recA^+$	CSH65	*leu, nal*A	
4) CSH52	*rec*A	CSH65	*leu, nal*A	
5) CSH52 (R68)	*rec*A	CSH65	*leu, nal*A	
6) CSH52 (R68.45)	*rec*A	CSH65	*leu, nal*A	

Strains	Genetic markers	Reference
CSH51	*ara*, Δ *(lac pro) str*A, *thi*, (φ80d *lac*⁺)	J. Miller (1972)
CSH52	Same as CSH51, but additonally *rec*A	J. Miller (1972)
CSH65	*leu, lac, nal*A, *str*A, *thi*	J. Miller (1972)

Plasmids		
R68	Ap, Km, Tc, Tra⁺	Haas and Holloway (1976)
R68.45	Ap, Km, Tc, Tra⁺, Cma⁺	Haas and Holloway (1976)

b) Objectives

1. To demonstrate mobilization of the chromosomal *leu* gene by plasmid R68.45 in *E. coli*.
2. To show that mobilization by plasmid R68.45 is *rec*A-independent.

c) Procedure

Bacterial crosses conducted to determine the plasmid transfer and the chromosomal mobilization frequency are summarized in Table 1. In addition, the genetic markers of strains and plasmids used in this experiment are listed.

Day 1

Start of E. coli Matings

Grow donor strains CSH51, CSH52, CSH51 (R68), CSH52 (R68), CSH51 (R68.45), and CSH52 (R68.45) and recipient strain CSH65 to log phase. Measure optical density of each strain and fill in Table 2.

Mix 0.5 ml of donor and recipient strain in a sterile Eppendorf tube in the following combinations:
1) CSH51   CSH65
2) CSH51 (R68)   CSH65
3) CSH51 (R68.45)   CSH65
4) CSH52   CSH65
5) CSH52 (R68)   CSH65
6) CSH52 (R68.45)   CSH65

Centrifuge mating mixtures in an Eppendorf centrifuge for 5 min and resuspend cells gently in 100 μl of saline buffer (PS).

Put 100 μl of the mating mixture on membrane filters (Sartorius SM 11325) attached to PA agar and incubate overnight at 37°C.

For control plate, 0.1 ml of each strain on selective agar A+Glu+Thi+Nx, respectively, PA+Tc+Nx.

Table 2. Calculation of plasmid transfer and chromosomal gene mobilization frequency

Donor	Recipient	Number of colonies on dilution step				
		A+Glu+Thi+Nx		PA+Tc+Nx		
		0	− 1	− 5	− 6	o.d. of donor
CSH51	CSH65					
CSH51 (R68)	CSH65					
CSH51 (R68.45)	CSH65					
CSH52	CSH65					
CSH52 (R68)	CSH65					
CSH52 (R68.45)	CSH65					

Day 3

Analysis of Exconjugants

Screen leu$^+$ colonies for presence of plasmids R68 or R68.45.

Pick exconjugant colonies onto PA plates and after incubation overnight, then replica plate onto PA+Ap, PA+Km, and PA+Tc.

Determine approximate frequency of plasmid transfer and of chromosomal mobilization per donor cell using the OD values obtained on day 1 (OD_{650} 0.1 $\triangleq$ 10^8 cells). Fill in the results in Table 2.

d) Results

Matings using the plasmid free donor strains CSH51 und CSH52 are control experiments in order to exclude the possibility of chromosome mobolization by a cryptic F plasmid. No exconjugants are expected in these matings. The donor strains CSH51 (R68) and CSH52 (R68) should transfer their plasmids up to 100% per donor cell. No exconjugants are expected when selection is made for mobilization of the chromosomal leu$^+$ marker. Using CSH51 (R68.45) and CSH52 (R68.45) as donors, plasmid transfer is expected to occur with the same frequency as with R68. In contrast to matings with R68, one expects those with R68.45 to show exconjugants on A+Glu+ Thi+Nx plates due to mobilization of the leu$^+$ marker. Since the Cma property of R68.45 is *rec*A independent, about the same mobilization frequencies should be found for the *rec*A$^+$ donor CSH51 (R68.45), as well as for the *rec*A donor CSH52 (R68.45). In our experience, chromosome mobilization with R68.45 occurs with a frequency of 10^{-5} to 10^{-6} per donor cell.

3. Experiment 2: Mobilization of *E. coli* Vector Plasmids

a) Introduction

Plasmid R68.45 can be used to mobilize *E. coli* vector plasmids such as pBR322, pBR325, or pACYC184. The mobilization process is presented schematically in Fig. 3. Vector plasmid mobilization occurs by transposition of IS*21* and cointegrate formation between R68.45 and the moblized plasmid in the donor cell. Willetts et al. (1981) showed that vector plasmids mobilized by R68.45 carry one copy of IS*21* after resolution of cointegrate in the recipient cell. In contrast to results found for cointegrate resolution with transposon Tn3 (McCormick et al. 1981), our recent experiments indicate that cointegrate resolution of IS*21* mediated cointegrates is *rec*A dependent (Rieß and Pühler, unpublished). The *E. coli* vector plasmid pACYC184 (Chang and Cohen 1978) is used to demonstrate cointegrate formation between R68.45 and the vector plasmid during the mobilization process. Donor strains carrying both the P-type and the vector plasmid will be constructed by transforming CSH51 (R68.45) and CSH51 (R68) with pACYC184 plasmid DNA. Mating the newly constructed donor strains with a recipient strain will lead to the transfer of R68.45 + pACYC184 cointegrates. Since the recipient strain S302.10 is *rec*A, no resolution of cointegrates is expected. The cointegrate state can be demonstrated either genetically, by mating with another recipient (CSH65:Nx^r) or physically, by an SDS-lysis procedure of transconjugant cells.

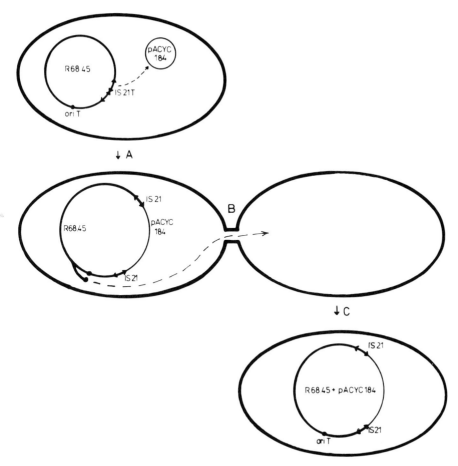

Fig. 3. Mobilization process of pACYC184 by R68.45. The mobilization of pACYC184 occurs via IS21 transposition to the vector plasmid and cointegrate formation in the donor strain CSH51 (R68.45) (pACYC184) (step A). Starting from the R68.45 specific origin of transfer replication site (oriT), the R6845 + pACYC184 cointegrate is transferred to the recipient strain S302.10 (step B) where the cointegrate is maintained since the recipient strain S302.10 is recA. Our results indicate that in the cointegrate state the vector plasmid DNA is bordered by one copy of IS21 at each junction with R68.45 (Rieß and Pühler, unpublished)

b) Objectives

1) To construct donor strains carrying the plasmids:
   pACYC184 and R68
   pACYC184 and R68.45.
2) To mobilize pACYC184 out of *E. coli* by R68.45.
3) To show that mobilization occurs via cointegrate formation.

c) Procedure

Bacterial crosses conducted to determine plasmid transfer and vector plasmid mobilization are summarized in Table 3. In addition, the genetic markers of strains and plasmids used in this experiment are listed.

Table 3. Crosses, strains, and plasmids

Crosses			
Donor		Recipient	
7) CSH51 (pACYC184)		S302.10	$recA$ Rf^r
8) CSH51 (R68.45)		S302.10	$recA$ Rf^r
9) CSH51 (R68) (pACYC184)		S302.10	$recA$ Rf^r
10) CSH51 (R68.45) (pACYC184)		S302.10	$recA$ Rf^r

Strains	Genetic markers	Reference
CSH51	ara, Δ (lac, pro) $strA$, thi, $(\phi 80d\ lac^+)$	J. Miller (1972)
CSH65	leu, lac, $nalA$, $strA$, thi	J. Miller (1972)
S302.10	Same as CSH51, but additionally $recA$ and Rf^r	R. Simon, Bielefeld

Plasmids		
R68	Ap^r, Km^r, Tc^r, Tra^+	Haas and Holloway (1976)
R68.45	Ap^r, Km^r, Tc^r, Tra^+, Cma^+	Haas and Holloway (1976)
pACYC184	Tc^r, Cm^r	Chang and Cohen (1978)

Day 1

Construction of Donor Strains

This is done by transforming the strains CSH51, CSH51 (R68), and CSH51 (R68.45) with pACYC184 plasmid DNA. The DNA of plasmid pACYC184 is prepared according to procedures described in this volume.

For competent cells, grow strains CSH51, CSH51 (R68), and CSH51 (R.6845) in 10 ml of RGMC medium to an optical density $OD_{650} = 0.6$.

Cool stocks in ice for 10 min.

Centrifuge 10 min 10^4 rpm at 0°C and resuspend in 5 ml 50 mM $CaCl_2$ 0°C.

Keep in ice for 20 min.

Centrifuge 10 min 6×10^3 rpm at 0°C and resuspend cells gently (!) in 1 ml 50 mM $CaCl_2$ 0°C. Hold competent cells in ice.

Prepare DNA solution by diluting 30 μl of pACYC184 DNA (100 ng/ml) to a final volume of 300 μl with TCM buffer.

Gently mix 200 μl of competent cells of each strain with 100 μl of DNA solution.

Keep in ice for 30 min.

Carry out heat shock for 2 min in a 42°C water bath by immediate plunging of cells from ice to 42°C.

Add 0.5 ml of RGMC medium and incubate at 37°C for 30 min to allow expression of transformants.

Plate 0.1 ml of each sample onto PA+Cm plates.

Spin down cells for 90 s in an Eppendorf centrifuge, pour off supernatant, resuspend cells in the rest and plate also onto PA+Cm plates.

For control, plate 0.1 ml of the nontransformed strains onto PA+Cm plates.

Incubate plates overnight at 37°C.

Day 2

Test of Transformants

Pick 50 colonies of each sample onto PA+Cm plates with sterile toothpicks.
Incubate overnight at 37°C.

Day 3

Test of Transformants continued

Screen the toothpicked colonies for presence of R68 and R68.45 encoded antibiotic resistances by replica plating onto PA+Ap, PA+Km, and PA+Tc plates.

Day 4

E. coli Matings

Carry out *E. coli* matings as described above for mobilization of chromosomal genes in the following combinations (see also Table 3):

Donor	Recipient
7) CSH51 (pACYC184)	S302.10
8) CSH51 (R68.45)	S302.10
9) CSH51 (R68) (pACYC184)	S302.10
10) CSH51 (R68.45) (pACYC184)	S302.10

For control, plate 0.1 ml of each strain onto selective agar PA+Cm+Rf.

Day 5

E. coli Matings Continued

Wash cells off the filters as described above.

Plate dilution steps $-1, -2, -3$ onto selective agar PA+Cm+Rf.

For determination of plasmid transfer frequency plate dilution steps -4 and -5 of matings 8, 9, and 10 onto selective agar PA+Tc+Rf.

Incubate plates overnight at 37°C.

Day 6

Screening of Transconjugants

Use sterile toothpicks to pick 50 Cm^r Rf^r colonies onto a PA+Cm plate.
After incubation at 37°C, replica plate onto PA+Ap, PA+Km, and PA+Tc.

Day 7

Determination of Plasmid Content of Transconjugants

Choose 20 clones for SDS lysis.
Use the CSH51 (R68.45) (pACYC184) donor strain as a standard.
Resuspend a loopful of bacteria in 100 µl 20% sucrose in TE in Eppendorf tubes.
Add 100 µl of 10 mg/ml lysozyme in H_2O, 5 min.
Add 100 µl 250 mM EDTA pH 8.5, 5 min.
Add 100 µl 2% SDS in H_2O, invert Eppendorf tubes until the lysate becomes clear.
Add 100 µl of TE-saturated phenol and shear lysate with a syringe by sucking up and down rapidly (5 ×).
Spin in an Eppendorf centrifuge for 3 min.
Take 50 µl of the supernatant, mix with 10 µl of 50% sucrose 0.1% bromphenol-blue and load sample on a vertical agarose gel (1% agarose in Tris-acetate buffer).
Separate DNA by gel electrophoresis (100 V, 2.5 h).
Stain gel with ethidium bromide (1 µg/ml) for 30 min.
Photograph gel under UV-light.

Transfer of an R68.45+pACYC184 Cointegrate to Another Recipient Strain

Choose one Ap^r, Km^r, Tc^r, Cm^r, Rf^r clone for mating with CSH65.
Carry out mating as described above.
For control plate, place donor strain S302.10 (R68.45+pACYC184) and recipient strain CSH65, separately, onto selective agar PA+Tc+Nx.

Day 8

Cointegrate Transfer Continued

Resuspend filter as described above and plate dilution steps -3, -4, -5 onto selective agar PA+Tc+Nx.

Day 9

Screening of Transconjugants

Use sterile toothpicks to pick 100 transconjugants parallel onto PA+Ap, PA+Km, and PA+Cm.
Incubate overnight at 37°C.

d) Results

Calculate plasmid transfer and vector plasmid mobilization frequencies using Table 4. Mating 6 only demonstrates that the vector plasmid pACYC184 is not self-transmissible. Mating 7 is conducted to determine plasmid transfer frequency of R68.45 alone. While in mating 9 no Cm^r transconjugants are expected, the plasmid transfer of R68 should occur up to 100% per donor cell. The same plasmid transfer frequency is expected for R68.45 in mating 10. Mobilization of pACYC184 should appear in

Table 4. Calculation of plasmid transfer and vector plasmid mobilization frequency

| | | Number of colonies on dilution step | | | | | |
| | | PA+Cm+Rf | | | PA+Tc+Rf | | |
Donor	Recipient	-1	-2	-3	-4	-5	o.d. of donor
I CSH51 (pACYC184)	S302.10						
II CSH51 (R68.45)	S302.10						
III CSH51 (R68+pACY184)	S302.10						
IV CSH51 (R68.45+pACYC184)	S302.10						

the range of 10^{-3} to 10^{-4} per donor cell. Since the recipient strain S302.10 is *rec*A, all Cm^r transconjugants should represent R68.45+pACYC184 cointegrates. Firstly, this can be confirmed by the SDS lysis procedure. The plasmid band corresponding to R68.45 in the donor strain should be enlarged and the plasmid band corresponding to pACYC184 should disappear. An agarose gel demonstrating this result is shown in Fig. 4. Secondly, all transconjugants of the cointegrate transfer experiment

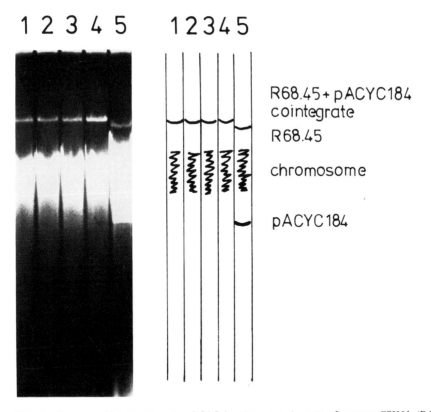

Fig. 4. Agarose gel electrophoresis of SDS lysed transconjugants of a cross CSH51 (R68.45) (pACYC184) × S302.10. The *lanes 1–4* show four independently isolated transconjugants, while the plasmid content of the donor strain CSH51 (R68.45) (pACYC184) is demonstrated in *lane 5*. Gel electrophoresis was carried out for 2.5 h at 100 V in Tris-acetate buffer

are expected to show Ap^r, Km^r, Tc^r, and Cm^r corresponding to the cointegrate state of the two plasmids. In a normal plasmid mobilization experiment (compare matings 8 and 10), about 1 of 1000 transconjugants is expected to show Cm resistance.

4. Materials

Media

PA:	Penassay Broth (Difco)	17.5 g/l
RGMC:	Rich medium	
	Tryptone	10.0 g/l
	Yeast Extract	1.0 g/l
	NaCl	8.0 g/l
	Glucose	1.0 g/l
	$MgCl_2$	1.0 g/l
	$CaCl_2$	0.3 g/l
A:	Minimal medium (J. Miller)	
	K_2HPO_4	10.0 g/l
	KH_2PO_4	4.5 g/l
	$(NH_4)_2SO_4$	1.0 g/l
	Na-citrate	0.5 g/l
	$MgSO_4$ 20%	1.0 ml/l
	Glucose 20%	10.0 ml/l

Amino acids added to a concentration of 40 µg/ml except thiamine (0.5 µg/ml)

Antibiotics added:

Tetracycline Tc	5 µg/ml
Streptomycin Sm	100 µg/ml
Rifampicin Rf	50 µg/ml (stock solution must be dissolved in methanol)
Ampicillin Ap	100 µg/ml
Kanamycin Km	50 µg/ml
Nalidixic acid Nx	100 µg/ml (stock solution must be dissolved in 0.1 N NaOH)

For solid media 13.5 g/l agar (Difco Bacto Agar) was added

Buffers

TE	10 mM Tris	
	1 mM EDTA	pH 7.5
TA	40 mM Tris	
	10 mM Na-acetate	
	1 mM EDTA	pH 7.8
TCM	10 mM Tris	
	10 mM $CaCl_2$	
	10 mM $MgCl_2$	pH 7.5

PS   Saline buffer
     7 g/l Na$_2$HPO$_4$
     3 g/l KH$_2$PO$_4$
     5 g/l NaCl   pH 7.0

Chemicals	Source
Agarose	Seakem, MCI Biochemicals, Rockland, Maine, U.S.A.
Ethidium bromide	Serva, Heidelberg

5. References

Beringer JE, Hopwood DA (1976) Chromosomal recombination and mapping in *Rhizobium leguminosarum*. Nature 264:291–293

Beringer JE, Hoggan SA, Johnston AWB (1978) Linkage mapping in *Rhizobium leguminosarum* by means of R-plasmid-mediated recombination. J Gen Microbiol 104:201–207

Burkhardt HJ, Rieß G, Pühler A (1979) Relationship of group P1 plasmids revealed by heteroduplex experiments: RP1, RP4, R68, and RK2 are identical. J Gen Microbiol 114:341–348

Chang ACY, Cohen SN (1978) Construction and characterization of amplifiable multicopy DNA cloning vehicles derived from the P15 cryptic miniplasmid. J Bacteriol 134:1141–1156

Datta N, Hedges RW (1972) Host ranges of R factors. J Gen Microbiol 70:453–460

Haas D, Holloway BW (1976) R factor variants with enhanced sex factor activity. Mol Gen Genet 144:243–251

Hamada SE, Luckey JP, Farrand SK (1979) R-plasmid-mediated chromsomal gene transfer in *Agrobacterium tumefaciens*. J Bacteriol 139:280–286

Holloway BW (1979) Plasmids that mobilize bacterial chromosome. Plasmid 2:1–19

Johnston AWB, Setchell SM, Beringer JE (1978) Interspecific crosses between *Rhizobium leguminosarum* and *R. meliloti*. Formation of haploid recombinants and of R-primes. J Gen Microbiol 104:209–218

Kondorosi A, Kiss G, Forrai T, Vincze E, Banfalvi Z (1977) Circular linkage map of *Rhizobium meliloti* chromosome. Nature 268:525–527

McCormick M, Wishart W, Ohtsubo H, Heffron F, Ohtsubo E (1981) Plasmid cointegrates and their resolution mediated by transposon Tn3 mutants. Gene 15:103–118

Miller JH (1972) Experiments in molecular genetics. Cold Spring Harbor Lab, Cold Spring Harbor, New York

Rieß G, Holloway BW, Pühler A (1980) R68.45, a plasmid with chromosome mobilizing ability (Cma) carries a tandem duplication. Genet Res Camb 36:99–109

Sistrom WR (1977) Transfer of chromosomal genes mediated by plasmid R68.45 in *Rhodopseudomonas sphaeroides*. J Bacteriol 131:526–532

Watson JM, Holloway BW (1978) Chromosome mapping in *Pseudomonas aeruginosa* strain PAT. J Bacteriol 133:1113–1125

Willetts NS, Crowther C, Holloway BW (1981) The insertion sequence IS21 of R68.45 and the molecular basis for mobilization of the bacterial chromosome. Plasmid 6:30–52

1.6 Characterization of Plasmid Relaxation Complexes

A. NORDHEIM[1] and K.N. TIMMIS[2]

Contents

1. General Introduction 64
2. Experiment 1: Demonstration of Plasmid DNA Relaxation Complexes 65
   a) Introduction 65
   b) Objective 65
   c) Procedure 66
   d) Results 66
3. Experiment 2: Rapid Method for Isolation of Relaxation Complexes 67
   a) Introduction 67
   b) Objective 67
   c) Procedure 68
   d) Results 68
4. Experiment 3: Localization of Relaxation Nick Sites on Plasmid Genomes 69
   a) Introduction 69
   b) Objective 70
   c) Procedure 70
   d) Results 71
5. Materials 72
6. References 73

1. General Introduction

Specific interaction of proteins with nucleic acids constitutes the mechanistic basis for almost all processes of genome function, including replication, recombination, and gene expression. Analysis in vitro of functional aspects of such interactions is greatly facilitated by the isolation from cells of intact, native nucleic acid-protein complexes. The experiments in this section describe the isolation and characterization of one type of specific DNA-protein complex, the so-called plasmid DNA relaxation complex.

Clewell and Helinski (1969) first described the isolation of ColE1 as a supercoiled plasmid DNA-protein complex after gentle lysis of plasmid-carrying bacteria with

---

1 Department of Biology, Massachusetts Institute of Technology, Cambridge, MA, USA
2 Department of Medical Biochemistry, University of Geneva, CH-1211 Geneva, Switzerland

nonionic detergents. This complex was termed relaxation complex because the torsionally stressed, supercoiled DNA molecule could be converted into one having a relaxed, open-circular configuration by treatment with various protein denaturing agents (SDS, proteases). Such treatments were demonstrated to activate the protein component of the complex, which introduced a site- and strand-specific, single stranded cut (nick) into the double stranded supercoiled plasmid genome. The formation of a covalent linkage between part of the protein component and the 5'-phosphoryl end of the nicked strand occurred simultaneously with, presumably as a consequence of, the nicking event (Blair and Helinski 1975, Lovett and Helinski 1975, Guiney and Helinski 1975).

A variety of plasmids, such as ColE2, R6K, RK2, pSC101, and RSF1010 have now been isolated as relaxation complexes and, in several instances, the relaxation nick sites were shown to be located at or close to origins of vegetative replication (Lovett et al. 1974, Nordheim et al. 1980). This finding led to the proposal that relaxation complexes might be involved in the initiation of plasmid replication (Lovett et al. 1974). Data obtained subsequently, however, demonstrated that relaxation complexes play an essential role in plasmid conjugal transfer (Warren et al. 1978). More recently, it was suggested that the proximity of relaxation nick sites and vegetative origins of replication may reflect a functional interaction of these determinants in the DNA synthesis associated with plasmid mobilization (Nordheim et al. 1980, Nordheim 1979, Covarrubias et al. 1981).

In the following experiments, the rapid detection, purification, and characterization of plasmid DNA-protein relaxation complexes are outlined.

2. Experiment 1: Demonstration of Plasmid DNA Relaxation Complexes

a) Introduction

Plasmid DNA-protein relaxation complexes can be isolated from log phase bacterial cultures by gentle lysis of cells with a nonionic detergent (e.g., Triton X-100), followed by removal of cell debris and chromosomal DNA by high-speed centrifugation. Plasmid DNA in the "cleared lysate" is purified by preparative centrifugation through a 15–50% neutral sucrose density gradient. The purified complexed supercoiled plasmid molecules can then be converted by treatment with SDS from a rapidly sedimenting supercoiled (CCC) species to a more slowly sedimenting open circular (OC) form. This relaxation is monitored by sedimentation of treated and untreated complexes through 20–30% analytical neutral sucrose density gradients.

b) Objective

Demonstration of pSC101 plasmid relaxation complexes.

c) Procedure

Isolation of Relaxation Complexes

Inoculate 40 ml TBM containing tetracycline (10 µg/ml) in a 250 ml Erlenmeyer flask with 0.4 ml of a fresh overnight TBM culture of *E. coli* strain C600 (pSC101).

Incubate the culture at 37°C with shaking until it reaches an absorbance of $OD_{590} = 0.2$.

Add 0.4 mCi ^{3}H-thymidine and 2.0 ml deoxyadenosine (5 mg/ml), and incubate until the culture reaches late log phase ($OD_{590} = 1.2$).

Harvest cells by centrifugation at 4°C for 5 min at 8000 rpm, wash twice with ice cold TE buffer and resuspend in 1 ml ice cold 25% sucrose in 0.05 M Tris-HCl, pH 8.0.

Add 0.2 ml lysozyme (5 mg/ml in 0.25 M Tris-HCl, pH 8.0), incubate 5 min on ice, add 0.4 ml 0.25 M EDTA, pH 8.0, incubate 5 min on ice, add 1.6 ml 0.2% Triton X-100 lysis mixture, and incubate further until solution becomes viscous (15–20 min).

Centrifuge at 2°C for 10 min at 15,000 rpm and carefully remove and store on ice the supernatant fluid (cleared lysate).

Layer cleared lysate on a 35 ml 15–50% linear neutral sucrose gradient, centrifuge in the SW27 rotor at 4°C for 15–18 h at 26,000 rpm.

Fractionate the gradient into 1 ml fractions, pipette 30 µl aliquots onto Whatman 3 MM filters, dry filters, and wash 3 × with 5% TCA at 4°C and 2 × with 95% ethanol.

Dry filters under an infrared lamp, place in scintillation vials, add toluene-based scintillation cocktail, count radioactivity, and pool fractions containing the fast sedimenting material, which represents complexed supercoiled pSC101 DNA (^{3}H-pS101-rc), keep on ice.

Induction of Relaxation

To two Eppendorf tubes in ice labeled A and B, add 150 µl ^{3}H-pSC101-rc, X µl ^{14}C-ColE1 or ^{32}P-ColE1 (representing ∼ 10^4 cpm) and 135-X µl of H_2O.

Keep A (no treatment) on ice.

To B, add 15 µl of 10% SDS in TE, mix gently, and incubate 15 min at 37°C.

Layer samples A and B separately on 4.8 ml 20–30% neutral sucrose gradients, centrifuge in SW 50.1 rotor at 15°C for 3 h at 45,000 rpm.

Drop gradients directly onto Whatman 3 MM filters (5 drops per filter, 30–40 fractions), dry filters, and wash and count as above.

d) Results

Plot the ^{3}H and ^{14}C or ^{32}P cpm determined against fraction number. Noncomplexed CCC ColE1 DNA has a sedimentation coefficient of 23S, whereas nicked OC ColE1 has a coefficient of 17S. Calculate S-values of the major ^{3}H peaks in the two gradients. Sedimentation profiles of the two samples are shown in Fig. 1. It can be seen that SDS treatment converts the 32S ^{3}H-pSC101 relaxation complex (A) into an 18 S form (B); the shift in sedimentation velocity is due to conversion of supercoiled plasmid-protein complex to a nicked, open circular DNA form.

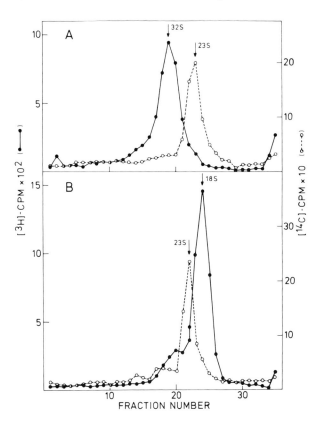

Fig. 1. Sucrose gradient (20–30%) sedimentation analysis of purified ^{3}H-pSC101 relaxation complex DNA *(filled circles)*, untreated *(A)* or treated with SDS *(B)*. Open circles represent protein-free, supercoiled ^{14}C-ColE1 marker DNA

3. Experiment 2: Rapid Method for Isolation of Relaxation Complexes

a) Introduction

Plasmid DNA-protein relaxation complexes can be rapidly purified by passage of cleared lysates through a column of Sepharose 2B, which excludes DNA and DNA-protein complexes and fractionates RNA and proteins. Elution of the former can be monitored by spotting aliquots of column fractions onto glass fiber filters, which are subsequently dried, stained with ethidium bromide (EtBr), and examined for EtBr-binding material (fluorescence) by illumination with UV. Plasmid DNA-protein complexes obtained by this method are relaxable, as will be shown by agarose gel electrophoresis, which separates supercoiled and open circular plasmid DNAs.

b) Objective

Rapid purification of pSC101 plasmid relaxation complex and demonstration of its relaxability by agarose gel electrophoresis.

c) Procedure

Isolation of Relaxation Complex

Prepare a cleared lysate from a 250 ml culture of unlabeled bacteria (omit ^{3}H-thymidine and 2-deoxyadenosine) grown to $OD_{590} = 1.2$, according to the instructions given in Experiment 1.

Pass the cleared lysate through a Sepharose 2B column (1.8 X 20 cm) equilibrated with TE and collect 1 ml fractions.

Spot 50 µl aliquots of the fractions onto GF-filters; dry filters and stain in ErBr solution (1 mg/ml) for 20 min. Illuminate filters with UV light to identify those containing nucleic acids and pool DNA-containing fractions (first fluorescing peak; referred to as pSC101-rc).

Induction of Relaxation and Analysis by Agarose Gel Electrophoresis

Treat 100 µl samples of pSC101-rc DNA with various agents known to induce relaxation [SDS (experiment 1), proteinase K (experiment 3), EtBr, etc.].

Add 1/5 vol Ficoll dye solution and apply to slots of a Tris-acetate buffered 0.65% agarose gel.

Apply to adjacent slots of the gel untreated pSC101-rc, noncomplexed CCC and OC pSC101 DNA.

Run the gel at 25 mA/50 V for 12–20 h, and strain in EtBr solution (1 µg/ml) for 15 min.

Photograph the gel.

d) Results

Figure 2 shows a typical electrophoretic separation of untreated and relaxed pSC101-rc DNA.

The purification of plasmid DNA-protein complexes by gel filtration and the electrophoretic analysis of treated and untreated complexes permit the rapid analysis of large numbers of plasmids for their existence as active relaxation complexes; no radioactive labeling of plasmid DNA is required. By adjustment of the dimensions of the Sepharose 2B column, larger volumes of cleared lysates can be processed, thus allowing preparative scale isolation of the complexes.

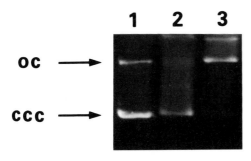

Fig. 2. Agarose gel electrophoresis of unrelaxed and SDS-relaxed pSC101 relaxation complex DNA. *Lane 1:* protein-free supercoiled (CCC) and open circular (OC) pSC101 DNA markers; *lane 2:* pSC101-rc; *lane 3:* SDS-treated pSC101-rc

4. Experiment 3: Localization of Relaxation Nick Sites on Plasmid Genomes

a) Introduction

One crucial piece of information required for the characterization of the functional roles of replicon-, strand-, and site-specific nucleases of the type found in relaxation complexes is the precise location of the relaxation nick site and its physical relationship to other DNA signals in its neighborhood. For this purpose, a nick labeling procedure has been developed which permits localization of the site on a plasmid restriction endonuclease cleavage map, generally to a DNA segment < 100 bp in length (Fig. 3).

In this procedure, complexed plasmid molecules are relaxed by treatment with EtBr or SDS and the protein that becomes covalently bound to the 5′-end of the nicked strand is removed by treatment with proteinase K. The nick is then radioactively labeled by extension of the free 3′-OH end of the nicked strand in a limited DNA synthesis/strand displacement reaction by means of DNA polymerase I (Klenow fragment). The plasmid DNA is then subjected to endonuclease cleavage analysis.

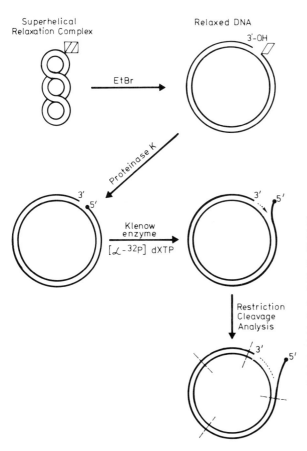

Fig. 3. The procedure used to label relaxation nick sites. The *box* on the supercoiled plasmid molecule represents relaxation proteins, one of which becomes covalently attached to the 5′-end of the nicked strand during the relaxation event. This and other proteins associated with the plasmid molecule after relaxation are removed by treatment with proteinase K. The free 3′-OH end of the nicked strand is then extended with radioactive nucleotides by DNA polymerase I (Nordheim et al. 1980) as described in the text

b) Objective

Identification of restriction endonuclease-generated DNA fragments that carry the relaxation nick of plasmid pSC101.

c) Procedure

Isolation of Relaxed Complex DNA

Inoculate 250 ml TBM containing tetracycline (10 µg/ml) with 2.5 ml of a fresh overnight culture of *E. coli* C600 (pSC101); incubate at 37°C with shaking until an absorbance of OD_{590} = 1.2 is reached.

Collect cells by centrifugation, wash cell pellet twice with cold TE buffer, and prepare a cleared lysate as described in Experiment 1.

Add 0.35 ml SDS (10%) and incubate at 37°C for 15 min.

Separate plasmid DNA from other cleared lysate components by gel filtration on Sepharose 2B, as described in Experiment 2.

Pool first fluorescent peak fractions, add 1/10 vol 3 M sodium acetate and 2.5 vol of absolute ethanol.

Store overnight at −20°C, collect precipitate by centrifugation, and resuspend in 250 µl TE.

Proteinase K Treatment

Add 3 µl proteinase K (10 mg/ml), incubate at 37°C for 15 min, phenol extract sample and ethanol precipitate DNA at −70°C for 5 min.

Collect precipitate and resuspend in 200 µl TE (referred to as pSC101-oc).

Nick Labeling Reaction

Dry down in an Eppendorf tube 20 µl (∼ 20 µCi) of α-^{32}P-dATP.

Add 50 µl pSC101-oc, 1.2 µl dCTP (0.5 mM), 1.2 µl dGTP (0.5 mM), 1.2 µl dTTP (0.5 mM), 6 µl DNA polymerase buffer (10 ×) and 3 units DNA polymerase I (Klenow fragment).

Incubate at 4°C for 30 min and then at 75°C for 10 min.

Pass sample through a 1 ml column of Sephadex G100 in a Pasteur pipette and collect the radioactive excluded fractions (referred to as ^{32}P-pSC101-oc).

Analysis of Nick Labeled DNA

Cleave nick labeled relaxed pSC101 DNA with *Hinc*II, *Hinc*II+*Bgl*I, and *Hinc*II+*Bgl*I + *Hae*II endonuclease combinations as follows: to 20 µl ^{32}P-pSC101-oc add 3 µl restriction buffer (10 ×) and 4 units restriction enzyme(s); incubate at 37°C for 30 min.

Add 10 µl of Ficoll dye solution.

Electrophorese samples at 50 mA/80 V for 4 h through a Tris-borate buffered 1.2% agarose gel.

Stain gel in EtBr solution (0.5 µg/ml) for 10 min and photograph.

Autoradiograph the gel overnight at room temperature using Kodak X-ray film. Align precisely the autoradiogram with the photograph of the stained gel and identify the labeled (nick-containing) fragments.

d) Results

Figure 4 shows a gel electrophoretic separation of ^{32}P-pSC101-oc DNA after cleavage with restriction endonucleases HincII (1), HincII/BglI (2), and HincII/BglI/HaeII (3), as visualized by EtBr staining (lanes labeled a) and autoradiography (lanes labeled b). The fact that in a given restriction digest, the radioactivity is confined to a single band demonstrates the high specificity of the nick labeling reaction. The relaxation nick of pSC101 is located in the second largest HincII fragment (lanes 1a, 1b), which is reduced in size by additional BglI cleavage (lanes 2a, 2b); this HincII/BglI fragment is not cleaved further by HaeII (lanes 3a, 3b; see Fig. 5). Such a series of multiple endonuclease cleavages permits the localization of relaxation sites on a plasmid genome and can provide useful information on the locations of additional restriction endonuclease cleavage sites in the vicinity of the nick. This greatly facilitates the development of strategies for DNA sequencing of plasmid relaxation regions. Note that precise localization and polynucleotide sequence determination of relaxation nick sites may be accomplished by a modification of the above experiment in which the nick labeling reaction is carried out on relaxed DNA and exonuclease III "gapped" relaxed DNA in the presence of dideoxynucleoside triphosphates (Nordheim 1979).

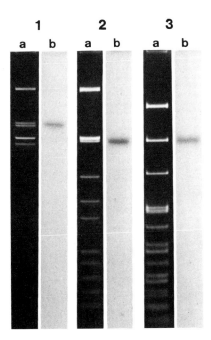

Fig. 4. Localization of the pSC101 relaxation nick site. Nick labeled ^{32}P-pSC101-oc DNA was cleaved with restriction endonucleases and the fragments thereby generated were separated by electrophoresis through a 1.2% agarose gel. The gel was stained with EtBr, photographed under UV illumination, and autoradiographed. The restriction endonucleases used were: HincII *(column 1)*, HincII/BglI *(column 2)*, and HincII/BglI/HaeII *(column 3)*. Within each column, the EtBr staining pattern *(lane a)* is aligned with the corresponding autoradiogram *(lane b)*

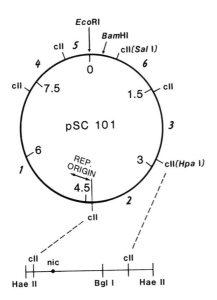

Fig. 5. Restriction endonuclease cleavage map of pSC101. cII, *Hin*cII cleavage site; nic, relaxation nick site

5. Materials

TBM (Tris buffered minimal medium, Clewell and Helinski 1969), contains per liter:
2.0 g	NH_4Cl
5.0 g	NaCl
0.37 g	KCl
0.01 g	$MgCl_2 \cdot 6H_2O$
0.02 g	Na_2SO_4
5 g	vitamin-free Casamino acids (Difco)
250 mg	thiamine
10 ml	20% solution of glycerol
100 ml	1 M Tris-HCl, pH 7.3

TE:
10 mM	Tris
1 mM	EDTA

Tris-borate buffer:
90 mM	Tris-HCl, pH 8.0
2.5 mM	EDTA
90 mM	boric acid

Tris-acetate buffer:
40 mM	Tris-HCl, pH 8.0
2.0 mM	EDTA
20 mM	CH_3COONa

DNA polymerase buffer (10 ×):
66 mM   Tris-HCl, pH 7.5
66 mM $MgCl_2$
66 mM   dithiothreitol

Ficoll dye solution:
20% Ficoll + 0.025% bromophenol blue in H_2O

6. References

Blair DG, Helinski DR (1975) Relaxation complexes of plasmid DNA and protein. I. Strand-specific association of protein and DNA in the relaxed complexes of plasmids ColE1 and ColE2. J Biol Chem 250:8785–8789

Clewell DB, Helinski DR (1969) Supercoiled circular DNA-protein complex in *E. coli:* Purification and induced conversion to an open circular DNA form. Proc Natl Acad Sci USA 62: 1159–1166

Covarrubias L, Cervantes L, Covarrubias A, Soberon Y, Vichido I, Blanco A, Kupersztoch-Portnoy YM, Bolivar F (1981) Construction and characterization of new cloning vehicles. V. Mobilization and coding properties of pBR322 and several deletion derivatives including pBR327 and pBR328. Gene 13:25–35

Guiney DG, Helinski DR (1975) Association of protein with the 5′ terminus of the broken DNA strand in the relaxed complex of plasmid ColE1. J Biol Chem 250:8796–8803

Lovett MA, Helinski DR (1975) Relaxation complexes of plasmid DNA and protein. II. Characterization of the protein associated with the unrelaxed and relaxed complexes of plasmid ColE1. J Biol Chem 250:8790–8795

Lovett MA, Guiney DG, Helinski DR (1974) Relaxation complexes of plasmids ColE1 and ColE2: Unique site of the nick in the open circular DNA of the relaxed complexes. Proc Natl Acad Sci USA 71:3854–3857

Nordheim A (1979) Charakterisierung von bakteriellen Plasmid DNA-Protein Relaxationskomplexen. Thesis, Free University Berlin

Nordheim A, Hashimoto-Gotoh T, Timmis KN (1980) Location of two relaxation nick sites in R6K and single sites in pSC101 and RSF1010 close to origins of vegetative replication: implication for conjugal transfer of plasmid deoxyribonucleic acid. J Bacteriol 144:923–932

Warren GJ, Twigg AJ, Sherratt DJ (1978) ColE1 plasmid mobility and relaxation complex. Nature 274:259–261

1.7 Use of the BAL31 Exonuclease to Map Restriction Endonuclease Cleavage Sites in Circular Genomes

J. FREY[1] and K.N. TIMMIS[2]

Contents

1. General Introduction 74
2. Experiment: Localization of *Hae*II Cleavage Sites on Plasmid pHSG415 75
   a) Introduction 75
   b) Procedure 76
   c) Results and Discussion 77
3. Materials 78
4. References 79

1. General Introduction

Restriction endonuclease cleavage sites can serve as covenient precise reference points on physical maps of DNA molecules. Moreover, a knowledge of the locations of sites of specific enzymes in relation to genetic determinants of interest greatly facilitates analysis and manipulation of the latter by gene cloning and DNA sequencing procedures. For these reasons, one of the first steps in the characterization of a specific DNA segment is the establishment of a restriction endonuclease cleavage map.

    A number of methods for the mapping of cleavage sites have been developed, all of which rely on the precise measurement of the lengths of DNA fragments, usually by gel electrophoresis, that are generated by cleavage with one or more restriction endonucleases. These include (1) complete digestion by enzyme A, of isolated DNA fragments generated by partial digestion with enzyme A (Danna et al. 1973, Griffin et al. 1974); (2) complete digestion by enzyme B, of DNA fragments generated by complete digestion with enzyme A (Lee and Sinsheimer 1974, Vereijken et al. 1975); (3) comparison of the digestion products obtained by cleavage of a genome, with those obtained by cleavage of a series of mutant derivatives of the genome that contain insertions and/or deletions at different locations (Allet et al. 1973, Smith et al. 1976); (4) electron microscopic localization of restriction endonuclease generated DNA fragments on a genome, by the formation of homoduplexes between the complete genome and purified fragments (Mulder et al. 1974, Brack et al. 1976); and

---

1 Department of Biochemistry, University of Geneva, CH-1211 Geneva 4, Switzerland
2 Department of Medical Biochemistry, University of Geneva, CH-1211 Geneva 4, Switzerland

(5) analysis of an entire range of partial digestion products of a DNA fragment radio-labeled at one end (Smith and Birnstiel 1976).

A very convenient method for the mapping of restriction endonuclease cleavage sites that was described recently depends upon the progressive exonucleolytic degradation of a linear DNA molecule, which is monitored by digestion with the restriction endonuclease whose cleavage sites are to be mapped. The sequential disappearance from the molecule ends of fragments generated by the restriction enzyme, following the exonucleolytic digestion reaction, reveals the order of fragments along the DNA molecule (Legerski et al. 1978). This method has the advantage that it is rapid and does not require radioactive labeling of DNA fragments.

The BAL31 exonuclease (Gray et al. 1975) is very convenient for this purpose because it (a) degrades linear DNA from both ends at a relatively constant and controllable rate; (b) is readily and irreversibly inactivated by removal of required Ca^{++} ions, through addition of EDTA (this latter feature enables digestion periods to be terminated precisely and subsequent digestions with endonucleases to proceed without the necessity for time-consuming phenol extractions); and (c) is a very stable enzyme that can be stored for months without loss of activity.

2. Experiment: Localization of *Hae*II Cleavage Sites on Plasmid pHSG415

a) Introduction

The pHSG415 plasmid (Fig. 1) is a pSC101-based cloning vector which has two features that provide it with a high degree of biological containment, namely (1) it exhibits temperature-sensitive replication and (2) it is deleted of the pSC101 *ori*T site that is necessary for plasmid mobilization (Hashimoto-Gohto et al. 1981). Both the vertical and horizontal transmission of this plasmid to new bacteria are thereby restricted. The pHSG415 plasmid contains single sites for endonucleases *Eco*RI, *Bst*EII, *Pst*I, *Hinc*II, *Bam*HI, *Hin*dIII, *Xma*I, and *Xho*I and four sites for *Hae*II. The *Hae*II sites on pHSG415 will be mapped relative to the *Bam*HI site by BAL31 digestion of *Bam*HI-cleaved plasmid DNA.

Ordinarily the experiment is carried out in two steps to determine (1) the location within the fragments to be ordered of the unique site at which the molecule will be linearized, by appropriate single and double enzyme digestions (in this case *Hae*II, and *Hae*II + *Bam*HI), and (2) the order of the fragments on the molecule, by BAL31 digestion of the linearized genome, followed by cleavage with a second enzyme (in this case, *Hae*II).

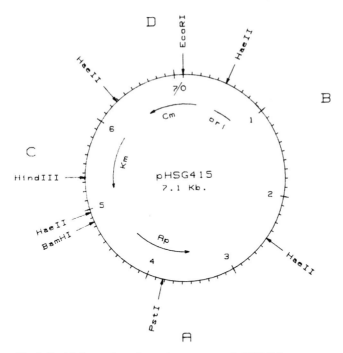

Fig. 1. Restriction endonuclease cleavage map of pHSG415

b) Procedure

Prepare the following digests in Eppendorf tubes:

	Tube 1	Tube 2	Tube 3
H_2O	12 µl	11 µl	12 µl
TMN × 10	–	–	6 µl
TM × 10	2 µl	2 µl	–
pHG415 DNA (∼ 100 µg/ml)	5 µl	5 µl	35 µl
HaeII (2 U/µl)	1	1	–
BamHI (2 U/µl)	–	1	7

Incubate tubes for 30 min at 37°C, then heat inactivate enzymes by placing tubes for 10 min in a water bath maintained at 70°C. Place tube 3 on ice.

Remove a 5 µl sample from tube 3 and add to 15 µl TBE (tube 4).

Add 3 µl stop solution to tubes 1, 2, and 4 and load these plus standards (e.g., λ-DNA digested with *Eco*RI + *Hin*dIII) onto a TBE-buffered 1% agarose gel.

Subject to electrophoresis at 80 V until the bromophenol blue has migrated 4/5 through the gel.

Stain gel with ethidium bromide (0.5 µg/ml in TBE) and photograph: the experiment may only be continued if sample 4 shows a single DNA band (complete digestion with *Bam*HI).

Label 10 Eppendorf tubes a–j, add 1 μl EDTA (0.1 M, pH 8.0) to each, and place on ice; transfer a 5 μl sample from tube 3 to tube a.

Add 5 μl NaCl (5 M), 5 μl CaCl$_2$ (140 mM), and 1 μl of EDTA (50 mM, pH 8.0) to tube 3 and pre-warm to 30°C.

Add 1 μl BAL31 exonuclease to tube 3 and incubate at 30°C.

At 1 min intervals transfer 5 μl samples from tube 3 to the series of tubes b–j to stop the digestion (the 1 min sample is transfered to tube b, the 2 min sample to tube c etc.).

Afterwards add to each tube 44 μl TMG, and 2 μl of HaeII (2 U/μl) to all samples, and incubate 1 h at 37°C.

Heat inactivate the endonuclease and analyse digestion products by electrophoresis through a 1% agarose gel.

c) Results and Discussion

Digestion of the pHSG415 plasmid with BamHI produces a single fragment of 7.1 kb, whereas digestion with HaeII (Fig. 2) produces 4 fragments with sizes of (A) 2.4 kb, (B) 2.0 kb, (c) 1.5 kb, and (D) 1.2 Kb. Double digestion of pHSG415 with HaeII + BamHI produces fragments having mobilties on a 1.0% agarose gel similar to those produced by HaeII alone, indicating that the BamHI cleavage site is located very close to (less than 300 bp from) a HaeII site.

Digestion of BamHI-cleaved pHSG415 DNA with BAL 31 for 1 min causes loss of HaeII fragments A and C and the appearance of two new diffuse bands on the gel which contain the degraded fragments. This shows that fragments A and C are adjacent in the plasmid and that the unique BamHI site is located close to the HaeII site

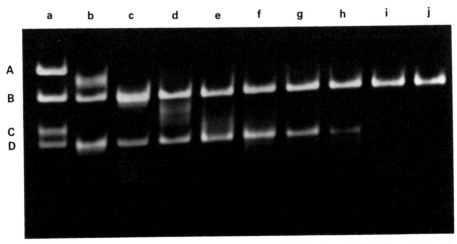

Fig. 2. HaeII digestion patterns of BAL31 degraded pHSG415 plasmid DNA linearized with BamHI. Track a contains DNA not treated with BAL31; lane b–j contain DNA treated with BAL31 for 1, 2, 3 minutes etc.

that forms the junction between these two fragments. While the degradation products of fragment C are no longer visible on the gel after 5 min of BAL31 digestion, those of fragment A disappear more slowly, appearing after 4 min of digestion as a diffuse band centered at a position on the gel corresponding to fragments 1.5 kb in size. Extrapolation of the digestion patterns indicates that fragment A is not completely degraded after 8 min of BAL31 digestion, although after 6 min, it is no longer visible as a discrete band. After 7 min of digestion, fragment D begins to disappear and after 8 min is no longer visible. This shows that fragments C and D are adjacent in the plasmid. Fragment B is not affected by the exonuclease after 9 min of digestion and must therefore be located on the side of the plasmid opposite to the BamHI cleavage site, between fragments A and D, as shown on the map of pHSG415 (Fig. 1).

The mapping of restriction enzyme sites on linear DNA molecules by means of the BAL31 exonuclease is rapid and simple and in many cases provides an unambiguous map in a single experiment. Ambiguities can, however, occur if two of the fragments generated in the final endonuclease digestion have a similar size and a symmetric location relative to the termini of the linear molecule being digested, because these will be degraded at the same time. Such ambiguities may be readily resolved either by one or more double digestions with the enzyme whose sites are to be mapped plus others, or by BAL31 digestion of the same molecule after linearization at a different site, which thus avoids the previous fragment symmetry.

The method can also be used successfully to map restriction sites on larger plasmids (Echarti 1981) by increasing the time of BAL31 digestion and the intervals between samples and/or by lowering the NaCl concentration in the reaction, which results in an increased activity of the nuclease.

3. Materials

10 × TM buffer (Tris-Cl, 100 mM, pH 7.5, $MgCl_2$, 100 mM)
10 × TMN buffer (Tris-Cl, 100 mM, pH 7.5, $MgCl_2$, 100 mM, NaCl, 500 mM)
TMG (Tris-Cl, 10 mM, pH 7.5; $MgCl_2$, 20 mM)
NaCl, 5 M
$CaCl_2$, 140 mM
EDTA, 50 mM, pH 8.0
EDTA, 100 mM, pH 8.0
pHSG415 DNA, 100 µg/ml, 50 µl
HaeII restriction endonuclease, 2 U/µl, 25 µl
BamHI restriction endonuclease, 2 U/µl, 25 µl
BAL31 exonuclease, 1.25 U/µl, 1 µl
TBE buffer (Tris, 0.09 M, boric acid, 0.09 M, EDTA 2 mM, pH 8.3)
Agarose, 2 g
Ethidium bromide, 10 mg/ml in H_2O
Gel standards, e.g., λ-DNA digested with HindIII + EcoRI (fragment sizes in kb:
    20.54, 5.13, 4.83, 4.27, 3.56, 2.04, 1.95, 1.62, 1.37, 0.89, 0.71, 0.47)

4. References

Allet B, Jeppesen PGN, Katagiri KJ, Delius H (1973) Mapping the fragments produced by cleavage of λ DNA with endonuclease RI. Nature 241:120–123

Brack C, Eberle H, Bickle TA, Yuan R (1976) A map of the sites on bacteriophage PM2 DNA for the restriction endonucleases *Hin*dIII and *Hpa*II. J Mol Biol 104:305–309

Danna KJ, Sack GH, Jr, Nathans D (1973) Studies of simian virus 40 DNA. VIII. A cleavage map of the SV40 genome. J Mol Biol 78:363–376

Echarti C (1981) Genetische Untersuchung des Kapsel-Polysaccharides K1 von *Escherichia coli*. Thesis, Freie Universität Berlin, Fachbereich Biologie (23)

Gray HB, Jr, Ostrander DA, Hodnett JL, Legerski RJ, Robberson DL (1975) Extracellular nucleases of *Pseudomonas* Bal31. I. Characterization of single-strand specific deoxyribonuclease and double-strand deoxyribonuclease activities. Nucleic Acids Res 2:1459–1492

Griffin BR, Fried M, Cowie A (1974) Polyoma DNA: A physical map. Proc Natl Acad Sci USA 71:2077–2081

Hashimoto-Gotho T, Franklin FCH, Nordheim A, Timmis KN (1981) Specific-purpose plasmid cloning vectors. I. Low copy number, temperature-sensitive, mobilization-defective pSC101-derived containment vectors. Gene 16:227–235

Lee AS, Sinsheimer RL (1974) A cleavage map of bacteriophage ∅X174 genome. Proc Natl Acad Sci USA 71:2882–2886

Legerski RJ, Hodnett JL, Gray HB, Jr (1978) Extracellular nucleases of *Pseudonomas* Bal31. III. Use of the double-strand deoxyribonuclease activity as the basis of a convenient method for the mapping of fragments of DNA produced by cleavage with restriction enzymes. Nucleic Acids Res 5:1445–1464

Mulder C, Arrand JR, Delius H, Keller W, Petterson U, Roberts RJ, Sharp PA (1974) Cleavage maps of DNA from Adenovirus types 2 and 5 by restriction endonuclease *Eco*RI and *Hpa*I. Cold Spring Harbor Symp Quant Biol 39:397–400

Smith DI, Blattner FR, Davies J (1976) The isolation and partial characterization of a new restriction endonuclease from *Providentia stuartii*. Nucleic Acids Res 3:343–353

Smith HO, Birnstiel ML (1976) A simple method for DNA restriction mapping. Nucleic Acids Res 3:2387–2398

Vereijken JM, Mansfield ADM van, Baas PC, Jansz HS (1975) *Arthrobacter luteus* restriction endonuclease cleavage map of ∅X174 RF DNA. Virology 63:221–233

1.8 IncP1 Typing of Plasmid DNA by Southern Hybridization

R. SIMON, W. ARNOLD, and A. PÜHLER[1]

Contents

1. General Introduction.. 80
2. Experiment 1: ^{32}P Labeling of the IncP1-Fragment of Plasmid R751 by Nick Translation. 82
   a) Introduction... 82
   b) Objectives... 82
   c) Procedure.. 82
   d) Results.. 84
3. Experiment 2: Southern Hybridization of Different Plasmid DNAs to a ^{32}P Labeled IncP1 Fragment... 85
   a) Introduction... 85
   b) Objectives... 86
   c) Procedure.. 86
   d) Results.. 88
4. Materials.. 89
5. References.. 91

1. General Introduction

In incompatibility tests, plasmids can be subdivided into different classes (see Chap. 1.4). This grouping is based on the fact that plasmids of one incompatibility class are unable to coexist in the same cell because they have homologous replication systems. In order to carry out incompatibility experiments, test plasmids are introduced by conjugation, transduction, or transformation into cells already harboring plasmids. For such experiments, it is important that the plasmids in question carry different selectable phenotypes. It is then possible to select for the presence of one or the other plasmid and to study their coexistence or their segregation into different daughter cells.

In this section, we present a method that uses Southern hybridization for incompatibility grouping. This method is based on a DNA probe specific for a single incompatibility group. Such a probe should hybridize to plasmid DNA of the same incompatibility class, but should show little, if any, hybridization to plasmid DNA of all

---

[1] Universität Bielefeld, Lehrstuhl für Genetik, Fakultät für Biologie, D-4800 Bielefeld 1, Fed. Rep. of Germany

the other classes. It should be mentioned that this incompatibility typing is independent of selectable phenotypes carried by the plasmids in question. Thus, it is also possible to determine the incompatibility group of cryptic plasmids.

The experiments described in this section will show that there is sequence homology between a specific DNA fragment responsible for incompatibility functions (IncP1) of plasmid R751 and plasmid RP4. The properties of plasmid RP4 and R751 are summarized in Table 1. Both plasmids belong to incompatibility group IncP1 and have a broad host range. In order to isolate a specific IncP1 probe for Southern hybridization, R751 DNA was digested by *Eco*RI restriction endonuclease and the resulting fragments cloned into the vector plasmid pACYC184. The hybrid plasmids were genetically analyzed and it was found that plasmid pWG1 carrying the smallest *Eco*RI fragment of plasmid R751 (0.9 kb) expresses incompatibility functions against the parent plasmid R751, as well as, against plasmid RP4. The map of plasmid pWG1 is shown in Fig. 1.

Table 1. Properties of plasmid RP4 and plasmid R751

R plasmid	Resistance markers	Additional markers	Incompatibility group	References
RP4	Ap, Km, Tc	Tra$^+$	IncP1	Barth and Grinter (1977)
R751	Tp	Tra$^+$	IncP1	Meyer and Shapiro (1980)

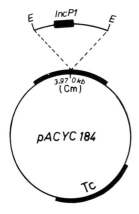

Fig. 1. Map of plasmid pWG1. Plasmid pWG1 consists of the vector plasmid pACYC184 and an *Eco*RI fragment of plasmid R751 carrying incompatibility function(s). This 0.9 kb IncP1 fragment is cloned into the single *Eco*RI site of pACYC184

In the first experiment of this section, the *Eco*RI insert of plasmid pWG1, we called the "IncP1-fragment", will be radioactively labeled with ^{32}P by nick translation. In the second experiment this probe will then be used for Southern hybridization. The plasmid DNA to be tested will be prepared in one case as a crude lysate according to the Eckhardt procedure. In another approach, we will use purified RP4 plasmid DNA digested with different restriction enzymes and separated by agarose gel electrophoresis. In both cases, the DNA will be denatured and transferred from the gel onto nitrocellulose filters. These filters will be hybridized with radioactively labeled IncP1 fragment DNA. Homology between the filter bound DNA and the probe can then be detected by autoradiography.

2. Experiment 1: ^{32}P Labeling of the IncP1-Fragment of Plasmid R751 by Nick Translation

a) Introduction

Procedures for preparing radioactively labeled DNA are extremely important in molecular biological studies, since they allow the production of labeled DNA probes for use in nucleic acid hybridization experiments. In principle, two approaches to the preparation of radioactively labeled DNA probes can be adopted. The first is to use in vivo labeling by supplying radioactive precursors of DNA to suitable bacterial or tissue culture systems. The second possibility is to prepare pure DNA and to label it in vitro, usually by using the nick translation reaction (Kelly et al. 1970) catalyzed by *E. coli* polymerase I (Rigby et al. 1977, Mackey et al. 1977). This method utilizes the ability of DNA polymerase I to carry out a repair type of replication dependent on nicks (single strand breaks) introduced by DNase I. Nucleotides are added to the 3'-hydroxyl side of the nick at the same time as existing nucleotides are removed from the other side of the nick by the 5' to 3' exonuclease activity of the enzyme DNA polymerase I. In consequence, the nick is "translated" along the DNA molecule in the 5' to 3' direction. If some of the nucleotides introduced during this process are radioactive (e.g., α-^{32}P dATP), the DNA will become radioactively labeled. Often it is important to use only a distinct DNA fragment for hybridization, e.g., to exclude the plasmid vector molecule that carries the fragment of interest. In this case, the fragment is cut out of the plasmid vector by restriction endonucleases. The mixture is then phenol extracted, and if necessary, the DNA fragments are ethanol precipitated. After the nick translation reaction, the fragments are separated by gel electrophoresis. The fragment of interest can then be isolated from the agarose gel by electrophoretic elution in a dialysis bag (McDonnell et al. 1977).

b) Objectives

1. To label pWG1 plasmid DNA with ^{32}P by nick translation.
2. To isolate the ^{32}P labeled IncP1 fragment of plasmid pWG1 by gel electrophoresis.

c) Procedure

Day 1

Labeling of pWG1 DNA with ^{32}P by Nick Translation

1. *Eco*RI restriction of plasmid pWG1 DNA

4 µg of pWG1 DNA in 50 µl *Eco*RI buffer is digested with *Eco*RI.
An aliquot of 2 µl is checked for complete digestion by gel electrophoresis.
After phenolization, ethanol precipitation and resuspension in water (final volume 40 µl), the DNA is ready for labeling.

2. Nick translation of EcoRI digested pWG1 DNA

Lyophilize 60 µl (α-^{32}P) dATP (2000–3000 Ci/mmol; 1 mCi/ml).
Suspend the dry radioactive dATP in 40 µl DNA solution.
Add dGTP, dCTP, and dTTP, in each case 8 µl from a 800 µM solution.
Add 7.4 µl 10 × nick translation buffer.
Start the reaction by adding 1 µl DNase I (1:10^6 dilution from a 1 mg/ml stock solution) and 1.5–2 µl DNA polymerase I.
Incubate 2–3 h at 18°C.
After incubation remove the proteins by phenol extraction.
Use Sephadex G50 column chromatography as described in Fig. 2, in order to separate labeled DNA from unincorporated nucleotides.
Concentrate the DNA by ethanol precipitation.
Measure the specific activity of the labeled DNA in a scintillation counter.

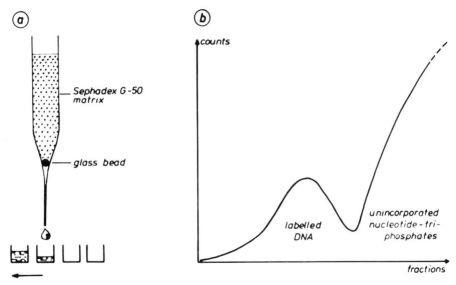

Fig. 2a, b. Separation of labeled DNA from unincorporated nucleotides. a Sephadex-G50 column and the elution of the column, b typical elution profile after measuring counts in a scintillation counter

Day 2

Separation and Isolation of the Labeled IncP1 Fragment

Some of the following steps are schematically described in Fig. 3.

Load the labeled DNA probe onto a horizontal agarose gel stained with ethidium bromide (1 µg/ml).
Run the gel in a dark room and check the separation of the fragments with a UV hand lamp.

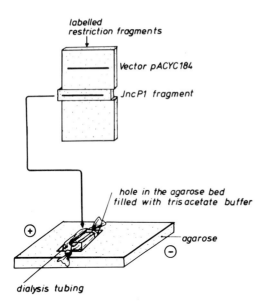

Fig. 3. Isolation of labeled DNA fragments from the agarose gel. For details see text

Cut out the fast moving IncP1 fragment.
Take a dialysis tube closed at both ends and cut it along the side.
Put the tube into a hole cut in the agarose bed of an horizontal gel.
Place the gel piece containing the DNA fragment into the dialysis tube and overlay with Tris-acetate buffer plus ethidium bromide.
Submit to electrophoresis for approximately 30 min; check the elution of the DNA with a UV lamp or a Geiger Müller Monitor.
Reverse the current during the last 5 min.
Collect the buffer containing most of the IncP1 fragment.
Phenolize the preparation.

d) Results

The elution profile of the Sephadex G50 column (Fig. 2) shows whether the nick translation step incorporated enough radioactivity into the DNA. After pooling the different DNA containing fractions, the specific radioactivity is mostly between 1×10^6 to 1×10^7 cpm/μg DNA. The separation and isolation of the labeled IncP1 fragment is presented in Fig. 3. The effectiveness of the separation can be checked by using an aliquot of the labeled IncP1 fragment DNA and repeating the agarose gel electrophoresis. Using autoradiography (see next experiment), the purity of the IncP1 fragment DNA can be determined.

3. Experiment 2: Southern Hybridization of Different Plasmid DNAs to a ^{32}P Labeled IncP1 Fragment

a) Introduction

In 1975, Southern developed a technique which provides a reliable way of detecting DNA containing complementary sequences to other DNA or RNA molecules. The different steps of this technique, namely, Southern blotting, hybridization, and autoradiography follow.

The DNA fragments or plasmids to be tested are separated by gel electrophoresis. They are denatured in situ by soaking the gel in alkali. Afterwards, the single stranded DNA molecules are transferred from the gel to a sheet of cellulose nitrate, retaining the original pattern. This transfer is achieved by laying the filter against the gel and by blotting solvent through it. The DNA molecules are transferred out of the gel by the flow of the high salt solution and trapped in the cellulose nitrate paper. The experimental set up of the blotting procedure is outlined in detail in Fig. 4. The time needed for the transfer of a particular DNA molecule depends on the gel concentration and the size of the molecule. Large fragments and DNA in the supercoiled form are not transferred very efficiently to nitrocellulose filters. Efficient transfer of such DNAs can be achieved after partial depurination with dilute acid and subsequent cleavage at the depurinated sites by alkali treatment (Wahl et al. 1979).

Denatured DNA fixed to nitrocellulose paper can then be hybridized to radioactively labeled RNA or DNA. Since labeled nucleic acids are mostly available only in small quantities, it is important to carry out hybridization in a volume as small as possible.

After hybridization, the DNA molecules forming hybrids with the radioactive probe can be detected as bands by autoradiography of the nitrocellulose sheet. An X-ray film is laid against the filter and stored at $-70°C$. The exposure time depends on the specific activity of the radioactive probe and the degree of sequence homology between the probe and the filter bound DNA (Laskey and Mills 1975).

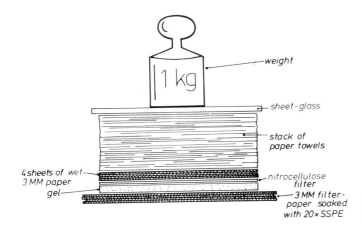

Fig. 4. Southern blotting procedure

b) Objectives

1. To prepare an Eckhardt gel of *E. coli* strains carrying the plasmids F, RP4, pPH1, or R751.
2. To restrict DNA of the plasmids RP4 and R751 and to separate the restriction fragments on agarose gels.
3. To transfer the DNA bands from the gels prepared in (1) and (2) to nitrocellulose filters.
4. To hybridize the filters with ^{32}P-labeled IncP1 DNA and to identify the hybridizing bands by autoradiography.

c) Procedure

Day 1

Preparation of the Eckhardt Gel

The following *E. coli* strains are used for the Eckhardt gel technique: (A) *E. coli* K12 (F); (B) *E. coli* C600 (RP4); (C) *E. coli* C600 (pPH1); and (D) *E. coli* C600 (R751). Additional *E. coli* or other strains carrying known or unknown plasmids can be included in this experiment. Experimental details for the Eckhardt lysis and gel technique can be found in Chap. 1.3 of this book.

Agarose Gel Electrophoresis with Restriction Fragments of RP4 and R751 DNA

RP4 and R751 DNA is purified according to methods described in this book. RP4 DNA is digested with the restriction endonucleases *Sma*I and *Pst*I, whereas R751 DNA is digested only with *Eco*RI. The restriction fragments of the different plasmids are separated by electrophoresis in a 1% agarose gel with Tris-acetate buffer. Again, it is possible to include additional plasmid DNAs of known or unknown origin in the experiment. It is recommended that both gels be stained for 30 min with EtBr (1 µg/ml), washed in water for about 1 h, and then photographed under UV light.

Pre-Treatment of the Gels for Blotting

The Eckhardt gel contains plasmid DNA mainly in CCC-form. For efficient transfer of this DNA, it is necessary to pre-treat the gel in dilute acid to fragment the molecules in the following way:

Submerge the gel in 0.25 M HCl and rock it gently for 5–15 min.
Decant the acid and rinse the gel briefly with water.

Subsequently both gels (the Eckhardt gel and the gel containing the restriction fragments) are treated in the same way:

Submerge the gels with denaturation solution (0.5 M NaOH and 1.5 M NaCl) and agitate gently for 15 min.
Decant alkali and repeat the last step.
Rinse with water briefly.
Add neutralization solution (1.5 M NaCl and 0.5 M Tris pH 7.5) and shake gently for 30 min.

Decant the neutralization solution and repeat this step.
Rinse the gel for 2 min in 2 × SSPE.

Blotting of the Gels

For the schematic drawing of the blotting procedure see Fig. 4.

Lay the gel on three Whatman 3 MM paper soaked with 20 × SSPE (the 3 MM paper should be somewhat larger than the gel).
Cut nitrocellulose filters to the size of the gels.
Wet the filter sheets with deionized water and then with 2 × SSPE.
Place the nitrocellulose (Schleicher & Schüll BA85) on top of the gel, carefully remove any bubbles between the gel and the nitrocellulose.
Cover the nitrocellulose filters with four sheets of 3 MM paper, wet with water (the 3 MM paper should have exactly the same size as the nitrocellulose filter).
A stack of 3–5 cm of paper towels (the same size) are placed on top and compressed with a 1 kg weight.
Let it blot for 2 h or longer.
After blotting, the nitrocellulose filter is rinsed for a few minutes in 2 × SSPE and dried between two sheets of 3 MM paper at 80°C in an oven for 2–4 h.

Day 2

Hybridization of the Filter Bound DNA with the Radioactive ^{32}P IncP1 Probe

Seal the filters in a plastic bag containing 10 ml of preincubation buffer. [The preincubation buffer contains: 4 × SSPE, 5 × Denhardts solution (Denhardt 1966), and 100 µg/ml calf thymus DNA. The calf thymus DNA is fragmented by autoclaving and freshly denatured by heating at 100°C for 5 min.]

Hybridization

Remove preincubation solution.
Prepare the following solution (4 ml/100 cm² NC filter): 4 × SSPE, 1 × Denhardt's solution, 0.3% SDS.
Heat the labeled DNA for 5 min at 100°C to denature the probe and quickly chill on ice.
Add hybridization solution and denatured labeled DNA to the filters, eliminate the air bubbles and seal the bag again.
Incubate at 65°C in a water bath for 6 h or overnight.

Washing

Remove hybridization solution from the bag (the probe can be stored at 4°C and used again several times).
Wash the filters 1–4 × for 5–15 min in 2 × SSPE and 0.1% SDS at 40–50°C with gentle agitation, check the filter for nonspecific binding of labeled DNA with the hand monitor (if necessary repeat washing).
After the final wash, the filters are dried at room temperature between 3 MM filter papers.

Autoradiography

Press X-ray film against the filter to maximize the band intensity.
Use intensifying screen.
Store at $-70°C$ for a suitable time (time can range from 1–14 days).
Develop X-ray film.

d) Results

Typical results for these experiments are shown in Figs. 5 and 6. In the hybridization experiment with plasmid DNA from the Eckhardt gel, it is clear that the *E. coli* F plasmid (lane 1) does not hybridize with the IncP1 probe, whereas all the other plasmids, i.e., RP4 (lane 2), pPH1 (lane 3), and R751 (lane 4) show a strong hybridization with the IncP1 fragment (Fig. 5). Thus, it is evident that IncP1 typing of plasmids can be carried out with the IncP1 probe derived from plasmid R751 since only the IncP1 plasmids hybridize with the probe and the F plasmid belonging to another incompatibility group (FII) fails to do so. In Fig. 6, the results of a hybridization experiment between the IncP1 probe and different restriction fragments of the plasmid R751 and RP4 are presented. No. 1 is a control experiment showing hybridization between the probe and the *Eco*RI fragments of plasmid R751. In No. 2, restriction fragments of the plasmid RP4 and their hybridization to the IncP1 probe are shown. To localize the region of homology on the RP4 plasmid, the restriction enzymes *Sma*I and *Pst*I were used to digest RP4 DNA. The IncP1-fragment of R751 hybridizes to the third *Sma*I and the second *Pst*I fragment of RP4 DNA (Fig. 6B). These two fragments overlap in the region where the origin of replication has been localized (Figurski 1979).

Thus, it can be concluded that the origin of replication of the RP4 plasmid and a region coding for incompatibility functions are in close proximity.

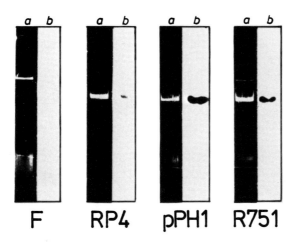

Fig. 5. Hybridization between the labeled IncP1-DNA sequence and intact DNA of the plasmids F, RP4, pPH1, and R751. *a* Agarose gel, stained with ethidium bromide and photographed under UV light; *b* autoradiogram

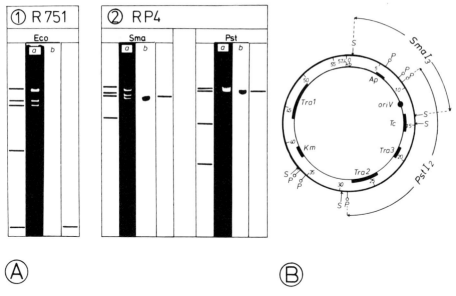

Fig. 6A, B. Hybridization between the labeled IncP1 sequence and restriction fragments of plasmid R751 and RP4. **A** Analysis of DNA fragments: *1* EcoRI fragments of plasmid R751, *a* agarose gel pattern; *2* SmaI and PstI fragments of plasmid RP4, *b* autoradiogram. **B** Restriction map of plasmid RP4: *S* SmaI; *P* PstI; *Ap* Ampicillin resistance; *Km* Kanamycin resistance; *Tc* Tetracycline resistance; *ori* origin of vegetative replication; *Tra* Transfer region

4. Materials

Strains and Plasmids

E. coli K 12 (F)
E. coli C600 (pPH1)
E. coli C600 (RP4)
E. coli C600 (R751)
E. coli C600 (pWG1)

Buffers

DNase Stock Solution:

50 mM Tris, pH 7.5
10 mM MgSO$_4$
 1 mM Mercaptoethanol
50% Glycerol
1 mg/ml DNase I
The stock solution can be stored at $-20°C$ for several months

DNase Dilution Buffer:

50 mM Tris, pH 7.5
10 mM $MgSO_4$
 1 mM Mercaptoethanol
50 µg/ml BSA (bovine serum albumine, Serva)

10 × Nick Translation Buffer:

500 mM Tris, pH 7.8
 50 mM $MgCl_2$
100 mM Mercaptoethanol
100 µg BSA/ml

Elution Buffer for the Sephadex G-50 Column Chromatography

10 mM Tris, pH 7.5
 1 mM EDTA

SmaI Restriction Buffer

20 mM KCl
10 mM Tris, pH 8
10 mM $MgSO_4$
 1 mM Mercaptoethanol

Restriction Buffer for the Other Endonucleases Used in this Experiment

50 mM NaCl
10 mM Tris, pH 7.4
10 mM $MgSO_4$
 1 mM Mercaptoethanol

20 × SSPE

20 mM EDTA
0.1 M NaOH
3.5 M NaCl
0.2 M NaH_2PO_4

50 × Denhardt's Solution

1% (w/v) BSA
1% (w/v) Ficoll 400 (Pharmacia)
1% (w/v) Polyvinylpyrrolidone (Sigma)

5. References

Barth PT, Grinter NJ (1977) A Tn7 insertion map of RP4. In: Bukhari AJ (ed) DNA insertion elements, plasmids, and episomes. Cold Spring Harbor Lab, Cold Spring Harbor, New York

Denhardt DT (1966) A membrane-filter technique for the detection of complementary DNA. Biochem Biophys Res Commun 23:641–646

Figurski DH, Helinski DR (1979) Replication of an origin-containing derivative of plasmid RK2 dependent on a plasmid function provided *in trans*. Proc Natl Acad Sci USA 76:1648

Kelly RB, Cozzarelli NR, Deutscher MP, Lehmann JR, Kornberg A (1970) Enzymatic synthesis of deoxyribonucleid acid XXXII. Replication of duplex deoxyribonucleic acid by polymerase at a single strand break. J Biol Chem 245:39–45

Laskey AR, Mills DA (1975) Quantitative film detection of ^{3}H and ^{14}C in polyacrylamid gels by fluorography. Eur J Biochem 56:335–341

McDonell MW, Simon MN, Studier FW (1977) Analysis of restriction fragments of T7 DNA and determination of molecular weights by electrophoresis in neutral and alkaline gels. J Mol Biol 110:119–146

Mackey JK, Brackmann KH, Green MR, Green M (1977) Preparation and characterization of highly radioactive in vitro labelled adenovirus DNA and DNA restriction fragments. Biochemistry 16:4478–4483

Meyer RJ, Shapiro JA (1980) Genetic organisation of the broad-host-range IncP1 plasmid R751. J Bacteriol 143:1362

Rigby PWJ, Dieckmann M, Rhodes C, Berg P (1977) Labelling deoxyribonucleic acid to high specific activity in vitro by nick translation with DNA polymerase I. J Mol Biol 113:237–251

Southern EM (1975) Detection of specific sequences among DNA fragments separated by gel electrophoresis. J Mol Biol 98:503–517

Wahl GM, Stern M, Stark GR (1979) Efficient transfer of large DNA fragments from agarose gels to diazobenzyloxymethyl-paper and rapid hybridisation by using dextran sulfate. Proc Natl Acad Sci USA 76:3683–3687

1.9 Molecular Epidemiology by Colony Hybridization Using Cloned Genes

M.A. MONTENEGRO, G.J. BOULNOIS, and K.N. TIMMIS [1]

Contents

1. General Introduction 92
2. Experiment: Screening of *E. coli* Isolates for the Presence of *tra*T Gene. 93
   a) Introduction. 93
   b) Procedure 1: Preparation and Characterization of the *tra*T DNA Probe 94
   c) Procedure 2: Screening of *E. coli* Isolates for the Presence of *tra*T Gene-Related
      Sequences 98
   d) Results and Discussion 99
3. Materials 101
4. References 102

1. General Introduction

Recently developed methods of gene analysis and manipulation have not only greatly accelerated the investigation of a large number of basic and applied biomedical problems, but are also being used increasingly as methods of diagnosis and epidemiology in medicine. This is especially true of combinations of gene cloning and nucleic acid hybridization techniques (Little 1981, Timmis 1981). The former provides a DNA probe for the polynucleotide sequence of interest and the latter involves the use of this probe to search for homolgous sequences in various preparations of nucleic acids.

Two nucleic acid hybridization procedures are particularly useful for this purpose: (1) colony hybridization (Grunstein and Hogness 1975), which determines the number of colonies containing nucleic acid sequences homologous to the probe; and (2) Southern blotting (Southern 1975), which determines the particular segment of a DNA preparation fractionated by gel electrophoresis containing sequences homologous to the probe. Colony hybridization and a modification of this technique, dot blot or touch blot hybridization, employing selected cloned genes of pathogenic microorganisms, is being increasingly used (1) in the analysis of microorganisms isolated from clinical specimens (and indeed of clinical specimens themselves), which could become a rapid and effective means of diagnosis, and (2) epidemiological studies, in which the prevalence of a particular microorganism in a large number of clinical

---

[1] University of Geneva, Department of Medical Biochemistry, CH-1211 Geneva, Switzerland

specimens, or the prevalence of a particular pathogenesis determinant in a large number of microorganisms, is determined. For example, colony hybridization has been widely used to examine the prevalence of enterotoxin genes in bacteria isolated from cases of diarrhea (e.g., Moseley et al. 1980); dot hybridization has been employed in the detection of rotaviruses in stools (Flores et al. 1983); and hybridization of tissue touch blots, derived from cutaneous lesions, with cloned kinetoplast DNA from *Leishmania*, has been proposed as a method of diagnosis of leishmaniasis (Wirth and Pratt 1982).

In the experiment described below, colony hybridization will be used to examine the prevalence of the *tra*T gene, which specifies a factor mediating bacterial resistance to complement, in a number of isolates of *E. coli*. Prior to this, however, the specificity of the *tra*T gene probe will be examined by Southern blotting.

2. Experiment 1: Screening of *E. coli* Isolates for the Presence of *tra*T Gene

a) Introduction

Bacterial pathogens generally cause disease either by production of one or more toxins that disturb the normal functioning of target cells, by invading the host and multiplying within a target tissue, or by a combination of both mechanisms. Invasive bacteria must not only be able to damage host tissues, but also be able to resist a battery of host defenses. The important first line defenses of higher animals against invasive pathogens are the lethal activities of complement and phagocytes (Mims 1982).

In recent years, considerable progress has been made in the identification of bacterial pathogenesis factors. Bacterial resistance to complement (serum) and phagocytes, for example, is known to be frequently mediated by cell surface components, such as outer membrane proteins (Moll et al. 1980), capsules (Gemski et al. 1980, Timmis et al. 1981), and lipopolysaccharides (M.M. Binns et al., submitted). One component analyzed in some detail is the *tra*T protein, a plasmid-encoded outer membrane protein of *E. coli*, that mediates resistance to serum and phagocytosis (Moll et al. 1980, Aguero et al. submitted). In a recent investigation, the *tra*T protein was found to increase the pathogenicity of an *E. coli* strain for an experimental infection model (Aguero et al. (1983).

Once a pathogenesis determinant has been identified, it is important to determine its prevalence in clinical isolates, that is, its epidemiological importance. In this experiment, the prevalence of the *tra*T gene in a collection of *E. coli* isolates will be examined by colony hybridization.

The isolates have not been obtained from clinical sources, which would not be appropriate for a class experiment, but from healthy individuals.

b) Procedure 1: Preparation and Characterization of the *tra*T DNA Probe

The *tra*T gene-containing *Eco*RI fragment E-7 of antibiotic resistance plasmid R6-5 (Fig. 1) has been cloned in the pACYC184 vector to produce hybrid plasmid pKT107 (Moll et al. 1980). The *Bst*EII endonuclease cleaves the pKT107 plasmid at two sites, about 700 bp apart. One site is located within the *tra*T gene and one is situated some 140 bp upstream of this gene. Thus, a small fragment that is highly specific for the *tra*T gene (Moll et al. 1980, Ogata et al. 1982) generates from this plasmid.

The small *Bst*EII fragment of pKT107 will be isolated after gel electrophoresis of *Bst*EII-cleaved DNA and will be labeled with ^{32}P by nick translation. The specificity of the probe will then be examined by Southern blotting, prior to it being used for colony hybridization.

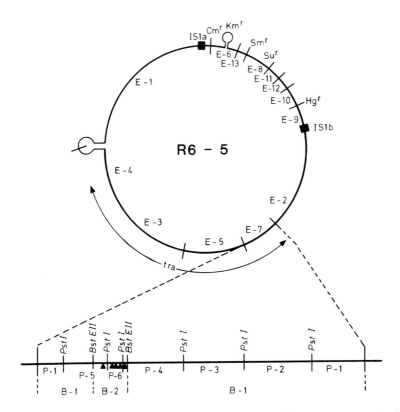

Fig. 1. Map of R6-5. *E, B,* and *P* restriction endonuclease fragments generated by *Eco*RI, *Bst*EII, and *Pst*I, respectively. The *diamond symbols* indicate the locations of Tn*3* insertions that inactivate the complement resistance determinant, *tra*T (Timmis et al. 1978, 1981)

Day 1

Label 6 Eppendorf tubes 1–6 and add the following components to each tube:

Tube	Plasmid	Volume (1 mg DNA/ml)	Restriction buffer (10 ×)	H_2O	Enzyme	Units
1	R386	1 µl	1 µl	8 µl	HindIII	2
2	R124	1 µl	1 µl	8 µl	HindIII	2
3	R6K	1 µl	1 µl	8 µl	HindIII	2
4	R6-5	1 µl	1 µl	8 µl	EcoRI	2
5	pKT107	1 µl	1 µl	8 µl	EcoRI	2
6	pKT107	20 µl	10 µl	70 µl	BstEII	30

Incubate the tubes for 1 h at 37°C.

Inactivate the endonucleases by holding the tubes at 70°C for 10 min.

To each tube add 1/5 volume of loading buffer.

Load all of the contents of tubes 1–5 and 2 µl of tube 6 into 6 adjacent slots of a 0.8% Tris-acetate buffered agarose gel containing 1 µg/ml ethidium bromide.

Store tube 6, containing the remainder of the BstEII digested pKT107 DNA, at −20°C overnight.

To a seventh slot, add 1 µg of λ-DNA digested with HindIII as a size standard.

Subject the gel to electrophoresis for 3–4 h at 80–90 V.

Examine the gel under UV illumination and if the enzyme digestions were complete, transfer the separated DNA fragments onto a nitrocellulose filter as follows:

Denature the DNA in situ by placing the gel in a solution of 0.5 M NaOH, 1.5 M NaCl and rocking the gel gently at room temperature for 30 min.

Rinse the gel with distilled water and neutralize it by placing it in a solution of 0.5 M Tris-HCl, 3 M NaCl, pH 7.0, for 30 min.

Transfer the gel into 20 × SSC and rock gently for 15 min.

Place the gel on several filter papers, pre-soaked in 20 × SSC.

Cut a nitrocellulose filter the same size as the gel, rinse in 2 × SSC and then 20 × SSC.

Place the washed filter on the gel, avoiding air bubbles.

Place on top of the nitrocellulose filter several sheets of Whatman 3 MM paper of the same dimensions, soaked in 20 × SSC.

Place on top of these several dry paper towels and compress them by applying a 1 kg weight.

Leave overnight to permit transfer of the DNA fragments from the gel to the nitrocellulose filter.

Day 2

Dismantle the blotting apparatus and wash the nitrocellulose filter in 2 × SSC for 15 min.

Allow the filter to air dry and then bake at 80°C for 4 h. This procedure irreversibly binds the DNA to the filter.

Load the BstEII-cleaved pKT107 DNA remaining in tube 6 into the slot of a 0.8% preparative agarose gel lacking ethidium bromide.

Include in the same gel 1 μg of λ-DNA cleaved with *Hin*dIII as molecular weight standard.

Subject the gel to electrophoresis at 80–90 V until the dye has migrated 3/4 of the length of the gel.

Stain the gel in a solution of 1 μg/ml ethidium bromide.

Visualize the DNA by UV illumination and cut out an agarose slice containing the 700 bp *Bst*EII fragment of pKT107.

Purify the DNA from the agarose using the method described in Chap. 4.4.

Collect the DNA by ethanol precipitation and redissolve in 25 μl of water. This procedure should yield about 500 ng of the probe DNA fragment, assuming 50% recovery of DNA from the agarose.

Day 3

Incubate the nitrocellulose filter for 3–4 h at 65°C in 10–20 ml of 4 × SSC containing 0.02% Denhardt solution in a sealed plastic bag. During the incubation, label the *Bst*EII fragment with ^{32}P by nick translation using the method of Jeffreys et al. (1983) as described below.

Mix in a tube placed on ice:

5   μl DNA
2.5   μl nick mix
1   μl dATP (100 μM)
1   μl dGTP (100 μM)
1   μl dTTP (100 μM)
11.5 μl H$_2$O
1   μl DNase (diluted to a final concentration of 8 ng/ml from a 1 mg/ml stock solution immediately before use)
1   μl DNA Polymerase I (5 units)
2   μl α-^{32}P-dCTP (3000 Ci/mmol)

Incubate at 15°C for 45 min.

Stop the reaction by addition of 25 μl quench mix.

Extract with 50 μl phenol, centrifuge for 1 min, separate and retain both phases.

Reextract the phenolic phase with 50 μl TE.

Pool the two aqueous phases and add 20 μl Na acetate (2 M, pH 5.6), 100 μl of high molecular weight salmon sperm DNA (1 mg/ml) and 500 μl cold absolute ethanol, and mix.

Remove the precipitated DNA with a Pasteur pipette and dissolve in 200 μl TE.

Repeat the ethanol precipitation and redissolve the DNA in a final volume of 500 μl TE.

An efficient labeling should yield at least 1×10^7 cpm/μg DNA.

Transfer the nitrocellulose filter to 10 ml of a solution consisting of 2 × SSC, 1 mM EDTA, 50 μg/ml denatured salmon sperm DNA, and 100 μl of the ^{32}P-labeled DNA, previously denatured by incubation at 100°C for 5 min. Save the remainder of the probe for the colony hybridization experiment (Procedure 2).

Allow hybridization to occur in a sealed plastic bag overnight at 65°C.

Day 4

Remove the nitrocellulose filter from the bag and place in a large beaker or bowl located in a water bath set at 65°C and containing 250 ml of washing solution I.

Agitate container from time to time and after 30 min, replace the washing solution with 250 ml of fresh solution and leave for a further 30 min.

Repeat the washing procedure with washing solution II (2 ×), followed by washing solution III (2 ×).

Dry the filter and expose to an X-ray film overnight at −70°C using an intensifying screen.

Day 5

Develop the X-ray film.

The autoradiograph of such a filter is shown in Fig. 2.

In addition to showing the expecting hybridization to the 6 kb EcoRI fragment of R6-5 (group Inc FII) and the small BstEII fragment of pKT107, the probe hybridized to large HindIII fragments of plasmids R386 (Inc FI) and R124 (Inc FIV),

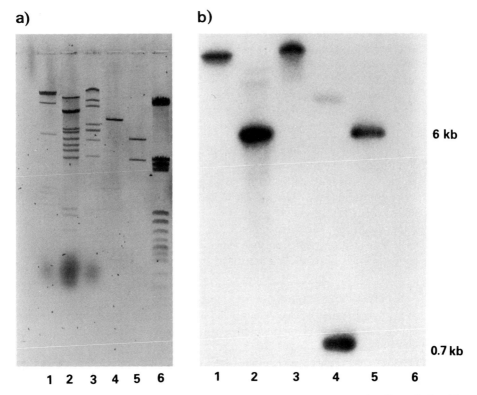

Fig. 2. Southern hybridization. *a* Ethidium bromide stained gel showing the digested plasmid DNAs, *b* autoradiograph of the nitrocellulose filter after hybridization to the *tra*T probe; *1* R386/HindIII; *2* R6-5/EcoRI; *3* R124/HindIII; *4* pKT107/BstEII; *5* pKT107/EcoRI; *6* R6K/HindIII

but not to any HindIII fragment of plasmid R6K (Inc X), nor to plasmids belonging to incompatibility groups A, C, Iα, Iγ, H2, L, N, O, Q, S, and W (Montenegro et al., in preparation). The probe DNA fragment is thus highly specific for the *tra*T gene.

c) Procedure 2: Screening of *E. coli* Isolates for the Presence of *tra*T Gene-Related Sequences

In this part, colony hybridization with the *tra*T gene probe will be used to examine the presence of *tra*T gene sequences in 45 isolates of *E. coli* obtained from the feces of healthy individuals. The 45 clones, plus appropriate controls, will be inoculated into separate wells of 96-well microtiter plates containing L-broth and cultured overnight. The following day, small quantities of the cultures will be transferred by means of a multipoint (48) inoculator from the microtiter plate to a filter on the surface of a nutrient agar plate. This plate will then be incubated overnight after which the colonies that develop will be lysed and their DNA denatured and bound to the filter by baking. The filter will then be used for hybridization with the radioactive *tra*T probe.

The advantages of inoculating wells of a microtiter plate, rather than the filter directly, are 3-fold: (1) relatively constant inocula are delivered to the filter, (2) several identical filters can be quickly prepared, enabling the screening of the same set of clones with several different probes, and (3) the original clones can be preserved economically at $-20°C$ or $-80°C$ in the microtiter plates, after addition of an equal volume of sterile glycerol.

Day 6

Distribute 0.1 ml quantities of L-broth into the wells of a 96-well microtiter plate.
Label one half of the plate A and the other half B, each half containing 48 wells that will correspond to one filter.
Inoculate the wells of both halves with the 45 given strains plus the control strains C600 (R6-5), C600 (pKT107), and C600, such that the two half plates constitute a duplicate series of cultures.
Incubate the microtiter plate overnight at $37°C$.
Clip a small V-shaped segment from two nitrocellulose filters, insert them individually between sheets of paper, and autoclave them in aluminium foil.

Day 7

Take two dry L-agar plates labeled A and B and draw an arrow on the rim of the agar-containing part of each plate; these orientation arrows will indicate the top of the plates (12 o'clock).
Carefully place the sterile nitrocellulose filters on the plates, such that the Vs are coincident with the arrows, avoiding air bubbles between the filters and the agar.
Inoculate the filters with the cultures in the microtiter plates using a multipoint inoculator.
Incubate the agar plates overnight at $37°C$.

Day 8

Remove the nitrocellulose filters from the agar plates and place colony side up on a filter paper soaked in 0.5 M NaOH for about 5 min to cause cell lysis and denaturation of DNA.

Neutralize the nitrocellulose filters by placing them on a sheet of filter paper soaked in 1 M Tris-HCl, pH 7.0, for 2 min and then on a second filter soaked in the same buffer for a further 2 min.

Finally, transfer them to a filter paper soaked in 1 M Tris-HCl, pH 7.0, 1.5 M NaCl and leave for 5 min.

Dry the nitrocellulose filters at 37°C and mark A and B with a ball point pen, before baking them at 80°C for 4 h.

Day 9

Pre-wash the filters with 10 ml of hybridization solution (50% formamide, 5 × SSC, 0.1% SDS, 0.02% Denhardt's solution) in a sealed plastic bag at 37°C for 3 h.

Add 0.5 ml calf thymus DNA to the radioactive probe (at least 5×10^5 cpm) and denature the DNAs by holding the tube at 100°C for 5 min.

Chill the probe on ice before adding to 10 ml of hybridization solution.

Remove the prewash solution from the plastic bag containing the filters and replace with hybridization solution containing the probe; seal the bag and incubate overnight at 37°C.

Day 10

Remove the filters from the plastic bag and rinse briefly with a small volume of hybridization solution, before washing them twice in 200 ml 5 × SSC, 0.1% SDS at 65°C for 30 min, and once in 200 ml 2 × SSC at room temperature for 30 min.

Allow the filters to dry, mount them on a piece of 3 MM paper and expose to an X-ray film overnight at –70°C using an intensifying screen.

Day 11

Develop the film and score *tra*T-positive and *tra*T-negative colonies.

d) Results and Discussion

An autoradiograph of such a colony hybridization filter is shown in Fig. 3. The control strains C600 (pKT107) and C600 (R6-5) gave strong positive signals, whereas no hybridization was detected with the plasmid-free strain C600. Thirteen of the 45 *E. coli* isolates were positive for *tra*T. The different intensities of signals observed for different strains may reflect different colony sizes, different copy numbers of the plasmids carrying these sequences, and/or different degrees of homology between probe and the *tra*T-related sequences detected.

The colony hybridization experiment described here was carried out in 50% formamide at 37°C, relatively stringent conditions that allow about 20% sequence mis-

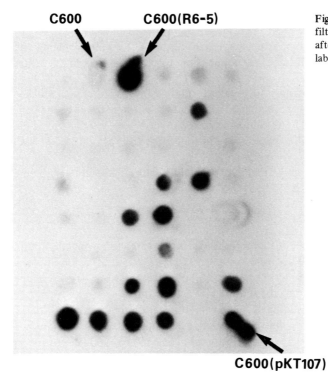

Fig. 3. Autoradiograph of a filter of lysed *E. coli* colonies after incubation with the ^{32}P-labeled *tra*T probe

match in hybrid molecules, depending on the G+C content of the DNA. We have found these conditions optimal for epidemiological screens of the type described above. However, where appropriate, the hybridization conditions can be easily adjusted by varying the formamide concentration, allowing a greater or lesser degree of mismatch (Howley et al. 1979) between the probe and related sequences in the colonies.

Colony hybridization, being highly sensitive and permitting the simultaneous analysis of hundreds of bacterial isolates, is a particularly useful method for epidemiological studies. The ease with which replica colony filters can be made enables the screen to be carried out with any number of existing probes. Filters carrying lysed colonies can be stored for several months at room temperature enabling them to be used at a later date with probes that are subsequently developed. Moreover, filters that have been used with a probe can be reused after washing in distilled water at 65°C for several hours.

3. Materials

Day 1

Purified DNAs of plasmids R386, R124, R6-5, R6K, and pKT107, each 1 mg/ml in TE.
Restriction enzymes *Eco*RI, *Hin*dIII, and *Bst*EII, and corresponding digestion buffers (as recommended by the supplier)
Tris-acetate electrophoresis buffer, 10 × (0.4 M Tris, 0.05 M Na acetate, 0.01 M EDTA, pH 7.9)
Materials for agarose gels
Materials for Southern blots: Denaturation solution: 0.5 M NaOH, 1.5 M NaCl; neutralization solution: 0.5 M Tris-HCl, pH 7.0, 3 M NaCl, 20 × SSC (1 × SSC = 0.15 M NaCl, 0.015 M Na citrate)
Loading buffer: 0.1% bromophenol blue, 20% Ficoll.

Day 2

Materials for purification of DNA from agarose gels (see Chap. 4.4)

Day 3

Nick mix: 0.5 M Tris-HCl, pH 7.8, 50 mM $MgCl_2$, 100 mM mercaptoethanol
Quench mix: 10 mM Tris-HCl, pH 7.5, 2% SDS, 50 mM EDTA
α-^{32}P-dCTP (specific activity 3000 Ci mmol, 10 mCi/ml, Amersham)
100 μM dATP, 100 μM dGTP, 100 μM dTTP
1 mg/ml DNase I
DNA polymerase I
2% Denhardt's solution: 2% BSA, 2% polyvinylpyrrolidon, 2% Ficoll 400
1 mg/ml salmon sperm DNA
0.25 M EDTA
2 M Na acetate, pH 5.6
Ethanol
20 × SSC
TE: 10 mM Tris-HCl, pH 7.6, 1 mM EDTA

Day 4

Washing solution I: 2 × SSC, 0.1% SDS, 20 mM phosphate buffer pH 7.3
Washing solution II: 1 × SSC, 0.1% SDS, 20 mM phosphate buffer pH 7.3
Washing solution III: 0.2 × SSC, 0.1% SDS, 20 mM phosphate buffer pH 7.3
X-ray films, casette with intensifying screen

Day 5

Materials for development of X-ray films

Day 6

Collection of isolates of *E. coli*
Control strains: C600, C600 (pKT107), C600 (R6-5)
L-broth
96-well microtiter plates
82 mm diameter, round nitrocellulose filters (0.45 μm pore size)

Day 7

L-agar plates
Multipoint inoculator

Day 8

0.5 M NaOH
1 M Tris-HCl, pH 7.0
1 M Tris-HCl, pH 7.0, 1.5 M NaCl

Day 9

20 × SSC
2% Denhardt's solution
10% SDS
1 mg/ml salmon sperm DNA
Formamide

Day 10

X-ray films, casette with intensifying screen

Day 11

Same as day 5

4. References

Aguero ME, De Luca AG, Timmis KN, Cabello FC (1983) A plasmid-encoded outer membrane protein, *tra*T, enhances resistance to phagocytosis and *E. coli* virulence (submitted)

Flores J, Boeggeman E, Purcell RH, Sereno M, Pérez I, White L, Wyatt RG, Chanock RM, Kapikian AZ (1983) A dot hybridization assay for detection of rotavirus. Lancet ii:555–559

Gemski P, Cross AS, Sadoff JC (1980) K1-antigen-associated resistance to the bactericidal activity of serum. FEMS Microbiol Lett 9:193–197

Grunstein M, Hogness DS (1975) Colony hybridization: a method for the isolation of cloned DNAs that contain a specific gene. Proc Natl Acad Sci USA 72:3961–3965

Howley PM, Israel MA, Law M-F, Martin MA (1979) A rapid method for detecting and mapping homology between heterologous DNAs. J Biol Chem 254:4876–4883

Jeffreys AJ, Boulnois GJ, Varley JM (1983) Analysis of DNA and RNA. In: Pritchard RH, Holland IB (eds) Basic cloning techniques

Little PFR (1981) DNA analysis and the antenatal diagnosis of hemoglobinopathies. In: Williamson R (ed) Genetic engineering, vol 1. Academic Press, London New York, pp 61–102

Mims CA (1982) The pathogenesis of infectious disease, 2nd edn. Academic Press, London New York

Moll A, Manning PA, Timmis KN (1980) Plasmid-determined resistance to serum bactericidal activity: a major outer membrane protein, the *tra*T gene product, is responsible for plasmid-specified serum resistance in *Escherichia coli*. Infect Immun 28:359–367

Moseley SL, Hug I, Alim ARMA, So, M, Samadpour-Motalebi M, Falkow S (1980) Detection of enterotoxigenic *Escherichia coli* by DNA colony hybridization. J Infect Dis 142:892–898

Ogata RT, Winters C, Levine RP (1982) Nucleotide sequence analysis of the complement resistance gene from plasmid R100. J Bacteriol 151:819–827

Southern EM (1975) Detection of specific sequences among DNA fragments separated by gel electrophoresis. J Mol Biol 127:502–517

Timmis KN (1981) Gene manipulation in vitro. In: Glover SW, Hopwood DA (eds) Genetics as a tool in microbiology. Soc Gen Microbiol Symp 31. Cambridge University Press, Cambridge, pp 49–109

Timmis KN, Cabello F, Cohen SN (1978) Cloning and characterization of *Eco*RI and *Hin*dIII restriction endonuclease-generated fragments of antibiotic resistance plasmids R6-5 and R6. Mol Gen Genet 162:121–137

Timmis KN, Manning PA, Echarti C, Timmis JK, Moll A (1981) Serum resistance in *E. coli*. In: Levy SB, Clowers RC, Koenig EL (eds) Molecular biology, pathogenicity, and ecology of bacterial plasmids. Plenum Press, New York London, pp 133–143

Wirth DF, Pratt DM (1982) Rapid identification of *Leishmania* species by specific hybridization of kinetoplast DNA in cutaneous lesions. Proc Natl Acad Sci USA 79:6999–7003

Chapter 2 Mutagenesis

2.1 Isolation of Suppressible Mutations by Hydroxylamine Mutagenesis in Vitro

R. EICHENLAUB [1]

Contents

1. General Introduction 106
2. Experiment: Isolation of Replication Defective Derivatives of pML31 107
    a) Objectives 107
    b) Strains 107
    c) Procedure 107
3. Materials 109
4. References 110

1. General Introduction

Bacterial plasmids are autonomously replicating mini chromosomes (replicons) which contain genetic information for many different properties, including the production of toxins and adhesion antigens, resistance to antibiotics, and restriction and modification of DNA. The analysis of these and other properties is greatly faciliated by the isolation of plasmid derivatives carrying mutations in relevant genes. Conditional plasmid mutants affected in plasmid maintenance have, for example, greatly facilitated investigations on the mechanism and control of plasmid replication. Treatment of plasmid-harboring bacteria with nitrosoguanidine was first used to obtain mutant plasmids defective in replication (Kingsbury and Helinski 1973, Collins et al. 1978), but this method did not prove to be useful for multicopy plasmids. Subsequently, the exposure in vitro of purified plasmid DNA to mutagens, such as hydroxylamine (HA), and introduction of the mutagenized DNA into bacteria by transformation, was found to be very effective for the isolation of mutants of low and high copy number plasmids (Hashimoto-Gotoh and Sekiguchi 1976, Humphreys et al. 1976, Eichenlaub 1979, Eichenlaub and Wehlmann 1980). The crucial element in this procedure is transformation, which enables the biological separation of individual plasmid DNA molecules and their subsequent propagation as pure clones; mutagenesis may be carried out equally well in vivo, provided that the mutagenized plasmid DNA is subsequently isolated and used to transform a non-mutagenized host.

---

1 Universität Hamburg, Institut für Allgemeine Botanik, Arbeitsbereich Genetik, Ohnhorststr. 18, D-2000 Hamburg 52, Fed. Rep. of Germany

Hydroxylamine preferentially reacts with cytosine, which is so modified that during DNA replication it pairs with adenine instead of guanine, leading to a transition-type mutations (Freese et al. 1961). If appropriate selection or screening methods are available, mutant plasmids exhibiting all or most potential phenotypes can be obtained.

2. Experiment: Isolation of Replication Defective Derivatives of pML31

a) Objectives

Plasmid pML31 (mini-F) DNA will be treated with HA and introduced into *E. coli* RH2358 $supF_{ts}$ by transformation. Specific selection and screening methods will then be employed to identify plasmids with thermosensitive and amber mutations affecting plasmid replication or antibiotic resistance.

b) Strains

E. coli RH2358 $supF_{ts}$ (thermosensitive suppressor tRNA) ara_{am} lac_{am} $galU_{K2am}$ $galE$ trp_{am} tsx_{am} (Delcure et al. 1977)
E. coli QD5003 $supF$
E. coli W3550 su^-
E. coli C600 (pML31) (Lovett and Helinski 1976; used for isolation of plasmid pML31 DNA).

c) Procedure

Day 1

Mix 10 µl of purified CCC plasmid DNA (1–2 µg) in TES buffer with 50 µl 0.1 M Na-phosphate, pH 6.0, and 40 µl 1 M HA, pH 6.0.
For the control, mix 10 µl of DNA with 90 µl of 0.1 M Na-phosphate.
Incubate both solutions for 30 min at 75°C or overnight (12–16 h) at 37°C.
Dialyze the DNA solutions at least 4 h against TES buffer (1 l, two changes) to remove HA, and transform strain *E. coli* RH2358 $supF_{ts}$, according to the procedure described in this volume.
Plate transformed cells on L-broth Km plates and incubate plates for 2 days at 25–28°C.

Day 3

Selection and Screening for Mutants

The control transformation with untreated plasmid DNA permits calculation of the proportion of plasmid DNA that "survives" mutagenesis, which should be between

1 and 10% (see Table 1). This level of survival guarantees a reasonable yield of mutant plasmids. A lower level provides higher yields, but the incidence of double mutations is also increased.

Replica plate Km^r clones on two L-broth plates and incubate one at 42°C and the other at 25°C for 1 and 2 days, respectively.

Note: All colonies/cultures to be cultivated at 42°C should be transferred to *pre-warmed* plates/broth, to avoid residual growth at lower temperatures.

Table 1. Relationship between mutation frequency and cell survival after HA treatment

Experiment	Survival	Clones tested	Mutants total	Mutants	
				rep^-	Km^-
1	30%	600	2	0	2
2	2%	550	18	4	14

Day 5

Clones not growing at 42°C contain plasmids mutated either in replication or in antibiotic resistance functions. Plasmids mutated in replication functions can be identified by testing for plasmid loss from actively dividing bacteria (segregation).

Culture clones at 42°C in antibiotic-free L-broth and maintain in log phase by periodic dilution with fresh pre-warmed medium.

At 1 h intervals, plate dilutions of cultures in parallel on L-broth and L-broth-Km plates.

Follow plasmid segregation for 10 generations (about 5 h).

Incubate plates for 2 days at 25°C, count colonies, and express maintenance of the plasmid (i.e., survival of the Km^r gene) as $\log (N/N_o)$, where N is the number of colonies obtained on L-broth Km and N_o is the number on L-broth plates.

A typical plot of plasmid segregation against time is given in Fig. 1. Plasmid mutants that segregate at 42°C are considered to be defective in DNA replication. Clones that exhibit antibiotic resistance at 25°C, but not at 42°C, and that do not lose the Km^r phenotype at the latter temperature, carry plasmids mutated in the antibiotic resistance gene.

It now remains to determine whether the mutant plasmids carry thermosensitive or amber mutations.

Prepare plasmid DNA from the strains harboring plasmid mutants by the procedure described in this volume (note: culture clones at 28°C), and use to transform *E. coli* strains QD5003 *sup*F and W3550 su^-.

Plate transformed cells on L-broth Km plates and incubate at 25–28°C and 42°C.

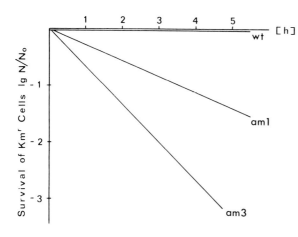

Fig. 1. Segregation kinetics of amber mutant derivatives of plasmid pML31 defective in replication. *wt* wild type; *am1* pML31am1; *am3* pML31am3

Although reversion frequencies are variable, plasmids containing amber mutations will tend to give very low transformation rates with an su^- host, compared to those obtained with a $supF$ host; such rates will be independent of the incubation temperature. In contrast, plasmids containing thermosensitive mutations will transform the two strains with similar frequencies, but these will be temperature-dependent, i.e., greatly reduced for bacteria plated at 42°C.

Mutant plasmids affected in replication can be further analyzed by measuring the incorporation of ^{3}H-thymidine into plasmid DNA at a nonpermissive temperature. This information may indicate whether mutants are defective in initiation of DNA replication, plasmid maintenance in dividing cells, or copy number control. Moreover, plasmids containing amber mutations can be used to identify specific gene products using the minicell or cell-free protein synthesis systems described in this volume.

3. Materials

Media

L-broth medium contains 10 g tryptone (Difco), 5 g yeast extract (Difco), and 5 g NaCl per liter.
L-broth-Km plates contain kanamycin at a concentration of 50 µg/ml.

Chemicals and solutions

Hydroxylamine 1 M, pH 6.0, in water
Sodium phosphate buffer 0.1 M, pH 6.0
DNA of plasmid pML31 in TES, DNA concentration 100–200 µg/ml
TES buffer (100 mM Tris-HCl, pH 7.5, 50 mM NaCl, 5 mM EDTA)

4. References

Collins J, Yanofsky S, Helinski DR (1978) Involvement of the DNA-protein complex in the replication of plasmid ColE1. Mol Gen Genet 167:21–28

Delcuve G, Cabezón T, Ghysen A, Herzog A, Bollen A (1977) Amber mutations in *Escherichia coli* essential genes: Isolation of mutants affected in the ribosome. Mol Gen Genet 157:149–153

Eichenlaub R (1979) Mutants of the mini-F plasmid pML31 thermosensitive in replication. J Bacteriol 138:559–566

Eichenlaub R, Wehlmann H (1980) Amber mutants of plasmid mini-F defective in replication. Mol Gen Genet 180:201–204

Freese E, Bautz E, Bautz-Freese E (1961) The chemical and mutagenic specificity of hydroxylamine. Proc Natl Acad Sci USA 47:845–855

Hashimoto-Gotoh T, Sekiguchi M (1976) Isolation of temperature-sensitive mutants of R plasmid by in vitro mutagenesis with hydroxylamine. J Bacteriol 127:1561–1563

Humphreys GD, Willshaw GA, Smith HR, Anderson ES (1976) Mutagenesis of plasmid DNA with hydroxylamine: Isolation of mutants of multi-copy plasmids. Mol Gen Genet 145:101–108

Kingsbury DT, Helinski DR (1973) Temperature-sensitive mutants for the replication of plasmids in *Escherichia coli*. I. Isolation and specificity of host and plasmid mutations. Genetics 74:17–31

Lovett MA, Helinski DR (1976) Method for the isolation of the replication region of a bacterial replicon: Construction of a mini-F' *Km* plasmid. J Bacteriol 127:982–987

2.2 Transposition of Tn*1* to the Phage P1 Genome: Isolation of Restriction Deficient Mutants

A. PÜHLER, V. KRISHNAPILLAI, and H. HEILMANN [1]

Contents

1. General Introduction 111
2. Experiment 1: Isolation of Restriction Deficient Mutants of Phage P1 Induced by Tn*1*
   Insertions 112
   a) Introduction 112
   b) Objectives 114
   c) Procedure 114
   d) Results 115
3. Experiment 2: Physical Mapping of *res*::Tn*1* Mutations in the P1 Genome 116
   a) Introduction 116
   b) Objective 119
   c) Procedure 119
   d) Results 120
4. Materials 122
5. References 123

1. General Introduction

The following experiments demonstrate how transposons can be used to induce mutations and how such mutations can be mapped by restriction analysis. In particular, the transposon Tn*1* of plasmid RP4 will be used to mutagenize the phage P1 genome.

The coliphage P1 was first described by Lennox (1955) as a phage capable of general transduction. Ikeda and Tomizawa (1968) have shown that P1 phage particles contain linear DNA molecules of $\approx$ 110 kb length. After lysogenization, the P1 prophage genome was found to be 12% shorter. It replicates (unlike λ) as an extrachromosomal replication unit and its DNA is packaged into the phage particle by a headful mechanism. The packaged DNA shows redundancy and is circularly permuted. This redundancy is important, since additional DNA can be inserted into the P1 genome without loss of viability, if the inserted DNA is shorter than the P1 redundancy. This property of P1 DNA plays a role in transposition experiments, e.g., when Tn*1* is transposed to the P1 genome.

---

[1] Lehrstuhl für Genetik, Universität Bielefeld, D-4800 Bielefeld, Fed. Rep. of Germany

Transposon Tn*1* played a key role in the discovery of transposition of antibiotic resistance genes. In 1974, Hedges and Jacob reported that the ampicillin resistance determinant of plasmid RP4 could be transposed to a variety of other replicons of different incompatibility groups. Plasmid RP4 (Datta et al. 1971) is well-known for its broad host range, which includes its transfer into enteric and soil bacteria from Pseudomonas. It carries genes conferring resistance against ampicillin, kanamycin, and tetracycline. The DNA element carrying the ampicillin resistance determinant of plasmid RP4 was initially called transposon TnA; it was later renamed Tn*1*. Transposition of Tn*1* was also found by the Erlangen group at the end of 1973. Upon transducing the RP4 plasmid with phage P1, we found in addition to RP4 transduction that the ampicillin resistant gene of the RP4 plasmid transposed onto the P1 genome; an event which was best explained by transposition (Pühler and Krauß 1977). It should be mentioned that such transposition events occur in a P1 lysogenic strain carrying the RP4 plasmid. The separation of RP4 and P1::Tn*1* genomes can be achieved very easily by inducing the lysogenic strain, harvesting the phage lysate, and reinfecting an *E. coli* cell culture. We used this system to mutagnize the P1 genome by Tn*1* insertions (Heilmann 1979) and were able to isolate and characterize mutants defective in the synthesis of the P1-encoded restriction endonuclease called *Eco*P1 (Meselson and Yuan 1968, Habermann 1974, Bächi et al. 1979). This enzyme is part of the P1 restriction/modification system. It is now known that the P1 modification activity methylates the marked adenine of the sequence AG$\overset{*}{A}$CC and that the P1 restriction activity cuts the DNA 25–27 base pairs from the site of methylation in the 3' direction, with a 2–4 base pair stagger between the cuts (Bächi et al. 1979).

In this article, the isolation of restriction deficient mutants *(res)* of phage P1 by transposon Tn*1* mutagenesis and the mapping of the *res*::Tn*1* mutations on the P1 genome are described.

2. Experiment 1: Isolation of Restriction Deficient Mutants of Phage P1 Induced by Tn*1* Insertions

a) Introduction

In this experiment the Rosner phage P1Cmts (Rosner 1972) is used. It carries a temperature sensitive (ts) repressor gene and is therefore inducible by a heat shock. In addition, transposon Tn*9* with a chloramphenicol resistance gene (Cm) is located on the phage genome. Thus, P1Cmts lysogenic cells can be recognized by forming colonies on agar plates containing chloramphenicol.

The scheme of the experiment is outlined in Fig. 1. An *E. coli* strain carrying the prophage P1Cmts and the resistance plasmid RP4 is heat induced and the phage progeny is harvested. It consists of the normal P1Cmts phage and a minor fraction of (P1Cmts)::Tn*1* phages that have received transposon Tn*1* by transposition from plasmid RP4. It should be noted that the phage progeny will also contain transducing particles, e.g., particles that are packed with chromosomal or RP4 DNA.

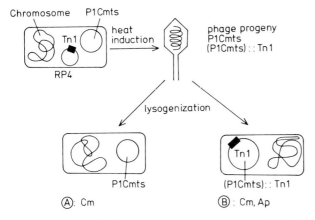

Fig. 1. Isolation of P1Cmts phages carrying transposon Tn*1*. Heat induction of an *E. coli* strain carrying the prophage P1Cmts and the resistance plasmid RP4 results in phage progeny containing mainly P1Cmts and only a few (P1Cmts)::Tn*1* phages. Separation of these phages can be obtained by a lysogenization step. Cells carrying P1Cmts (A) are chloramphenicol-resistant *(Cm)*, whereas cells carrying (P1Cmts)::Tn*1* (B) are chloramphenicol *(Cm)*, and ampicillin resistant *(Ap)*

The phage progeny is now used to lysogenize (or to transduce) an *E. coli* strain. Lysogenic cells, carrying the prophage P1Cmts or (P1Cmts)::Tn*1* can be distinguished by their antibiotic resistance markers. (P1Cmts)::Tn*1* lysogenic cells are resistant to chloramphenicol and ampicillin. In addition, RP4 transducants can be isolated. They show resistance to ampicillin, kanamycin, and tetracycline.

In order to detect restriction deficient (P1Cmts)::Tn*1* phages, the high plating efficiency of P1-unmodified λ-phages on (P1Cmts)::Tn*1* lysogens is used as a test system. The plating efficiencies (e.o.p.) of P1-unmodified λ-phages on different *E. coli* strains are summarized in Table 1. The low e.o.p. value can be explained by the restriction system of phage P1. Since P1-unmodified λ-phages do not carry the P1 modification, they are restricted in a P1 lysogenic cell. On the other hand, after inactivation of the restriction system of phage P1 (*res* mutants) the e.o.p. value is as high as that of a nonlysogenic strain. For this experiment, the special mutant phage λ*vir* is used. λ*vir* is a virulent phage unable to lysogenize *E. coli* (Jacob and Monod 1961).

Table 1. Efficiency of plating (e.o.p.) of unmodified λ-phages [a] on different nonlysogenic and P1 lysogenic *E. coli* C600 strains

Test strain	P1 restriction phenotype	e.o.p. of P1-unmodified λ-phages [a]
E. coli C600	Res⁻	1
E. coli C600 (P1Cmts)::Tn*1*	Res⁺	10^{-4}
E. coli C600, P1Cmts, *res*::Tn*1*	Res⁻	1

[a] "P1-unmodified λ" denotes that λ phage was grown on *E. coli* C600 and does not carry the P1 modification

b) Objectives

1. To insert the ampicillin transposon Tn*1* into the genome of phage P1.
2. To isolate restriction deficient mutants of phage P1 induced by transposon Tn*1*.

c) Procedure

Day 1

Production of a P1 Lysate by Heat Induction of a Lysogenic E. coli Strain

10 ml of RGMC medium are inoculated with *E. coli* C600 (RP4) (P1Cmts) and grown at 30°C until a cell density of 6×10^8 cells per ml or an OD_{580} of 0.6 is reached.

For heat induction, add 23 ml of fresh RGMC medium heated to 50°C. This will give a temperature shift from 30°C to 44°C. The induced cell suspension is kept in a water bath at 44°C for 30–45 min. After this time, the OD_{580} will have dropped to approximately 0.05.

After lysis, add 3 drops of chloroform and pellet the cell debris (10 min, $12,000 \times g$). Collect the supernatant. Add again 3 drops of chloroform. Such a P1 lysate is stable for several months.

Titrate the phage lysate by plating 0.1 ml aliquots of the serial dilutions (10^{-4}, 10^{-6}, 10^{-8}) onto an *E. coli* C600 bacterial lawn. The titer of the lysate should be 10^8 to 10^{10} pfu/ml.

Day 2

Lysogenization of E. coli with a P1Cmts Lysate

Mix 0.5 ml of the P1Cmts lysate (T = 10^9 pfu/ml) and 0.5 ml of an o/n culture of *E. coli* C600 in 5 mM $CaCl_2$ and incubate for 30 min at 32°C.

Plate 0.1 ml of dilution step −1, −2, −3 (decimal) onto Pa agar plates containing ampicillin (PA + Ap).

Plate 0.1 ml of dilution step −3, −4, −5 onto LB agar plates containing chloramphenicol (LB + Cm).

As a control, streak the uninfected C600 culture and the P1Cmts lysate separately onto PA + Ap and LB + Cm.

Incubate the plates overnight at 32°C.

Day 3

Toothpick Isolation

Record the number of colonies grown on different plates.

Transfer 100 single colonies each from PA + Ap and LB + Cm onto PA plates by toothpick isolation.

Incubate the plates overnight at 32°C.

Day 4

Replica Plating

Replica plate the master plates onto PA + Ap, PA + Km, PA + Tc, and LB + Cm plates. Incubate the plates o/n at 32°C.

Day 5

Test for res::Tn1 Mutants of P1Cmts Phages

Record the results for each master plate individually and compute the frequency of RP4 transduction and the occurrence of (P1Cmts) lysogens and (P1Cmts)::Tn1 lysogens.

Note: *E. coli* strains carrying the different extrachromosomal elements RP4, P1Cmts, and P1Cmts::Tn1 can be recognized by their antibiotic resistance pattern: *E. coli* C600 (RP4) is Ap, Km, Tc; *E. coli* C600 (P1Cmts) is Cm; and *E. coli* C600 ((P1Cmts)::Tn1) is Cm, Ap.

Resuspend each of the (P1Cmts)::Tn1 lysogens in 1 ml PS and put a drop of this suspension onto a PA plate. After drying, add one drop of a P1-unmodified λvir lysate (T = 10^8 pfu/ml). The P1-unmodified λvir lysate is produced by growing the λvir phage on an *E. coli* C600 host using the confluent lysis procedure. Up to 10 strains can be tested on 1 plate. As a control, include a C600 nonlysogenic strain for P1 on each test plate. Incubate overnight at 32°C.

Day 6

Isolation of P1Cmts, res::Tn1 Lysogens

P1-unmodified λvir phage multiplies on P1Cmts, *res*::Tn1 lysogens at the same frequency as the control. Record their occurrence and stock the strains from the master plate.

d) Results

Typical results which can be obtained in the lysogenization experiment are shown in Table 2. Cm-r cells appear at a frequency of 10^{-1}. They are due to lysogenization by P1Cmts, amp-r cells are less frequent. They can be explained by lysogenization with (P1Cmts)::Tn1 or by transduction of the RP4 plasmid. The prophage or plasmid type of these amp-r colonies is shown in Table 3. 81% carry the prophage (P1Cmts)::Tn1. Thus, the majority of the amp-r colonies are due to transposition of transposon Tn1 onto the P1 genome. 10% have plasmid RP4 and 5% contain RP4 as well as P1Cmts. Evidently, these strains were produced by a double infection. It is interesting to note that 4% show the phenotype (P1ts)::Tn1. We assume that they have lost the gene for chloramphenicol resistance.

Table 2. Frequency of antibiotic resistance markers after lysogenization with a P1Cmts lysate obtained by heat induction of *E. coli* strain C600 (P1Cmts) (RP4)

Antibiotic resistance marker	Selective medium	Lysogens or transductants per ml	Frequency [a]
cam-r	LB + Cm	5×10^7	10^{-1}
amp-r	PA + Ap	5×10^3	10^{-5}

[a] For frequency calculation a recipient titer of 5×10^8 cells/ml was used

Table 3. Resistance pattern of primarily amp-r lysogens

Resistance pattern	Phage or plasmid type	Frequency in % [a]
amp-r	(P1ts)::Tn*1*	4
amp-r, cam-r	(P1Cmts)::Tn*1*	81
amp-r, kam-r, tet-r	RP4	10
amp-r, kam-r, tet-r, cam-r	P1Cmts, RP4	5

[a] Altogether 164 amp-r colonies were tested

3. Experiment 2: Physical Mapping of res::Tn*1* Mutations on the P1 Genome

a) Introduction

In the preceeding experiment, restriction deficient mutants of phage P1Cmts were isolated. The mutations were induced by insertion of transposon Tn*1* into the P1Cmts genome. In this experiment, the mutational sites of these P1Cmts, res::Tn*1* mutants will be mapped by restriction analysis. The experimental steps include large scale production of the phage, concentration and purification of the lysates, and isolation and restriction analysis of phage DNA.

In order to evaluate the fragment pattern of different P1Cmts, res::Tn*1* DNAs, it is necessary to know the P1 restriction map. The first restriction map of the P1 genome was published by Bächi and Arber (1977). It contains restriction sites for the endonucleases *Eco*RI, *Bam*HI, and *Pst*I (Fig. 2). The lengths of the *Eco*RI and *Bam*HI fragments of the P1 genome are summarized in Table 4. The restriction map of phage P1Cmts is very similar, except that the *Eco*RI fragment 4 carries the chloramphenicol transposon Tn*9* in tandem arrangement and the *Eco*RI fragment 7 has an additional IS*1* element. Fig. 3 shows the map of transposon Tn*9*. Tn*9* is flanked by two IS*1* elements in the same orientation. It is important to note that the only *Pst*I sites of the P1 genome occur within the IS*1* elements. In Fig. 4, the *Eco*RI and *Pst*I restriction map of the DNA fragment containing the Tn*9* tandem is shown. From the restriction map, it can be concluded that the first Tn*9* transposon is integrated into the P1 genome via IS*1* homology. The tandem structure is then the result of an intramolecular

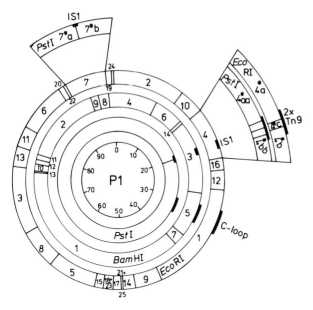

Fig. 2. Restriction maps of the genomes of phage P1 and phage P1Cmts. The *Eco*RI, *Bam*HI, and *Pst*I restriction map of the phage P1 are shown. The fragments are numbered according to their position in the gel and their lengths are summarized in Table 4. The restriction map of phage P1Cmts differs as indicated in the *Eco*RI fragments 4 and 7: *Eco*RI fragment 7 carries an additional IS*1* element, whereas *Eco*RI fragment 4 contains two Tn*9* transposons in tandem arrangement. The maps of IS*1* and Tn*9* are shown in more detail in Fig. 3. The exact location of the Tn*9* tandem in the altered *Eco*RI fragment 4 of the P1Cmts genome is indicated in Fig. 4. The lengths of the numbered fragments of the P1Cmts genome are also included in Table 4. The P1Cmts map was constructed by Heilmann (1979)

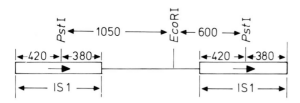

Fig. 3. Map of the chloramphenicol transposon Tn*9*. Transposon Tn*9* is flanked by two IS*1* elements in the same orientation. These elements carry two *Pst*I sites. In addition, there is one *Eco*RI site in the cam-r gene. The lengths of the fragments are given in bp. The map was constructed by Heilmann (1979), using gel electrophoresis for fragment sizing. The measurement are in agreement with published data, e.g., with the IS*1* sequence determined by Ohtsubo and Ohtsubo (1978) and by Johnsrud (1979)

Table 4. Lengths of EcoRI and BamHI fragments of the P1 and P1Cmts genome

Fragment numbers [a]	Lengths of fragments in kb after digestion with		
	EcoRI	BamHI	EcoRI/PstI
1	14,55	34,65	
2	9,90	13,65	
3	8,85	11,55	
4	6,45	8,40	
4*a	5,03		
4*b	3,53		
4*c	1,65		
4*aa			4,05
4*bb			2,40
5	6,30	7,35	
6	6,30	5,03	
7	5,63	1,77	
7*a			3,60
7*b			2,78
8	5,33	1,76	
9	3,45	1,29	
10	3,30	1,23	
11	2,93	0,80	
12	2,70	0,71	
13	2,70	0,29	
14	1,65	0,29	
15	1,47		
16	1,29		
17	1,07		
18	1,04		
19	0,90		
20	0,62		
21	0,39		
22	0,35		
23	0,30		
24	0,15		
25	0,08		

[a] The fragment numbers are those indicated in Fig. 1. "*" denotes fragments obtained when P1Cmts DNA is digested

amplification step. The information still required is the restriction map of transposon Tn1. This map is shown in Fig. 5. It is important to note that Tn1 does not carry an EcoRI site. Thus, integration of Tn1 into an EcoRI fragment of the P1Cmts genome results in an enlargement of this fragment to the size of Tn1. By this method, it is possible to identify EcoRI fragments that carry res::Tn1 mutations. Since Tn1 contains one BamHI site, BamHI digestions can be used for more precise measurement of Tn1 insertions.

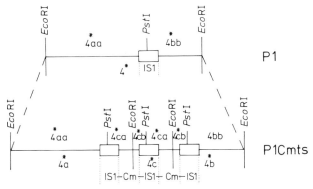

Fig. 4. The tandem transposon Tn9 of the phage P1Cmts. The tandem transposon Tn9 contains three IS*1* elements and two chloramphenicol resistance genes. The lengths of restriction fragments obtained after *Eco*RI and/or *Pst*I digestions were determined by Heilmann (1979): *4*a*, 5,03; *4*b*, 3,53; *4*c*, 1,65; *4*aa*, 4,05; *4*bb*, 2,40; *4*ca*, 1,05; *4*cb*, 0,60

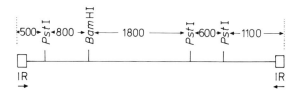

Fig. 5. Restriction map of the ampicillin transposon Tn*1*. The restriction map of the ampicillin transposon Tn*1* was described by Priefer et al. (1981). Tn*1* carries one *Bam*HI and three *Pst*I sites. Distances in bp were determined by gel electrophoresis and electron microscopy (Simon 1980). They are in general agreement with sequence data for the ampicillin transposon Tn*3* (Heffron et al. 1979)

b) Objective

To map *res*::Tn*1* mutations on *Eco*RI and *Bam*HI fragments of phage P1Cmts.

c) Procedure

Day 1

Large Scale Production of P1Cmts, res::Tn1 Phages

Heat induce P1Cmts, *res*::Tn*1* lysogenic strains in 100 ml of RGMC medium. Use the same method as described in the preceeding experiment.

After removal of cells by low speed centrifugation, add chloroform (1 ml) and spin down the cell debris (5 min, 10,000 rpm). Add chloroform again (1 ml).

Titrate the lysate. Normally, titers of 10^9 pfu/ml or more should be obtainable.

Day 2 and 3

Concentration and Purification of the P1Cmts, res::Tn1 Lysates

Spin down the lysate at high speed (40 min, 50,000 rpm) and resuspend the pellet in 0.5 ml SM buffer overnight.
Treat the concentrated lysate with DNase/RNase (10 μg/ml) for 60 min at 37°C.
Titrate the lysate. Titers of 10^{11} pfu/ml should be obtainable.
Purify and concentrate the lysate by CsCl step gradient centrifugation in a swingout rotor.
CsCl steps: 1.6 g/ml, 1.5 g/ml, 1.4 g/ml.
Overlay these CsCl steps with the phage lysate and centrifuge for 60 min at 40,000 rpm. A blue phage band should be visible in the CsCl steps after centrifugation.
Isolate the phage band by puncturing the tube with a syringe and dialyze against SM buffer.

Day 4

Isolation of Phage DNA by Phenol Treatment

Mix the purified and concentrated lysate with an equal volume of TE buffer saturated with phenol in Eppendorf tubes.
Shake the tube for 2 min by hand. After centrifugation for 2 min in the Eppendorf centrifuge, collect the aqueous phase containing the DNA.
Repeat this procedure twice.
Residual phenol is extracted from the DNA solution by treatment with an equal volume of either (2 ×) or by dialysis against TE buffer (change buffer 2 ×).

Day 5

Restriction Analysis of P1Cmts, res::Tn1 DNA

P1Cmts and P1Cmts, *res*::Tn*1* DNA is restricted by *Eco*RI and *Bam*HI by the usual procedure.

The fragments obtained are separated on agarose gels. By comparing the fragment pattern of P1Cmts and P1Cmts, *res*::Tn*1* DNA, the *Eco*RI and *Bam*HI fragments that carry the Tn*1* insertion indicating the location of a *res* gene, can be determined.

d) Results

It should be mentioned that it is sometimes difficult to get a high phage titer by heat induction of a P1Cmts lysogenic strain. Optimal growth of the strain at 32°C and violent shaking of the culture after the heat shock treatment, followed by slow shaking when the cells start to lyse, can help to overcome this problem.

After isolation and restriction of P1Cmts, *res*::Tn*1* DNA fragment patterns can be obtained identical to those shown in Fig. 6. Tn*1* insertions leading to restriction deficient phage strains map in the *Eco*RI fragments 2 and 24. Note that (*Eco*RI fragment 2)::Tn*1* and *Eco*RI fragment 1 and that (*Eco*RI fragment 24)::Tn*1* and *Eco*RI

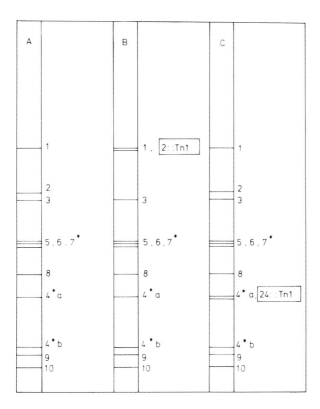

Fig. 6. Analysis of P1Cmts, res::Tn*1* DNA by *Eco*RI restriction. In lane *A*, the *Eco*RI fragments of P1Cmts DNA separated on agarose gel are shown. The numbers marking the different DNA bands correspond to those shown on the restriction map in Fig. 2. Lane *B* and lane *C* show the analysis of 2 types of res::Tn*1* mutants. In *B*, Tn*1* is inserted into *Eco*RI fragment 2, whereas in *C* *Eco*RI fragment 24 carries the Tn*1* insertion

fragment 4*a, form double bands (Fig. 6). If the same analysis is repeated with *Bam*HI digestions, the *Bam*HI fragment 4 is found to carry the Tn*1* insertions.

The experiment described so far shows only the clustering of res::Tn*1* mutations. The question of how many genes are involved in P1 restriction cannot be answered. Additional experiments are necessary to solve this problem. One way is to determine the gene products encoded by the DNA segment marked by res::Tn*1* mutations. This type of analysis has been done by Heilmann et al. (1980). It has been found that there is only one *res* gene, ≈ 1,5 kb long and coding for a polypeptide with a molecular weight of 110,000. The location of the *res* gene on the *Eco*RI map of P1Cmts and the direction of its transcription are presented in Fig. 7. It should be noted that Tn*1* insertions into the *res* gene result in truncated polypeptides. Analysis of the molecular weights of these truncated polypeptides coupled with the precise measurement of the Tn*1* insertions enabled Heilmann et al. (1980) to determine the coding region of the *res* gene on the P1Cmts genome.

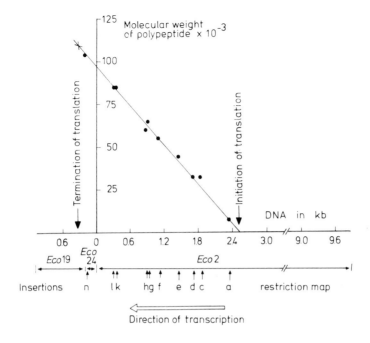

Fig. 7. Map of the *res* gene of phage P1. The *abscissa* gives kb of DNA and the *ordinate* molecular weight of the polypeptides ($\times 10^{-3}$). In addition, part of the *Eco*RI map of phage P1Cmts with some transposon insertion sites *(a–n)* are indicated. Each *point* on the graph represents the molecular weight of a truncated gene.product. The *arrow* shows the direction of transcription of the *res* gene

4. Materials

Strains. Phages, and Plasmids

E. coli C600	thr, leu, B_1, gal, lacY, supE, tonA
P1Cmts	(Rosner 1972)
λvir	(Jacob and Monod 1961)
RP4	Ap, Km, Tc (Datta et al. 1971)

Media

PA	Pennassay broth, 17.5 g per liter
LB	Luria broth per liter: 10 g Bactotryptone 5 g Yeast extract 5 g NaCl adjust to pH 7.4 with NaOH

RGMC   per liter:
                 10 g Bactotryptone
                 1 g Yeast extract
                 8 g NaCl
                 1 g glucose
                 1 g MgCl$_2$
                 0.3 g CaCl$_2$
                 adjust to pH 7.6 with 12% Tris

                 Antibiotics added:
                 100 µg/ml ampicillin
                 50 µg/ml kanamycin
                 10 µg/ml tetracycline
                 100 µg/ml chloramphenicol

Buffers

TE: 1 mM EDTA, 10 mM Tris/HCl, pH 7.5
SM: 100 mM NaCl, 10 mM MgSO$_4$, 20 mM Tris/HCl, pH 7.5
PS: 8.5 g NaCl per liter

5. References

Bächi B, Arber W (1977) Physical mapping of *Bgl*II, *Bam*HI, *Eco*RI, *Hin*dII and *Pst*I restriction fragments of bacteriophage P1 DNA. Mol Gen Genet 153:311–324

Bächi B, Reiser J, Pirotta V (1979) Methylation and cleavage sequences of the *Eco*P1 restriction-modification enzyme. J Mol Biol 128:143–163

Datta N, Hedges RW, Shaw EJ, Sykes RB, Richmond MH (1971) Properties of a R factor from *Pseudomonas aeruginosa*. J Bacteriol 108 (3):1244–1249

Habermann A (1974) The bacteriophage P1 restriction endonuclease. J Mol Biol 89:545–563

Hedges RW, Jacob AE (1974) Transposition of ampicillin resistance from RP4 to other replicons. Mol Gen Genet 132:31–40

Heffron F, McCarthy BJ, Ohtsubo H, Ohtsubo E (1979) DNA sequence analysis of the transposon Tn3: Three genes and three sites involved in transposition of Tn3. Cell 18:1153–1163

Heilmann H (1979) Molekulare Analyse des Restriktionsgens des Phagen P1, Dissertation, Univ Erlangen

Heilmann H, Pühler A, Brukardt HJ, Reeve JN (1980) Transposon mutagenesis of the gene encoding the bacteriophage P1 restriction endonuclease: Co-linearity of the gene and gene product. J Mol Biol 144:387–396

Ikeda H, Tomizawa J (1968) Prophage P1, an extrachromosomal replication unit. Symp Quant Biol Cold Spring Harbor 33:791–798

Jacob F, Monod J (1961) Genetic regulatory mechanism in the synthesis of proteins. J Mol Biol 3:318

Johnsrud L (1979) DNA sequence of the transposable element IS1. Mol Gen Genet 169:213–218

Lennox ES (1955) Transduction of linked genetic characters of the host by bacteriophage P1. Virology 1:190–206

Meselson M, Yuan R (1968) DNA restriction enzyme from *E. coli*. Nature 217:1110–1114

Ohtsubo H, Ohtsubo E (1978) Nucleotide sequence of an insertion element, IS1. Proc Natl Acad Sci USA 75:615–619

Priefer UB, Burkardt HJ, Klipp W, Pühler A (1981) ISR1: An insertion element isolated from the soil bacterium *Rhizobium lupini.* Cold Spring Harbor Symp Quant Biol 45:87–91

Pühler A, Krauss G (1977) Transposition of the ampicillin resistance gene from the RP4 factor to the bacteriophage P1 and to the *Escherichia coli* fertility factor. In: Mutsuhashi S, Rosival L, Krĉméry V (eds) Plasmids 3rd Int Symp Antibiotic Resistance. Avicenum Praque. Springer, Berlin Heidelberg New York, pp 151–160

Rosner JL (1972) Formation, induction and curing of bacteriophage P1 lysogens. Virology 48: 679–689

Simon R (1980) Inkompatibilität und Replikation des Resistenzplasmids RP4. Dissertation, Univ Erlangen

2.3 In Vivo Genetic Engineering: Use of Transposable Elements in Plasmid Manipulation and Mutagenesis of Bacteria Other than *E. coli*

R. SIMON[1]

Contents

1. General Introduction 125
   a) Transposable Elements 125
   b) P-type Plasmids 126
   c) Mutagenesis and Physical Mapping of Plasmid Borne Genes 127
   d) Transposon Mutagenesis of Bacteria Other than *E. coli* 127
2. Experiment 1: Insertion of Bacteriophage Mu Genome into Plasmid RP4 129
   a) Introduction 129
   b) Objectives 129
   c) Procedure and Results 130
3. Experiment 2: Tn5 Transposition into the Self-Transmissible Plasmid RP4-Km::Mu . . 132
   a) Introduction 132
   b) Objectives 132
   c) Procedure and Results 132
4. Experiment 3: Tn5 Transposition into Mobilizable *E. coli* Vectors 133
   a) Introduction 133
   b) Objectives 134
   c) Procedure and Results 134
5. Experiment 4: Transposon Mutagenesis of *Rhizobium* Using Broad Host Range "Suicide Plasmids" and Mobilizable *E. coli* Vectors 135
   a) Introduction 135
   b) Objectives 136
   c) Procedure and Results 136
6. Materials 138
7. References 139

1. General Introduction

a) Transposable Elements

Recently a number of bacterial drug resistance genes capable of translocation between chromosomal, phage, or plasmid DNA have been described. They are called transposable elements or transposons and are defined as discrete DNA segments which can insert into multiple sites of a genome (Campbell et al. 1977). Transposons have not

---

[1] Lehrstuhl für Genetik, Fakultät für Biologie, Universität Bielefeld, D-4800 Bielefeld 1, Fed. Rep. of Germany

yet been shown to exist autonomously. They need to reside in a functional bacterial replicon. Several transposons and their genetical and physical characteristics are summarized in Table 1. The experimental part of this chapter will give some examples which demonstrate how spontaneous transposition events from one replicon to another can be obtained. The potential use of transposons as genetic tools will be discussed later.

Table 1. Transposable elements

Transposable element	Plasmid origin	Resistance markers [a]	Size (bp)	Reference
Tn*1*	RP4	Ap	4957 [b]	Hedges and Jacob (1974)
Tn*5*	JRG7	Km/Nm	5400	Berg et al. (1975)
Tn*7*	R483	Tp, Sm	14,000	Barth et al. (1976)
Tn*9*	pSM14	Cm	2638	Gottesmann and Rosner (1975)
				Alton and Vapnek (1979)
Tn*10*	R100	Tc	9300	Foster et al. (1975)
				Kleckner et al. (1975)

[a] Abbreviations: *Ap* = Ampicillin; *Km* = Kanamycin; *Nm* = Neomycin; *Tp* = Trimethoprim; *Sm* = Streptomycin; *Cm* = Chloramphenicol; *Tc* = Tetracycline
[b] Tn*1*, Tn*2*, and Tn*3* are very similar; the complete sequence of Tn*3* is available (Heffron and McCarthy 1979)

The temperate bacteriophage Mu has also been described as a "giant" transposon under the cloak of a virus (Bukhari 1976). It has all the genes needed for its replication and morphogenesis and an intricate system to control all these various functions. Upon infection of *E. coli* K12, Mu can replicate either by lytic phage multiplication, or Mu DNA can become stably integrated into the host DNA. The Mu integration can be described as transposition from the linear viral DNA form into random sites on the host genome. Mu insertion causes a very stable polar mutation in a given operon. This property has been widely used in *E. coli* genetics. In this chapter, a method for the introduction of mutations into plasmid borne genes by Mu insertion is described.

For a comprehensive introduction into the biology of transposable elements, the reader is referred to the collection of recent papers in *DNA Insertion Elements, Plasmids and Episomes,* A.J. Bukhari, J.A. Shapiro, S.L. Adhya, Cold Spring Harbor Laboratory (1977) and *Movable Genetic Elements,* Cold Spring Harbor Symposia on Quantitative Biology, vol XLV (1981).

b) P-Type Plasmids

In Gram-negative bacteria, a few particular plasmids have been identified, which can be transferred and are able to replicate stably in many different species (Datta and Hedges 1972, Olsen and Shipley 1973, Beringer 1974). This wide host range group of plasmids comprises the incompatibility class P in *E. coli* (Datta et al. 1971). The best studied member of these promiscuous plasmids is the plasmid RP4. Fig. 1 shows the

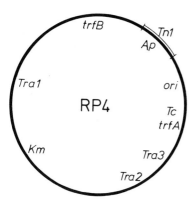

Fig. 1. Genetic map of plasmid RP4. Summary of data taken from Thomas et al. (1979) and Barth (1979). Abbreviations: *Ap, Km, Tc:* see Table 1; *trf* transacting replication function; *ori* origin of replication; *Tra* transfer functions

genetic map of this plasmid. RP4 has been shown to be identical to plasmids RP1, R68, and RK2 (Burkhardt et al. 1979). RP4 and its derivatives habe been proven to be very useful tools in genetic studies of many different Gram-negative bacteria. For example, a "broad host range"cloning vector system has been derived from the RK2 plasmid and used for constructing a gene library of *Rhizobium meliloti* (Ditta et al. 1980). The ability of RP4 derivatives to promote host chromosomal gene transfer in different bacterial species has been widely used to map bacterial chromosomes (for review see Holloway 1979). Here we will describe the use of RP4 in transposon mutagenetic experiments of bacteria other than *E. coli*.

c) Mutagenesis and Physical Mapping of Plasmid Borne Genes

Perhaps the widest application of transposons in genetic studies or strain manipulations is "insertional mutagenesis", which is used to obtain stable mutants. Insertion of a transposon into a given gene leads to non-leaky polar mutations (Kleckner 1977). In addition, the introduction of a resistance gene into a DNA molecule provides it with a readily identifiable genetic and physical marker. Transposon induced mutations of genetic loci of a transmissible or mobilizable plasmid can be isolated in simple mating experiment. The precise insertion of the transposon can then be easily located. For example, Tc- and Km-resistance in RP4 have been located by restriction analyses of RP4::Tn7 hybrid plasmids (Barth and Grinter 1977, Barth et al. 1978). In addition, very accurate mapping results can be obtained by heteroduplex experiments with transposon- or Mu-marked plasmid DNA.

d) Transposon Mutagenesis of Bacteria Other than *E. coli*

Methods for transposon mutagenesis are well established for the enteric bacteria *E. coli* and Salmonella (Kleckner 1977). There is also increasing interst in extending the applicability of transposon mutagenetic technology to other Gram-negative bacteria of economic and scientific importance. In working with transposons in strains other than *E. coli*, the first problem to overcome is the introduction of the transposon into this particular strain, i.e., to find a suitable vector molecule. As mentioned above, plasmids of the Inc-P class can be transferred to many Gram-negative bacteria. This very promiscuity makes P-type plasmids useful for transposon transfer outside *E. coli.*

Once a transposon has been introduced into a particular host, the second problem is the elimination of the vector, since transposition events can only be found if the resistance gene of the transposon is retained after the plasmid that had carried it is eliminated. It has been reported that P-type plasmids that carry an inserted Mu genome have a much reduced ability to become stably established in *Rhizobium* or *Agrobacterium* strains (Boucher et al. 1977, Denarié et al. 1977, Van Vliet et al. 1978, Beringer et al. 1978). Taking advantage of this unstable "plasmid suicide effect", the transposon Tn5 has been inserted into the genome of different Rhizobia (Beringer et al. 1978). The application of the hybrid plasmid (RP4-Km::Mu)::Tn5 which was constructed in experiment 1 and 2, for Tn5 mutagenesis of *Rhizobium* will be demonstrated in a later section.

However, the elimination of RP4::Mu vectors is not always equally efficient. In some bacterial strains, e.g., in *Rhizobium meliloti,* the elimination is very inefficient. Therefore, we offer an alternative procedure for transposon mutagenesis for bacteria other than *E. coli.* Fig. 2 describes the essential features of the system. It is assumed that pBR325 is not able to replicate in hosts other than *E. coli.* Therefore, pBR325-Mob plasmids should be the ideal vectors for introduction of transposons into strains for which this assumption is true. We will demonstrate random transposon mutagenesis of *Rhizobium* with the mobilization system that is described. In addition, we will show how mobilizable *E. coli* vectors can be used for site-specific transposon mutagenesis of *Rhizobium.* A specific DNA fragment of *R. meliloti* carrying the gene *nif*H (Ruvkun 1980) has been cloned into pBR325-Mob, then Tn5 has been inserted into *nif*H (Experiment 3 will demonstrate this procedure). If the resulting plasmid

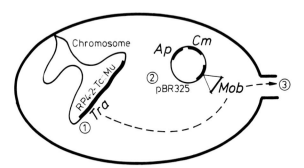

Fig. 2. Mobilization of *E. coli* plasmid vectors. *1* A derivative of the broad host range plasmid RP4 is integrated in the chromosome of a *rec*A⁻ *E. coli* strain. The genes for Tc- and Ap-resistance have been inactivated, but the Km-resistance and the transfer genes are still functional. This *E. coli* strain is called the mobilizing strain. *2* The RP4-specific Mob-site (origin of conjugal transfer) has been inserted into the plasmid pBR325. The Mob-site is located on a small *Sau*3A fragment from RP4 which has been cloned into the *Bam*HI site of pBR325. *3* The resulting plasmid pBR325-Mob (pSUP201) can be mobilized by the "trans"-acting transfer functions of the integrated RP4 plasmid. Due to the broad host range of the conjugation system probably any Gram-negative bacterium can act as recipient

pBR325-Mob-*nif*H::Tn5 is transferred into *R. meliloti*, homologous recombination between the wild type *nif*H gene in the genome and the mutated *nif*H::Tn5 region on the plasmid can be expected. This event allows us to isolate specifically *nif*H::Tn5 *R. meliloti* transconjugants simply by selection for the Tn5-mediated resistance (neomycin), the vector plasmid being lost automatically due to its inability to replicate in *Rhizobium*. A similar procedure using the same host *R. meliloti* has been published (Ruvkun 1981). In this system, the broad host range vector pRK290 has been used for introduction of the *nif*H::Tn5 DNA. But this vector has to be displaced after the recombination event by introducing another IncP plasmid which may pose problems in subsequent analysis. Provided an appropriate cloned DNA fragment is available, the site-specific transposon mutagenesis procedure described here is of great value. It will overcome the problems of random transposon mutagenesis, which is often very time-consuming to analyze, especially if markers, which are difficult to test, are mutated.

2. Experiment 1: Insertion of Bacteriophage Mu Genome into Plasmid RP4

a) Introduction

The principle of the method is outlined in Fig. 3.

1. In the first step, an *E. coli* strain carrying RP4 infected with the bacteriophage Mu_{c^+} (c^+ denotes wild type gene for heat stable repressor). Upon lysogenization, the Mu genome becomes inserted at random into either (a) chromosomal or (b) plasmid DNA of the host.

2. A mixture of Mu-lysogenized cells is used then as a donor culture in a mating experiment. The recipient strain in this mating experiment is Sm^r and carries a Mu_{cts} prophage in the chromosome (the cts gene codes for a thermosensitive Mu reproessor molecule). Transconjugants resulting from this mating are selected on a medium containing Tc and Sm. The RP4::Mu_{c^+} hybrids can then easily be found by incubating these plates at elevated temperature.

3. RP4 transconjugants will lyse at 45°C due to heat induction of the Mu_{cts} prophage in the recipient chromosome. But if RP4::Mu_{c^+} has been introduced, the recipient cells can survive at 45°C, because the heat stable repressors provided by Mu_{c^+} are transdominant over the thermosensitive ones of the resident prophage.

About 1% of the RP4::Mu transconjugants should be Km-sensitive, indicating Mu insertion into the Km-resistance determinant.

b) Objectives

1. Integration of bacteriophage Mu into the transmissible plasmid RP4.
2. Isolation of RP4 derivatives carrying the Mu insertion in the Km-resistance gene (RP4-Km::Mu).

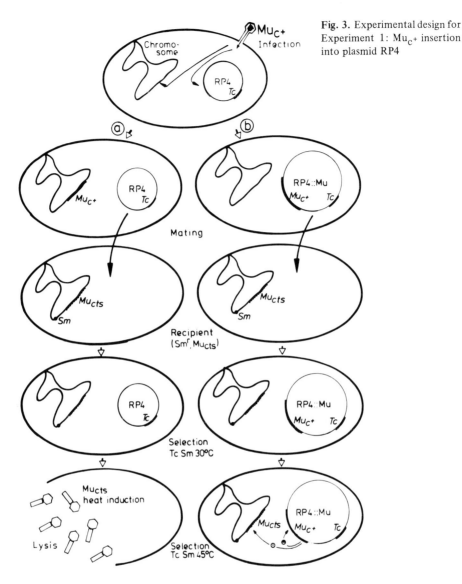

Fig. 3. Experimental design for Experiment 1: Mu_{c^+} insertion into plasmid RP4

c) Procedure and Results

Day 1

Lysogenization of C600 (RP4) with Mu_{c^+} Phages

Spread 0.1–0.3 ml of an o/n broth culture of C600 (RP4) onto RGMC plates. Spot serial decimal dilutions of a Mu_{c^+} lysate onto the bacterial lawn. Incubate overnight at 37°C.

Day 2

Mating Experiment and Selection of Transconjugants

Scrape off from a spot with a turbid lysis from the plate.

Suspend the cells in PA medium ($5 \times 10^7 - 10^8$ cells/ml).

Grow this Mu-lysogenized donor culture to early log phase. For control, grow a Mu-free culture of the same strain. The mating is performed as follows:

Mix 2×10^8 donor cells with 5×10^8 recipients (S53-30) in a sterile Eppendorf tube. Spin down for 20 sec, pour off the supernatant. Resuspend the cells as gently as possible in about 50 μl of PA medium.

Spread the mixture onto sterile millipore filters on pre-warmed PA plates. (The plates should be prepared several days before use; dried and checked for sterility.) Incubate the mating mixture for 1–2 h at 32°C.

Prior to plating on selection plates, suspend the cells by shaking the filter vigorously in a small volume of buffer containing 200 μg/ml Sm. Plate serial decimal dilutions of this suspension parallel onto PATcSm medium. Incubate one set of plates overnight at 32°C and the other at 45°C.

Day 3

Results of the Mating and Preparation of Master Plates

Examine the plates incubated at (a) 32°C for plasmid transfer; (b) 45°C for Mu_{c+} insertions into RP4.

The number of heat resistant transconjugants from the control mating (without Mu_{c+} being involved) should be at least 100 × less than the mixture obtained from the mating with the Mu_{c+} lysogenized donor.

Prepare master plates by picking 200 or more colonies (containing the putative RP4::Mu_{c+}) from the PATcSm 45°C plates and transferring them onto the same medium. Incubate overnight at 37°C.

Day 4

Replica Plating

The clones are tested for drug resistance by replica plating the master plates onto PAKmSm, PAApSm medium and incubating overnight at 37°C.

Day 5

Isolation of Km-Sensitive RP4::Mu Plasmids

Screen replicas for Km-sensitive clones, presumed to contain RP4-Km::Mu derivatives. Verify the Mu_{c+} insertion and the Tra^+ phenotype of the plasmids as follows:

Grow log phase cultures of the Km-sensitive clones and perform two filter mating experiments in parallel with the recipients S 380-0 and S 380-2, respectively. The filter mating is carried out as described before with two exceptions: (a) the incubation temperature is 37°C and (b) the cells are resuspended after the mating in buffer containing 100 μg/ml Nx.

Spread serial dilutions of the mating mixture onto PATcNx medium.

Day 6

Results of the Second Mating

Compare the transfer frequencies of the two parallel matings. Strain S 390-0 is a Mu-sensitive and S 380-2 a Mu-immune recipient. The transfer of RP4::Mu leads to zygotic induction in S 380-0 strain and therefore about a 100 × less transconjugants are obtained than from the cross using the S 380-2 recipient.

3. Experiment 2: Tn5 Transposition into the Self-Transmissible Plasmid RP4-Km::Mu

a) Introduction

The Km-sensitive derivative of RP4 constructed in experiment 1 should now be loaded again by Tn5-insertion to confer Km-resistance. This experiment demonstrates the simplicity with which transmissible plasmids can be marked with transposons. The transposon Tn5 is carried in the chromosome in strain S605. The plasmid RP4-Km::Mu is introduced into this strain by standard techniques (conjugation or P1-transduction). Spontaneous Tn5 transposition onto the plasmid can occur. This event is detected simply by transferring the plasmid into a new strain and selecting for Km-resistance (Km for Tn5). This procedure works for any chromosomally located transposon. The advantage of using Tn5 lies in its high transposition frequency.

b) Objectives

1. The RP4-derivative RP4-Km::Mu is to be labeled with Tn5.
2. This mating experiment allows the evaluation of Tn5 transposition frequency in *E. coli*.

c) Procedure and Results

Day 1

Mating Experiment and Selection of Transconjugants

Subculture an overnight culture of S605 (RP4-Km::Mu) by 1:10 dilution into prewarmed PA medium. Grow cells for 2 h at 37°C.

A filter mating experiment is carried out as described above, Mix 5×10^8 donor cells (S605 RP4-Km::Mu) and 10^9 recipients (S51-12) and incubate the mixture on filters for 2 h at 37°C.

Resuspend the cells in buffer containing 200 µg/ml Sm. Spread serial dilutions parallel onto PATcSm and PAKmSm plates and incubate them overnight at 37°C.

Day 2

Results of the Mating

Examine the plates for (a) plasmid transfer (PATcSm) and (b) Tn5 insertion into RP4-Km::Mu (PAKmSm).

The transposition frequency can be calculated by comparing the number of RP4 transconjugants with the number of Km-resistant recipients.

The new (RP4-Km::Mu)::Tn5 plasmids should be tested further for their Tra$^+$ phenotype if one plans to use them in *E. coli* × *Rhizobium* matings as described in the last experiment.

4. Experiment 3: Tn5 Transposition into Mobilizable *E. coli* Vectors

a) Introduction

In the following two experiments, the practical use of the mobilizable *E. coli* vector pBR325-Mob (pSUP201) will be demonstrated. Here, we will insert Tn5 (a) into the vector itself and (2) into a specific site of a DNA fragment, which has been cloned in the pSUP201 vector. The principle of the method is described in Fig. 4. Donor

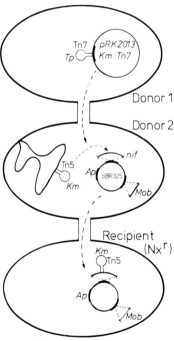

Fig. 4. Experimental design for Experiment 3: Tn5-insertion into mobilizable *E. coli* vectors by a double donor mating experiment

strain (donor 2) contains the Tn5 in its chromosome and the vector plasmid. During its growth, spontaneous Tn5 transposition occurs onto the vector. The mobilizing plasmid pRK2031Km::Tn7[2] is then introduced by short-term mating with a second donor strain (donor 1). The RP4 specific Tra-functions are expressed from this plasmid and as a result, the mobilization of the pBR325-Mob plasmids into a Nx-resistant recipient can take place. After selection of transconjugants on plates containing Nx, Ap, and Km, the desired plasmids with (pBR325-Mob)::Tn5 can be isolated. The probability that Tn5 transposes to the mobilizing plasmid pRK2013-Km::Tn7 is very low, since the mating time is very short. The new plasmid (pBR325-Mob)::Tn5 will be used for random Tn5 mutagenesis in *Rhizobium meliloti* in the last experiment. Using the same technique, we will also isolate cells carrying Tn5 insertions in the *R. meliloti nif*H gene which had previously been cloned into pBR325-Mob.

b) Objectives

1. In a double donor mating experiment, we will isolate insertions of Tn5 into the mobilizable *E. coli* vectors pBR325-Mob and pBR325-Mob-*nif*H.
2. The location of Tn5 insertion will be mapped for restriction analysis.

c) Procedure and Results

Day 1

Double Donor Mating Experiment and Selection of Transconjugants

Prepare log phase culture cells of the donor strains (a) S605 (pBR325-Mob) and (b) S605 (pBR325-Mob-*nif*H) in PAApKm medium (Km: 50–75 µg/ml; the high concentration of Km increases the probability of finding Tn5 transpositions into the vector plasmid since cells containing Tn5 on multicopy plasmids have a growth advantage in this medium. Wash the donor cultures in PA medium to remove the antibiotics.

Mix 2×10^8 cells of the donor strains with the same number of the second donor C600 (pRK2013Km::Tn7) in a sterile Eppendorf tube, spin down for 30 s, resuspend the cells gently in about 50 µl medium and spread the mixture onto millipore filters on pre-warmed PA plates. Incubation time: 30–60 min at 37°C.

Add 2×10^9 recipients (S380-2, about 50 µl of a concentrated cells suspension) to the filters and incubate for further 2–3 h.

Resuspend the cells by shaking the filters in buffer with 100 µg/ml Nx. Plate serial dilutions onto the following selection plates: PAApNx, PAApKmNx, and IsoTpNx.

Controls: plate both donors separately without recipient cells and the first or the second donor alone with the recipient.

Incubate overnight at 37°C.

---

[2] pRK2013 contains the Tra-region of RK2 cloned onto *Col*E1 (Figurski and Helsinki 1979). The Km gene in this plasmid is inactivated by Tn7 insertion

Day 2

Results of the Mating, Clone Purification

Examine the selection plates as follows:
1. Colonies on IsoTpNx indicate the transfer of pRK2013-Km::Tn7;
2. colonies on PAApNx indicate mobilization of the vectors;
3. clones growing on PAApKmNx are very likely to contain Tn5 insertion derivatives of the mobilzed vectors.

Several clones from the PAApKmNx medium should be purified on the same medium containing 50–75 µg/ml Km. The elevated Km concentration accelerates the segregation of clones which no longer contain plasmids without Tn5 integration.

Day 3

Analysis of Plasmids Containing Tn5 Insertions

For location of the Tn5 insertion in the new plasmids prepare plasmid DNA and analyse it with appropriate restriction endonucleases as described in Chapters 1.2 and 1.3.
In the following experiment, we will use plasmids which have been isolated as described above.

5. Experiment 4: Transposon Mutagenesis of *Rhizobium* Using Broad Host Range "Suicide Plasmids" and Mobilizable *E. coli* Vectors

a) Introduction

Here we demonstrate the introduction of transposons into the *Rhizobium* strains *R. leguminosarum* and *R. meliloti*. We will use the plasmids which have been constructed in the preceding experiments. As mentioned earlier, most of the RP4::Mu plasmids that are transferred into *Rhizobium* do not become stably established. Therefore, RP4::Mu::Tn5 can act as a vector for Tn5 mutagenesis. A disadvantage of this system is that elimination of RP4::Mu is not always complete. Sometimes stably replicating deletion-derivatives do arise from the vector which then can serve as a donor for Tn5 transposition. The use of the pBR325-Mob plasmids eliminates this problem. These plasmids are mobilized by an RP4 derivative (RP4-2-Tc::Mu), which is integrated in the donor host chromosome (see Fig. 2). Excision of the integrated mobilizing plasmid, which still carries the Mu genome, occurs at a frequency of about 10^{-5}. Because of the low excision frequency and the plasmid "suicide effect" due to the Mu genome, the excised mobilizing plasmids do not interfere with the transposon carrier plasmids in *Rhizobium*. Both (RP4-Km::Mu)::Tn5 and (pBR325-Mob)::Tn5 are used for random Tn5 insertions into the *Rhizobium* genome. Transconjugants from these mating experiments can be screened for auxotrophic or any other mutation. The phenomenon of "homogenotization" or marker exchange will be demonstrated experimentally. The principle of this method is to transfer a specific DNA fragment, previously mutated with Tn5 in *E. coli*, back into the original host. As shown in Fig. 5, there are two ways of Tn5 rescue in the recipient cell, where the vector itself does not replicate. One way is by normal Tn5 transposition into randomly distributed sites of the genome. This would happen with the same frequency as in the case of (pBR325-Mob)::Tn5 (about 10^{-5}). The second way of integrating Tn5 into the host DNA is by homologous recombination resulting in transfer of the mutated *nif*H::Tn5 sequence into the genome. This marker exchange reaction occurs with a much higher frequency (about 10^{-3}) than the nonspecific normal transposition and can therefore be easily identified.

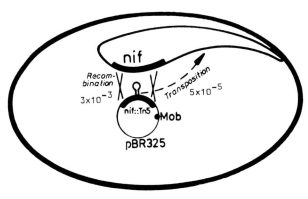

Fig. 5. Tn5 transposition and site-specific Tn5 mutagenesis with plasmid pBR325-Mob-*nif*H::Tn5 in *R. meliloti*

b) Objectives

1. To demonstrate the plasmid suicide effect after transfer of RP4::Mu into *Rhizobium*.
2. Identification of randomly distributed Tn5-insertions in the *Rhizobium* genome by use of two different transposon mutagenesis systems: (a) RP4::Mu and (b) mobilizable *E. coli* vectors as Tn5-donor plasmids.
3. Isolation of *R. meliloti nif*H::Tn5 mutants by site specific Tn5 mutagenesis.

c) Procedure and Results

Mating Experiment

Grow early log phase cultures of the following donor strains in PA medium:

1. S49-20 (RP4)
2. S49-20 (RP4::Mu)
3. S49-20 (RP4-Km::Mu)::Tn5
4. SM10
5. SM10 (pSUP2011 = (pBR325-Mob::Tn5)
6. SM10 (pSUP2111-101 = pBR325-Mob-*nif*H::Tn5)

(The medium for growth of donor 5 and 6 should contain 100 μg/ml Ap to ensure the maintenance of the vector plasmids.)

Use 2×10^8 donor cells for the mating experiment.

Grow late log or stationary phase cultures of the recipients *R. meliloti* 2011 (Sm) and *R. leguminosarum* 897 (Sm) in TY broth. Use about 10^9 recipient cells for the mating.

Mix the donor and recipient cultures, concentrate the suspension by centrifugation, and resuspend it carefully in a small volume of RGMC medium.

Spread the mixtures onto millipore filters on pre-warmed RGMC plates. Incubate for 3 or more h at 32°C.

For control, grow the parent strains separately in the same way.

Selection of Transconjugants

Suspend the mating mixtures in about 5 ml of TY medium containing 200 μg/ml Sm. Incubate these suspensions for 1–2 h at 32°C. Prepare serial decimal dilutions in sterile water.

Spread onto selection medium: for *R. meliloti* recipients 100 μg/ml TYSm + Nm; for *R. leguminosarum* recipients 25 μg/ml TYSm + Km. Incubate for at least 3 days at 32°C.

Results of the Mating Experiments

Examine the plates in the following way:

Donor 1 is the positive control of the whole system. RP4 plasmids should be transferred at a high frequency (at least 10^{-3} per recipient).

Donor 2: RP4::Mu plasmids should show a much reduced transfer frequency as compared to RP4.

Donor 3: the Tn5 transposition from (RP4-Km::Mu)::Tn5 plasmids into the host genome is indicated by a significantly higher yield of Km/Nm resistant transconjugants relative to the RP4::Mu control.

Donor 4: strain SM10 carrying the Mob region should not give rise to Km/Nm resistant transconjugants.

Donor 5 mobilizes the (pBR325-Mob)::Tn5 vectors. Km/Nm resistant recipients can only result from random Tn5 transpositon events in the *Rhizobium* cells.

Donor 6 mobilizes the (pBR325-Mob)*nif*H::Tn5 vectors. In *R. meliloti* recipients, the *nif*H/*nif*H::Tn5 marker exchange reaction is indicated by about a 10–100 × higher yield of Nm resistant colonies as compared to random mutagenesis with donor 5. The *R. leguminosarum* recipient does not show this effect probably due to the lack of sufficient homology between the transferred *R. meliloti* specific *nif*H::Tn5 region and its own corresponding DNA sequence.

6. Materials

Bacterial Strains

All *E. coli* strains are derivatives of C600 (thi, thr, leu, su$_{III}$, lac) or CSH 51 [thi, Δ (lac, pro), ϕ80dlac, Sm^r]

E. coli strains

C600 (RP4)	Ap, Km, Tc
S53-30 CS	CSH51 Mu$_{cts}$
S380-0	C600 Nx^r
S380-2	C600 Nx^r, Mu$_{c+}$
S605 (RP4-Km::Mu)	C600 *met*::Tn5, Ap, Km, Tc
S51-12	XSH51 *his*::Mu$_{c+}$
S605 (pSUP201)	C600 *met*::Tn5, pBR325-Mob, Ap, Cm, Km
S605 (pSUP2110)	C600 *met*::Tn5, pBR325-Mob-*nif*H, Ap, Km
S49-20 (RP4)	C600 *rec*A, Ap, Km, Tc
S49-20 (RP4::Mu)	C600 *rec*A, Ap, Km, Tc
S49-20 (RP4-Km::Mu)::Tn5)	C600 *rec*A, Ap, Km, Tc
SM10	C600 *rec*A, RP4-2-Tc::Mu integrated [3], Km
SM10 (pSUP2011)	SM10 + (pBR325-Mob)::Tn5, Km, Ap, Cm
SM10 (pSUP2111-101)	SM10 + pBR325-Mob-*nif*H::Tn5, Km, Ap

Rhizobium strains

R. leguminosarum 897	*Rhizobium leguminosarum*, phe, trp, Sm^r
R. meliloti 2011	*Rhizobium meliloti* Sm^r

Media and Antibacterial Drugs

PA   1000 ml H$_2$O contains
         17.5 g Penassay Broth (Difco)

RGMC 1000 ml H$_2$O contains
         10 g Bacto tryptone (Difco)
         1 g Yeast extract (Difco)
         8 g NaCl
         1 g Glucose
         0.3 g CaCl$_2$
         1 g MgCl$_2$

TY   1000 ml H$_2$O contains
         5 g Bacto tryptone (Difco)
         3 g Yeast extract (Difco)
         0.3 g CaCl$_2$

Agar media contain 1.5% Difco Agar

Iso   1000 ml H$_2$O contains 31.4 g "Iso-Sensitest"-Agar (Oxoid)

---

[3] See Fig. 2

Antibiotics are prepared freshly and sterilized by filtration

Antibiotics		Concentration (µg/ml)	
		PA	TY
Ampicillin	(Ap)	100	–
Kanamycin	(Km)	25	25
Neomycin	(Nm)	–	100
Tetracycline	(Tc)	5	5
Trimethoprim[4]	(Tp)	100	–
Streptomyccin	(Sm)	200	200
Nalidixic acid	(Nx)	100	–

7. References

Alton NK, Vapnek D (1979) Nucleotide sequence analysis of chloramphenicol resistance transposon Tn9. Nature 268:864–869

Barth PT (1979) RP4 and R300B as wide host-range plasmid cloning vehicles. In Timmis KN, Pühler A (eds) Plasmids of medical, environmental and commercial importance. Elsevier/North-Holland Biomedical Press, Amsterdam New York Oxford, pp 399–401

Barth PT, GrinterNJ (1977) Map of plasmid RP4 derived by insertion of transposon C. J Mol Biol 113:455

Barth PT, Datta N, Hedges RW, Grinter NJ (1976) Transposition of a deoxyribonucleic acid sequence encoding trimethoprim and streptomycin resistance from R483 to other replicons. J Bacteriol 125:800

Barth PT, Grinter NJ, Bradley DE (1978) Conjugal transfer system of plasmid RP§: analysis by transposon 7 insertion. J Bacteriol 113:43

Berg DE, Davies J, Allet B, Rodraix J (1975) Transposition of R factor genes to bacteriophage λ. Proc Natl Acad Sci USA 72:3628

Beringer JE, Beynon JL, Buchanan-Wollaston AV, Johnston AWB (1978) Transfer of the drug-resistance transposon Tn5 to Rhizobium. Nature 276:633

Beringer JG (1974) R factor transfer in *Rhizobium leguninosarum.* J Gen Microbiol 84:188

Boucher C, Bergeron B, Bertalmio MB, Dénarié J (1977) Introduction of bacteriophage Mu into *Pseudomonas solanacearum* and *Rhizobium meliloti* using the R factor RP4. J Gen Microbiol 98:253–263

Bukhari AI (1976) Bacteriophage Mu as a transposition element. Annu Rev Genet 10:389

Burkardt HJ, Rieß G, Pühler A (1979) Relationship of group P1 plasmids revealed by heteroduplex experiments: RP1, RP4, R68 and RK2 are identical. J Gen Microbiol 114:341

Campbell A, Berg D, Botstein D, Lederberg E, Novick R, Starlinger P, Szybalski W (1977) Nomenclature of transposable elements in prokaryotes. In: Bukhari AI, Shapiro JA, Adhya SL (eds) DNA insertion elements, plasmids and episomes. Cold Spring Harbor Lab, Cold Spring Harbor, pp 15–22

Datta N, Hedges RW (1972) Host ranges of R factors. J Gen Microbiol 70:453

Datta N, Hedges RW, Shaw EJ, Sykes RB, Richmond MH (1971) Properties of an R factor from *Pseudomonas aeruginosa.* J Bacteriol 108:1244–1245

---

4 Tp can be autoclaved and should be used in Iso-Sensitest-Agar and not in PA Broth

Dénarié J, Rosenberg C, Bergeron B, Boucher C, Michel M, de Bertalmio MB (1977) Potential of RP4::Mu plasmids for in vivo genetic engineering of Gram-negative Bacteria. In Bukhari AI, Shapiro JA, Adhya SL (eds) DNA insertion elements, plasmids and episomes. Cold Spring Harbor Lab, Cold Spring Harbor New York, pp 507–520

Ditta G, Stanfield S, Corbin D, Helinski DR (1980) Braod host range DNA cloning system for Gram-negative bacteria: construction of a gene bank of *Rhizobium meliloti*. Proc Natl Acad S Sci USA 77:7347

Figurski DH, Helinski DR (1979) Replication of an origin-containing derivative of plasmid RK2 dependent on a plasmid function provided in trans. Proc Natl Acad Sci USA 76:1648

Figurski D, Meyer R, Miller DS, Helinski DR (1976) Generation in-vitro of deletions in the broad host range plasmid RK2 using phage Mu insertions and restriction endonucleases. Gene 1:107

Foster TJ, Howe GB, Richmond KMV (1975) Translocation of the tetracycline resistance determinant from R100-1 to the *Escherichia coli* K12 chromosome. J Bacteriol 124:1153

Gottesmann M, Rosner JL (1975) Acquisition of a determinant for chloramphenicol resistance by coliphage λ. Proc Natl Acad Sci USA 72:5041

Hedges RW, Jacob AE (1974) Transposition of ampicillin resistance from RP4 to other replicons. Mol Gen Genet 132:31

Heffron F, McCarthy BJ (1979) DNA sequence analysis of the transposon Tn*3*: three genes and three sites involved in transposition of Tn*3*. Cell 18:1153–1163

Heffron F, Sublett R, Hedges RW, Jacob A, Falkow S (1975) Origin of the TEM beta lactamase gene found on plasmids. J Bacteriol 122:250

Holloway BW (1979) Plasmids that mobilize bacterial chromosome. Plasmid 2:1–19

Kleckner N (1977) Translocatable elements in prokaryotes. Cell 11:11–13

Kleckner N, Chan R, Tye B, Botstein D (1975) Mutagenesis by insertion of a drug-resistance element carrying an inverted repetition. J Mol Biol 97:561

Olsen RH, Shipley P (1973) Host range and properties of the Pseudomonas R factor R1822. J Bacteriol 113:772

Ruvkun GB, Ausubel FM (1980) Interspecies homology of nitrogenase genes. Proc Natl Acad Sci USA 77:191

Ruvkun GB, Ausubel FM (1981) A general method for site-directed mutagenesis in prokaryotes. Nature 289:85

Thomas CM, Stalker D, Guiney D, Helinski DR (1979) Essential regions for the replication and conjugal transfer of the broad host range plasmid RK2. In: Timmis KN, Pühler A (eds) Plasmids of medical, environmental and commercial importance. Elsevier/North-Holland Biomedical Press, Amsterdam New York Oxford, pp 375–386

Van Vliet F, Silva B, van Montagu M, Schell J (1978) Transfer of RP4::Mu plasmids to *Agrobacterium tumefaciens*. Plasmid 1:446–455

2.4 Generation of Deletion Mutations in Vitro with the BAL31 Exonuclease

J. FREY [1], M. BAGDASARIAN [2], and K.N. TIMMIS [3]

Contents

1. General Introduction 141
2. Experiment: Isolation of Plasmid Deletion Derivatives 143
   a) Introduction 143
   b) Objective 144
   c) Strains 144
   d) Procedure 145
   e) Results and Discussion 146
3. Materials 148
4. References 149

1. General Introduction

Mutations are heritable alterations in genetic material. They are responsible for the enormous variation and variability of life forms and thus, in combination with natural selection, for phylogenetic evolution. Mutations have of course been of tremendous importance to geneticists and molecular biologists in the genetic and biochemical analysis of life processes.

Mutations may be of four basic types: (1) *point* mutations, the replacement of one nucleotide by another; (2) *insertions*, the addition of one or more nucleotides at a specific point in a genome; (3) *deletions*, the removal of one or more nucleotides from a specific point in the genome; and (4) genome rearrangements, inversion or transposition of genomic segments. All types of mutations have proven to be of considerable utility in the analysis of gene structure and function and experiments involving the first two have been presented earlier in the chapter. In this section, a strategy for generating deletion mutations will be described.

Because deletion mutations involve at net loss of genetic information they are generally non-revertible, i.e., are stably inherited over many generations. This stability is of paramount importance for experiments requiring long-term maintenance of

---

1 Department of Biochemistry, University of Geneva, CH-1211 Geneva, Switzerland
2 Max-Planck-Institute for Molecular Genetics, D-1000 Berlin-Dahlem, Fed. Rep. of Germany
3 Department of Medical Biochemistry, University of Geneva, CH-1211 Geneva, Switzerland

mutant phenotypes and for those in which conditions prevail that are strongly selective for revertants. Deletion mutations are also of great utility for the physical mapping of genes since the sites of deletions may often be localized readily by one or more simple physical techniques.

Deletion mutations may occur naturally or be generated by a variety of genetic manipulations. Of particular use to the molecular geneticist are in vitro procedures for the deletion of segments of DNA fragments that have been cloned in plasmid or bacteriophage genome vectors (for review, see Timmis 1981). These include:

1. Transformation of DNA genomes cleaved at single random sites: intracellular nuclease attack of the molecule termini prior to their rejoining in vivo results in the introduction of deletions of varying lengths at random sites in a small proportion of the population of transformed genomes (Shenk et al. 1976). The efficiency of deletion production can be greatly increased by exonuclease digestion in vitro of the cleaved DNA molecules prior to transformation (Shenk et al. 1976, Heffron et al. 1977).
2. Transformation of DNA genomes cleaved at a single specific site (e.g., by an appropriate restriction endonuclease): this results in the introduction of deletions of varying lengths at the cleavage site in a small proportion of transformed genomes (Lai and Nathans 1974, Murray and Murray 1974). Again, this proportion can be greatly increased by exonuclease digestion in vitro of the cleaved DNA (Carbon et al. 1975, Covey et al. 1976).
3. As for the exonuclease digestion in vitro version of (2), but followed by joining in vitro of deletion termini by DNA ligase, prior to transformation: this greatly increases the frequency of transformation and hence, the overall yield of deletion derivative genomes.
4. Deletion of specific restriction endonuclease-generated DNA segments by gene cloning procedures: this results in the introduction of deletions with precisely defined termini at specific sites (Lai and Nathans 1974, Timmis et al. 1975).

The versions of Methods 1 and 2 which lack the exonuclease step are very inefficient for deletion formation, whereas the version with the exonuclease step give a very low yield of transformed genomes. Methods 3 and 4 give much higher yields because functional genomes are reformed in vitro prior to transformation and are very efficient for deletion formation. Method 4 is used in an experiment in the gene cloning section. Method 3 is used here to produce deletion derivatives of a plasmid cloning vector that are deleted of one or more mobilization genes, i.e., genes which are undesirable in containment vectors.

Several exonucleases, including exonuclease III, which digests DNA from its 3' ends, λ exonuclease which digests DNA from its 5' ends, and S1 nuclease, which removes single stranded termini, have been employed both singly and in combinations to produce deletions in linearized DNA. More recently, however, the BAL31 enzyme, which is a single strand specific endonuclease that also possesses a potent double strand exonuclease activity (Gray et al. 1975, Legerski et al. 1978), has been extensively used. BAL31 exonucleolytic digestion of linear double stranded DNA proceeds relatively slowly in high salt at 30°C ($\sim$ 300 base pairs/min/DNA terminus) and is irreversibly stopped by chelation of Ca^{++} in the digestion buffer. Thus, the extent of deletion formation by this enzyme may be carefully regulated. The new DNA termini generated by BAL31 digestion may be joined in vitro by the action of DNA ligase as efficiently as those produced by restriction endonucleases that generate blunt ends.

2. Experiment: Isolation of Plasmid Deletion Derivatives

a) Introduction

The RSF1010 replicon (Fig. 1) is a small multicopy antibiotic resistance plasmid (streptomycin/sulphonamide) that exhibits an extraordinarily broad host range specificity and is able to replicate in, and be stably propagated by, a wide range of Gram-negative bacteria. These properties make the RSF1010 plasmid a highly attractive replicon for constructing vectors for genetic engineering in bacteria other than *E. coli;* a number of such vectors have been developed (Bagdasarian et al. 1981, Bagdasarian and Timmis 1982).

Plasmid RSF1010 and a number of its derivatives are efficiently transferred (mobilized) by conjugation from one host bacterium to another, if fertility functions are provided by a coexisting conjugative plasmid, such as F, RP4, or ColI in the donor (Barth and Grinter 1974). This feature of the plasmid is very convenient for genetic engineering in bacteria that are poorly transformable. It is possible to clone in a readily transformable host, such as *E. coli* or *Pseudomonas putida,* and thereafter

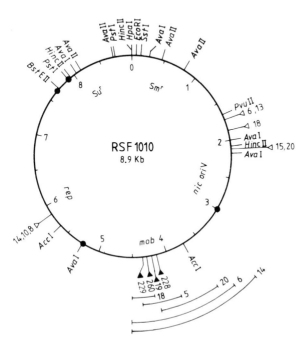

Fig. 1. Physical and genetic map of RSF1010. Abbreviations: Su^r, Sm^r, resistance to sulphonamide and streptomycin, respectively; *mob, nic, oriV, rep,* determinants for plasmid mobilization, relaxation nick site, origin of vegetative replication and a positive replication factor, respectively; *triangles* represent the insertions of Tn3 transposon; *filled triangles* are the insertions of Tn3 that result in *mob* phenotype, the *lines outside of the circle* represent the deletions of the *mob* region of the plasmid; *filled circles* are RNA polymerase binding sites

to transfer recombinant plasmids into the desired host by conjugation. For the cloning of DNA segments which may encode a hazardous product, however, a high level of physical and/or biological containment must be maintained and mobilization-deficient cloning vectors must be employed.

Mobilization of RSF1010, like that of other small plasmids, requires a DNA site, *ori*T, the origin of transfer replication, and the products of one or more mobilization *(mob)* genes. The plasmid relaxation nick site, which is thought to be synonymous with *ori*T, has been located by the method described in Chap. 1.6 (Nordheim et al. 1980), whereas the plasmid segment containing the *mob* genes has been localized by transposition mutagenesis with Tn*3* as described in Chap. 2.3 (Bagdasarian et al. 1982). Plasmid pKT260 is an RSF1010 derivative that contains a Tn*3* element in a *mob* gene (Fig. 2). Because it is known that transposon mutant derivatives may revert by excision of the transposon, and because transposons promote replicon fusion, which can also facilitate conjugal transfer of the RSF1010 plasmid, it is desirable to isolate a mutant derivative of RSF1010 that is deleted of the *mob* gene(s). In the following experiment, the BAL31 enzyme will be used to delete the Tn*3* element and flanking sequences [*mob* gene(s)] from pKT260.

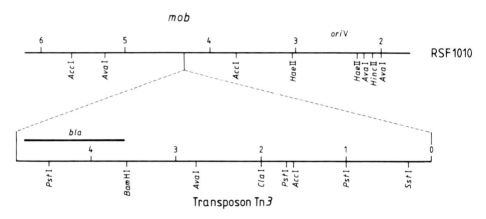

Fig. 2. Map of *mob* region of the *mob*⁻ derivative RSF1010::Tn*3*, plasmid pKT260. *bla*, gene for β-lactamase

b) **Objective**

Generation of Mob⁻ deletion derivatives of the pKT260 plasmid.

c) **Strains**

SK1592 *E. coli* K-12 F⁻ *end*A, *gal*, *hsd*R4, *hsd*M⁺, *sbc*B15, thi, Tl^r (Kushner 1978); SK1592 (pKT260); SK1592 (RSF1010).

d) Procedure

Day 1

Deletion Formation in Vitro

Digestion of pKT260 DNA with BamHI

To 100 µl of pKT260 DNA (10 µg) in TE (10 mM Tris-HCl, pH 8.0, 1 mM EDTA) in an Eppendorf tube add 11 µl TMN buffer × 10 and 15 units of *Bam*HI endonuclease and incubate for 60 min at 37°C.
Prepare a 0.8% agarose gel buffered with TBE.
Remove a 1 µl sample and add to 10 µl of H_2O and 3 µl of stop solution.
Subject to agarose gel electrophoresis for about 2 h at 60 V (bromophenol blue should have migrated halfway through the gel).
Stain gel in TBE containing 0.5 µg/ml ethidium bromide for 30 min and examine on a UV box to determine if cleavage of the pKT260 DNA is complete. If not, add more *Bam*HI to the DNA and continue the digestion.
Remove 2 µl sample, treat with *Ava*I as indicated below in parallel with samples of BAL31-treated DNA.

BAL31 Digestion

Precipitate *Bam*HI-cleaved pKT260 DNA by addition of 1/10 vol of 3 M Na acetate, pH 5.5, and 0.7 vol of isopropanol.
Incubate the solution at −20°C for 15 min, and centrifuge in an Eppendorf centrifuge for 10 min. Wash pellet twice with 80% ethanol at room temperature to remove salt, dry, and resuspend in 110 µl of BAL buffer in an Eppendorf tube.
Prewarm tube to 30°C; add 2 µl of BAL31 enzyme (1200 units/ml) and incubate at 30°C.
Remove 25 µl samples 10, 12, 14, and 16 min after addition of BAL31 and stop reaction in each sample by addition of 5 µl of ice cold 0.2 M EDTA and 80 µl of ice cold H_2O.
Precipitate DNA in each sample with isopropanol, wash 2 × with 80% ethanol and resuspend in 10 µl TM buffer. In order to monitor the extent of BAL31 digestion of the pKT260 DNA in each sample, remove 2 µl from each tube, treat with *Ava*I endonuclease (1 unit), and analyse by agarose gel electrophoresis.

Ligation

To the remainder of each sample add 0.5 µl ATP (10 mM), 0.5 µl DTT (40 mM) and 1 µl of T4 DNA ligase (100 units/ml). Incubate 2 h at 14°C, add 90 µl of TM buffer (this dilution step promotes intramolecular ligation, i.e., recircularization) and continue the incubation overnight at 14°C.
Inoculate a colony of SK1592 into 5 ml L-broth and incubate overnight at 37°C.

Day 2

Transformation

Transform SK1592 with half of each sample and with RSF1010 as a control DNA. Select transformants on nutrient agar plates containing streptomycin (50 µg/ml).

Day 3

Identification of Deletion Derivatives by Plasmid Phenotype

From each transformation, pick 50 transformant clones to ampicillin and streptomycin plates, in order to identify streptomycin resistant (Sm^r) ampicillin sensitive (Ap^s) clones that must contain plasmids deleted of all or part of the Ap^r gene.

Day 4

Identification of Deletion Derivatives by Plasmid Size

Select 5 Sm^r Ap^s colonies from each transformation plus colonies containing RSF-1010 and pKT260 (1 of each type); as controls, make mini-preparations of the plasmids they contain, and analyze by electrophoresis through an 0.8% agarose gel in TBE. Each plasmid preparation should be analyzed (a) untreated; (b) treated with *Sst*I (TM buffer), and (c) treated with *Ava*I (TMN buffer). Run all samples of one type together, i.e., run all (a) samples in adjacent gel slots, etc.

Stain and photograph the gel and calculate the approximate sizes of the plasmids and the extents of their deletions. Alternatively, plasmids may be analyzed according to the method of Eckhardt (1978).

Purify by single colony isolation on Sm agar, the clones containing the two smallest plasmids, which should preferably be smaller than RSF1010 (8.9 kb).

Day 5

Inoculate a colony from each of the cultures into 200 ml M9 medium + 0.5% casamino acids supplemented with thiamine (10 μg/ml) and streptomycin (50 μg/ml), and incubate overnight at 37°C.

Day 6–8

Prepare cleared lysates of the two cultures and purify the plasmid DNAs by centrifugation in CsCl-EtBr gradients.

Analyze these DNAs, RSF1010, and pKT260 control DNAs by cleavage with the following restriction endonucleases: (a) none; (b) *Sst*I (TM); (c) *Acc*I (TMN); (d) *Ava*I (TMN); and (e) *Acc*I + *Ava*I (TMN). Map the endpoints of the deletions of the plasmids constructed.

e) Results and Discussion

The *Ava*I digestion patterns of the pKT260 DNA samples digested with BAL31 for different periods of time are shown in Fig. 3. *Ava*I digestion of pKT260 DNA previously linearized by *Bam*HI treatment produces 7 DNA fragments having sizes of 4.85, 2.8, 2.2, 1.75, 1.2, 0.8, and 0.2 kb. After a 10 min digestion with BAL31, the 0.8 and 2.2 kb fragments which flank the *Bam*HI site have disappeared, and after

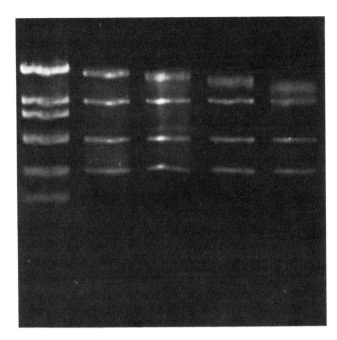

Fig. 3. Agarose gel electrophoresis analysis of BAL31 digested pKT260 DNA. The samples shown represent, from left to right, BamHI linearized pKT260 DNA digested with BAL31 for 0, 10, 12, 14 and 16 min, respectively, and subsequently digested with AvaI (see text)

longer digestions the 4.85 kb fragment becomes shortened. The most suitable period of digestion for removal of the Tn3 element and flanking sequences in pKT260 is generally 12–16 min, depending upon the batch of BAL31 enzyme.

After ligation and transformation approximately 300 transformants per half sample are normally obtained. About 50% of these are $Sm^r Ap^r$, i.e., clones of bacteria transformed with plasmids that had either not been cleaved with BamHI or had not been deleted of a substantial number of nucleotides from the BamHI site. The remainder of the clones are $Sm^r Ap^s$ and contain plasmid deletion derivatives. Such derivatives exhibit deletions of varying sizes that because of the small number tested, do not necessarily reflect the time of BAL31 digestion used in their construction. Mapping of the deletion endpoints of the smallest derivatives should reveal that most or all of the Tn3 element and some of the adjacent RSF1010 DNA (i.e., mob gene region) has been deleted.

Although it has been reported that BAL31 degrades double stranded linear DNA at the same velocity from both ends, most of the deletion derivatives of RSF1010 obtained will have deletions that are asymmetric with respect to the BamHI site. This presumably reflects the presence of a determinant essential for plasmid replication or maintenance relatively close to the Tn3 insertion in RSF1010 which limits the extent of deletions on one side of the insertion.

The method described is highly effective in producing random sized deletions from a given restriction endonuclease cleavage site. In the form presented, it also permits conversion of transposon mutants, which may be revertable, to deletion mutants, which are not. The Tn3 element is particularly convenient in this regard, since it has a number of single cleavage sites for enzymes that cleave DNA infrequently (BamHI, BstEII, AccI, ClaI, and SstI).

3. Materials

Day 1

Plasmid pKT260 DNA, 10 μg in 100 μl TE
Agarose for gel electrophoresis (1 g)
Isopropanol (2 ml)
Na acetate 3 M (1 ml)
Ethanol 80% (4 ml)
BAL buffer (20 mM Tris-Cl, pH 8.0, 600 mM NaCl, 12 mM $MgCl_2$, 12 mM $CaCl_2$, 1 mM EDTA; 1 ml)
BAL31 enzyme (2.5 units)
*Bam*HI restriction enzyme, 15 units
*Ava*I restriction endonuclease, 5 units
Stop solution (see 3.2)
ATP 10 mM
DTT 40 mM
T_4 ligase
L-broth
0.2 M EDTA, pH 8.0
TMNx10 buffer (see 1.7)
TMN buffer
TM buffer (see 1.7)
TBE buffer (see 1.7)

Day 2

Solutions for transformation
12 L-broth agar plates containing Sm (50 μg/ml)
RSF1010 DNA

Day 3

3 L-broth agar plates containing Ap (50 μg/ml)
3 L-broth agar plates containing Sm (50 μg/ml)
Sterile toothpicks

Day 4

Solutions and agarose for mini-preparations of plasmids and their analysis by agarose gel electrophoresis
*Sst*I, *Ava*I endonucleases
Plates with colonies of strains containing RSF1010 and pKT260
L-broth agar plates containing Sm (50 μg/ml)
L-broth agar plates with colonies of SK1592 (RSF1010) and SK1592 (pKT260)

Day 5

2 × 200 ml M9 medium containing 0.5% casamino acids, 10 µg/ml thiamine and 50 µg/ml Sm

Day 6–8

Solutions for preparation of cleared lysates
CsCl and ethidium bromide for gradients
Phenol, isopropanol, ethanol for purification of plasmid DNA from gradients
*Sst*I, *Acc*I, and *Ava*I restriction enzymes + buffers
Agarose + buffer for gel

4. References

Bagdasarian M, Timmis KN (1982) Host:vector systems for gene cloning in *Pseudomonas*. Curr Top Microbiol Immunol 96:47–67

Bagdasarian M, Bagdasarian MM, Lurz R, Nordheim A, Frey J, Timmis KN (1982) Molecular and functional analysis of the broad host range plasmid RSF1010 and construction of vectors for gene cloning in Gram-negative bacteria. In: Mitsuhashi S (ed) Bacterial drug resistance. Jpn Sci Soc Press Tokyo, pp 183–197

Bagdasarian M, Lurz R, Rückert B, Franklin RCH, Bagdasarian MM, Frey J, Timmis KN (1981) specific-purpose plasmid cloning vectors. II. Broad host range, high copy number, RSF1010-derived vectors, and a host vector system for gene cloning in *Pseudomonas*. Gene 16:237–247

Barth PT, Grinter NG (1974) Comparison of the deoxyribonucleic acid molecular weights and homologies of plasmids carrying linked resistance to streptomycin and sulfonamides. J Bacteriol 120:618–630

Carbon J, Shenk TE, Berg P (1975) Biochemical procedure for production of small deletions in simian virus DNA. Proc Natl Acad Sci USA 72:1392–1396

Covey C, Richardson D, Carbon J (1976) A method for deletion of restriction sites in bacterial plasmid deoxyribonucleic acid. Mol Gen Genet 145:155–158

Eckhardt T (1978) A rapid method for the identification of plasmid deoxyribonucleic acid in bacteria. Plasmid 1:584–588

Gray Jr, Ostander DA, Halnett JL, Legerski RJ, Robberson P (1975) Extracellular nucleases of *Pseudomonas* BAL31. I. Characterization of single-strand specific deoxyribonuclease and double strand deoxyribonuclease activities. Nuclei Acids Res 2:1459–1492

Heffron F, Bechinger P, Champoux JJ, Falkow S (1977) Deletions affecting the transposition of an antibiotic resistance gene. Proc Natl Acad Sci USA 74:702–706

Kushner SR (1978) An improved method for transformation of *Escherichia coli* with ColE1 derived plasmids: In: Boyer HW, Nocosia S (eds) Genetic Engineering. Elsevier/North Holland, Amsterdam, pp 17–23

Lai CJ, Nathans D (1974) Deletion mutants of simian virus 40 generated by the enzymatic excision of DNA segments from the viral genome. J Mol Biol 89:179–193

Legerski RJ, Halnett JL, Gray Jr (1978) Extracellular nucleases of *Pseudomonas* BAL31. III. Use of the double strand deoxyribonuclease activity as the basis of a conventional method für the mapping of fragments of DNA produced by cleavage with restriction enzymes. Nucleic Acid Res 5:1445–1464

Murray NE, Murray K (1974) Manipulation of restriction targets in phage λ to form receptor chromosomes for DNA fragments. Nature 251:476–481

Nordheim A, Hashimoto-Gotoh T, Timmis KN (1980) Location of two relaxation nick sites in RK6 and single sites in pSC101 and RSF1010 close to origins of vegetative replication: Implications for conjugal transfer of plasmid DNA. J Bacteriol 144:923–932

Shenk TE, Carbon J, Berg P (1976) Construction and analysis of viable deletion mutants of simian virus 40. J Virology 18:669–671

Timmis KN (1981) Gene manipulation in vitro. Symp Soc Gen Microbiol 31:49–109

Timmis KN, Cabello F, Cohen SN (1975) Cloning, isolation and characterization of replication regions of complex plasmid genomes. Proc Natl Acad Sci USA 72:2242–2246

Chapter 3 Gene Cloning

3.1 Gene Cloning: An Introduction

K.N. TIMMIS[1]

Contents

1. General Introduction 152
2. References 153

1. General Introduction

Genetic recombination is the fundamental biological process of exchange of genetic information between different chromosomes. Its constant activity during evolution has enabled the large number of spontaneously occurring genetic changes that improve the fitness of individuals for their particular environments to accumulate in the most successful members of each biological species. Recombination has provided much of the experimental basis for the science of genetics and has been exploited to great social benefit by plant and animal breeders and to great scientific benefit by classical and molecular geneticists.

Ordinary or generalized recombinational events commonly involve the reciprocal exchange of genetic material and require DNA sequence homology in the regions of exchange. Nonreciprocal recombinational events, e.g., involving insertion of segments of DNA at chromosomal sites that exhibit little or no sequence homology, have also been documented in many biological systems. A feature of this type of recombination is that the inserting DNA fragments are often discrete segments (insertion sequences, transposons), which have special repeated sequences at their termini and which may encode specific enzyme systems that accomplish the insertion reactions.

Nonreciprocal recombination usually involves insertion of DNA segments originating in the same organism as that in which the recombination takes place, or originating from a parasite of that organism (e.g., tumor viruses, mammalian cells; Ti plasmid segment of *Agrobacterium,* plant cells). Recombination between unrelated species of organisms having little DNA sequence homology is apparently rare and experimentally difficult to accomplish by classical procedures. However, it has long been apparent that great benefits could be derived from *inter*generic, as well as *intra*generic, genetic manipulations.

---

1 Department of Medical Biochemistry, University of Geneva, CH-1211 Geneva, Switzerland

Genetic recombination essentially consists of the breakage and joining of DNA molecules. The development of gene cloning procedures now permits DNA obtained from a wide variety of prokaryotic and eukaryotic sources to be cut in vitro at precisely defined locations and the DNA fragments thereby generated to be coupled enzymatically to a self-replicating genetic element, known as a cloning vector or vehicle (either a plasmid or virus genome). Hybrid molecules generated in this fashion are introduced into a suitable host cell in which they are propagated (Cohen 1975). Thus, the host cells containing a hybrid molecule can serve as "cellular factories" for selective amplification of the cloned DNA segment and, in some instances, the gene product(s) specified by the cloned DNA. Moreover, they can serve as a well-defined genetic background in which to study expression of cloned DNA fragments.

DNA manipulation procedures are applied in the biomedical sciences to obtain basic information about fundamental biological processes, and in the applied sciences to obtain a variety of biological products that are of medical, agricultural, and commercial importance and that are otherwise expensive or unobtainable in large quantity. Exceptional achievements accomplished thus far with these techniques have clearly demonstrated their extraordinary usefulness (for recent reviews see Wu 1979, Setlow and Hollaender 1980, Timmis 1981).

Gene cloning experiments are usually carried out in 5 steps: (1) generation of DNA fragments that are suitable for cloning (in its simplest form, this involves treating the DNA sample with a site-specific restriction endonuclease that generates DNA fragments having termini with an appropriate polynucleotide sequence); (2) linkage of the fragments to be cloned to a vector plasmid or virus genome by treatment of the DNA fragment/vector mixture with DNA ligase; (3) introduction of the DNA products of ligation into a suitable host cell system, such as *E. coli* bacteria, by transformation, transfection, or infection; (4) detection of transformed clones containing required hybrid DNA species; and (5) isolation and characterization of the hybrid DNA species in those clones. The following experiments will demonstrate (1) some parameters of the ligation of different types of DNA fragment, (2) the use of plasmid and bacteriophage λ-vectors for gene cloning, (3) the construction of gene banks or libraries using cosmid vectors, and (4) the selective cloning of replication origins.

2. References

Cohen SN (1975) The manipulation of genes. Sci Am 233:24–35

Setlow JK, Hollaender A (eds) (1980) Genetic engineering, principles and methods, vol 2. Plenum Press, New York

Timmis KN (1981) Gene manipulation in vitro. In: Glover SW, Hopwood DA (eds) Soc Gen Microbiol Symp 31, Genetics as a tool in microbiology. Cambridge Univ Press, Cambridge, pp 49–109

Wu R (ed) (1979) Methods in enzymology, vol 68. Recombinant DNA. Academic Press, New York

3.2 Ligation of Cohesive-Ended and Flush-Ended DNA Fragments

G. BRADY[1] and K.N. TIMMIS[2]

Contents

1. General Introduction 154
2. Experiment 1: Ligation of Cohesive-Ended and Flush-Ended DNA Fragments 155
   a) Introduction 155
   b) Objective 155
   c) Procedure 1 156
   d) Procedure 2 156
   e) Results and Discussion 157
3. Materials 159
4. References 159

1. General Introduction

DNA fragments having 3'-hydroxyl and 5'-phosphate termini can be covalently joined via phosphodiester linkages by the action of DNA ligase. Because the substrate components in the reaction are DNA ends, the kinetics of joining reflect the concentration of these ends rather than the concentration of DNA per se. The tendency toward either *inter*molecular joining or *intra*molecular joining (circle formation) depends upon the concentration of DNA ends and upon the length of DNA fragments (Dugaiczyk et al. 1975). A high concentration of DNA ends favors *inter*molecular joining, whereas a low concentration of ends favors circle formation. Because of the intrinsic rigidity of duplex DNA, circle formation is not possible with fragments that are less than 200 base pairs long. Efficiency of circle formation increases above this length but falls off rapidly with fragments longer than 20 kb (Dugaiczyk et al. 1975).

The initial reaction in the joining of two ends of one or two double stranded DNA fragments is the formation of a single phosphodiester linkage in one strand. Complete covalent joining of the two ends requires the formation of a second phosphodiester bond on the opposite strand. The initial reaction requires contact between the two ends to be joined and DNA ligase. This tricomponent reaction is very inefficient. Formation of the second phosphodiester bond is, however, efficient because the ends are permanently held together in the precise configuration required for ligation.

---

1 Deutsches Krebsforschungszentrum, D-6900 Heidelberg, Fed. Rep. of Germany
2 Department of Medical Biochemistry, University of Geneva, CH-1211 Geneva, Switzerland

Contact between fully base-paired (i.e., blunt or flush-ended) DNA fragments occurs by random collision. Therefore, high concentrations of ends and of DNA ligase are important for optimal joining (Sgaramella et al. 1970). The primary contact between DNA fragments having single stranded, self-complementary (cohesive) 3'- or 5'-extensions also occurs by random collision. However, base-pairing of the cohesive termini stabilizes the initial contact without covalent linkage, in the same way that formation of the initial phosphodiester linkage establishes a stable contact between flush-ended DNA fragments. Because of this, efficient ligation can occur even when the concentrations of DNA ends and DNA ligase are low. However, the temperature of the ligation reaction must be maintained close to the T_m of the cohesive termini in order to permit stable base-pairing of these ends (Dugaiczyk et al. 1975).

In gene cloning experiments involving plasmid vectors, the yield of recombinant plasmids obtained is a measure of the proportion of circular heterologous molecules generated during ligation (only circular molecules efficiently transform *E. coli* bacteria). The formation of heterologous molecules through *inter*molecular joining is favored by high concentrations of DNA ends, whereas circle formation through *intra*molecular joining is favored by low concentrations of DNA ends. In order to obtain a high ratio of hybrid circles to nonhybrid circles, compromise reaction conditions are usually employed that favor slightly the formation of nonproductive linear hybrid molecules. In those instances where the identification of clones containing desired hybrid molecules is problematic, or where the quantities of DNA fragments to be cloned is low, the formation of nonhybrid circles of vector plasmids can be greatly reduced by first removing their terminal 5'-PO_4 groups through treatment with alkaline phosphatase (see Chap. 3.3), or by "tailing" the vector fragments with one nucleotide homopolymer (e.g., poly G) and the fragments to be cloned with a complementary homopolymer (e.g., poly C).

2. Experiment 1: Ligation of Cohesive-Ended and Flush-Ended DNA Fragments

a) Introduction

In the following experiment, the influence of DNA termini concentration and DNA ligase concentration upon the efficiency of ligation and the types of products found, will be examined. In procedure 1, pBR322 plasmid DNA will be linearized with endonuclease *Hin*dIII, which generates DNA fragments having cohesive termini; the resulting molecules will be treated at different concentrations with DNA ligase. In procedure 2, pBR322 DNA will be cleaved into two fragments with endonuclease *Hin*cII, which generates DNA fragments having flush-ends; the fragments will be ligated at several different concentrations of DNA and enzyme.

b) Objective

To examine some parameters of DNA ligase reactions.

c) Procedure 1

Day 1

In a sterile Eppendorf tube, pipette

H_2O	54 µl
MRB × 10	7 µl
pBR322 (1 mg/ml)	7 µl
*Hin*dIII (10 U/µl)	2 µl

Mix briefly on vortex, spin a few sec in Eppendorf centrifuge and incubate at 37°C for 60 min.

Inactivate endonuclease by holding tube at 70°C for 10 min, then pipette 10 µl samples into 6 tubes on ice labeled *a, b, c, d, e, f*.

Add ligation components as follows:

- a   1   µl 10 mM ATP +   1   µl 40 mM DTT
- b   1.5 µl 10 mM ATP + 1.5 µl 40 mM DTT +   3 µl MRB* + 2.0 µl T4 ligase (Miles; 20 M/ml)
- c   2.5 µl 10 mM ATP + 2,5 µl 40 mM DTT + 10 µl MRB* + 2.0 µl T4 ligase
- d   5   µl 10 mM ATP +   5   µl 40 mM DTT + 30 µl MRB* + 2.0 µl T4 ligase
- e   10 µl 10 mM ATP + 10   µl 40 mM DTT + 70 µl MRB* + 2.0 µl T4 ligase

* **Note:** Always distinguish solutions at normal strength (here) from stock solutions at higher strengths.

Incubate tubes overnight at 14°C.

Day 2

To tube *f* add 1 µl pBR322 DNA. Make volumes of all tubes up to 100 µl with MRB, heat to 70°C for 10 min to destroy ligase and melt base-paired termini.

Transfer 40 µl samples from each tube to fresh tubes containing 4 µl stop solution.

Analyse these by electrophoresis through a TBE buffered 0.8% agarose gel containing ethidium bromide, final concentration 0.1 µg/ml.

Use the remaining samples in the original tubes for analysis by transformation of *E. coli* C600 (use 4 µl and 40 µl samples and select transformants resistant to 50 µg/ml ampicillin) and electron microscopy (score 100 molecules as monomer, dimer, etc. and as linear or circle). Which conditions favor circle formation; which conditions produce the most transformants?

d) Procedure 2

Day 1

In a sterile Eppendorf tube, pipette

H_2O	51 µl
MRB × 10	7 µl
pBR322 DNA (1 mg/ml)	7 µl
*Hinc*II (4 U/µl)	5 µl

Mix briefly on a vortex, spin a few sec in Eppendorf centrifuge and incubate at 37°C for 60 min.

Inactivate endonuclease by holding tube at 70°C for 10 min, then pipette 10 µl samples into 6 tubes on ice labeled g, h, i, j, k, l.

Add ligation components as follows:

g	1 µl ATP + 1 µl DTT	
h	1 µl ATP + 1 µl DTT	+ 2.0 µl T4 ligase (Miles; 100 U/ml)
i	1 µl ATP + 1 µl DTT	+ 2.0 µl T4 ligase (Miles; 20 U/ml)
j	5 µl ATP + 5 µl DTT + 30 µl MRB	+ 2.0 µl T4 ligase (Miles; 100 U/ml)
k	5 µl ATP + 5 µl DTT + 30 µl MRB	+ 2.0 µl T4 ligase (Miles; 20 U/ml)
l	10 µl ATP + 10 µl DTT + 65 µl MRB	+ 2.0 µl T4 ligase (Miles; 100 U/ml)

Treat as in procedure 1.

e) Results and Discussion

Electrophoretic analysis of the products of ligation of the two experiments are shown in Figs. 1 and 2. Electron microscopic analysis of the ligation reactions provides a quantitative evaluation of their products and, additionally, facilitates the identification of the various DNA forms seen on the gels.

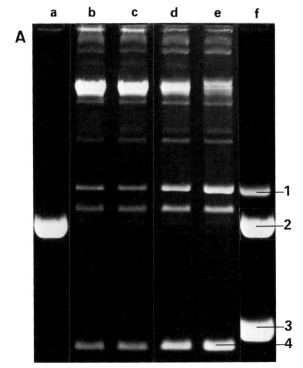

Fig. 1. Agarose gel electrophoresis of the DNA ligation mixtures of procedure 1. *1* open circular (nicked, relaxed) monomer; *2* linear monomer; *3* supercoiled covalently-closed circular monomer; *4* nonsupercoiled covalently closed circular monomer ligation product (in fact is supercoiled during electrophoresis due to the presence of ethidium bromide)

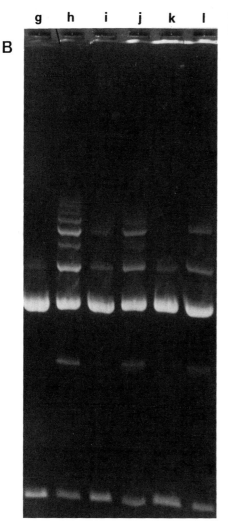

Fig. 2. Agarose gel electrophoresis of the DNA ligation mixtures of procedure 2

In the case of the ligation of cohesive-ended DNA fragments (Fig. 1), it can be seen clearly that as DNA concentration increases, the efficiency of ligation increases (high molecular weight, polymeric molecules predominate, i.e., both ends of each DNA fragment have a high probability of being joined to other fragments) and the probability of circle formation is low. At low DNA concentrations, the tendency toward circle formation increases and the tendency to form multimers decreases. Analysis of the products of ligation by transformation should indicate that the highest frequency of transformation results from the highest proportion of circular monomers, i.e., from the reaction with the lowest DNA concentration.

In the case of ligation of flush-ended DNA fragments (Fig. 2), it can be seen that the most effective ligation occurs under conditions of highest DNA and enzyme concentrations and that, in this experiment, the most important factor for efficient ligation is the enzyme concentration. (However, note that the inclusion of T4 RNA ligase can stimulate flush end ligation up to a factor of 20-fold; Sugino et al. 1977).

3. Materials

MRB × 10 (100 mM Tris-Cl, pH 7.4, 100 mM $MgCl_2$, 500 mM NaCl)
MRB 1/10 MRB × 10
TBE (45 mM boric acid, 45 mM Tris, 1.25 mM EDTA, pH 8.5)
ATP (10 mM)
DTT (40 mM)
pBR322 DNA (1 mg/ml)
*Hin*dIII (10 U/μl)
*Hin*cII (4 U/μl)
T4 ligase (Miles, 20 U/ml and 100 U/ml)
Stop solution [20% Ficoll (Pharmacia), 0.025% bromophenol blue, 0.035% xylene cyanol (optional), 0.8% sodium dodecyl sulphate, in TBE]
Agarose
Ethidium bromide stock solution (5 mg/ml)
Competant cells of *E. coli* K-12 C600
Nutrient agar plates containing ampicillin (50 μg/ml)

4. References

Dugaiczyk A, Boyer HW, Goodman HM (1975) Ligation of *Eco*RI endonuclease-generated DNA fragments into linear and circular structures. J Mol Biol 96:171–184

Sgaramella V, van de Sande JH, Khorana HG (1970) Studies on polynucleotides. C. A novel joining reaction catalyzed by the T4-polynucleotide ligase. Proc Natl Acad Sci USA 67:1468–1475

Sugino A, Goodman HM, Heyneker HL, Shine I, Boyer HB, Cozzarelli NR (1977) Interaction of bacteriophage T4 RNA and DNA ligases in joining of duplex DNA at base-paired ends. J Biol Chem 252:3987–3994

3.3 Cloning with Plasmid Vectors

K.N. TIMMIS[1] and J. FREY[2]

Contents

1. General Introduction 160
2. Experiment 1: Cloning of DNA Fragments of Plasmid R6-5 Using the pBR322 Vector . 162
   a) Introduction. 162
   b) Procedure 1 163
   c) Procedure 2 164
   d) Results and Discussion 166
3. Materials 169
4. References 169

1. General Introduction

The overriding consideration in any gene cloning strategy is the generation of hybrid DNA molecules containing the determinant of interest and their identification in transformant clones (Timmis 1981). As reported in Chap. 3.1, the ligation of DNA fragments to linear vector molecules usually produces a high proportion of nonhybrid recircularized vector molecules that transform host cells with high efficiencies. Thus, in any cloning strategy, the means of identifying or excluding recircularized vector molecules must be considered.

The simplest situation is that in which the desired fragment specifies a directly selectable function, e.g., resistance to an antibiotic (selection of transformant clones on media containing the corresponding antibiotic) or a component of a pathway that can be selected by complementation of a defective pathway in a mutant host cell (Franklin et al. 1981, Long et al. 1982). Where direct selection of desired DNA sequences is possible, simple cloning procedures can be used, although specific strains may need to be employed as host cells (i.e., in the cases above, antibiotic sensitive cells and mutants specifically defective in the desired determinant, respectively).

Another relatively straightforward situation is the cloning of all possible DNA fragments presented to the vector, i.e., the construction of a "bank" or "library" of the DNA molecule(s) being cloned. In this case, if the number of fragments to be cloned is relatively small, it suffices to be able to distinguish transformants carrying

---

1 Department of Medical Biochemistry, University of Geneva, CH-1211 Geneva, Switzerland
2 Department of Biochemistry, University of Geneva, CH-1211 Geneva, Switzerland

hybrid molecules from those carrying nonhybrid vector molecules. This may be readily accomplished by the use of a vector whose phenotype changes when fragments are inserted into it, i.e., whose cloning sites are located in genes that specify readily scored phenotypes, namely, "insertional inactivation" sites (Timmis 1981, Timmis et al. 1974, 1978).

If, however, the number of fragments to be cloned is large, it is imperative to use a host:vector system or experimental protocol which ensures that all transformant clones obtained contain hybrid molecules. Examples of vectors that permit the direct selection of hybrid molecules are pKN80 and pKY2289 (Schumann 1979, Ozaki et al. 1980. see Timmis 1981 for discussion). Procedures that permit the direct selection of hybrid molecules generally prevent vector recircularization or include a biological selection for molecules larger than the vector. The former includes (1) the removal of the 5'-phosphate groups of the vector molecule by treatment with alkaline phosphatase (Ullrich et al. 1977); (2) "tailing" of the vector and fragments to be cloned with complementary homopolymer blocks (e.g., tailing of vector molecules with poly dG and of the fragments to be cloned with poly dC; Roychoudhury et al. 1976); and (3) use of vector DNA molecules having different termini to clone fragments also having the same distinct termini. The latter includes the use of the λ-phage genome packaging system in conjunction with vectors (λ genome vectors, see Chap. 3.5; cosmids, see Chap. 3.6) that are themselves too small to be packaged (Hohn and Hinnen 1980).

Generally, all of the above procedures are applicable when the DNA to be cloned is available in reasonable quantities and a molar ratio of vector:DNA to be cloned of $\leqslant 1$ can be arranged. Where the amount of available DNA to be cloned is very small, it becomes necessary to accept a ratio of $\geqslant 1$, in order that a high proportion of the fragments become linked to vector molecules. In this case, it is imperative to prevent *intra*molecular linkage of, or *inter*molecular linkage between vector molecules.

These are some aspects of general cloning strategies. In addition, many specific strategies have been developed for particular cloning problems that have arisen. For further information, the reader should consult recent reviews (Setlow and Hollaender 1980, Timmis 1981). In any case, the beginner should be aware that additional biochemical steps which may facilitate the cloning of desired DNA fragments constitute additional stages in the experimental protocol at which losses of DNA can occur and problems can arise.

In the following experiments, three procedures will be followed, namely, the direct selection of transformant clones carrying desired fragments, the use of insertional inactivation to identify clones carrying hybrid molecules, and the use of alkaline phosphatase to select clones carrying hybrid molecules. The DNA to be cloned will be plasmid R6-5, a well-studied conjugative plasmid that specifies resistance to chloramphenicol, fusidic acid, kanamycin, streptomycin/spectinomycin, sulphonamide, and mercury salts (Fig 1).

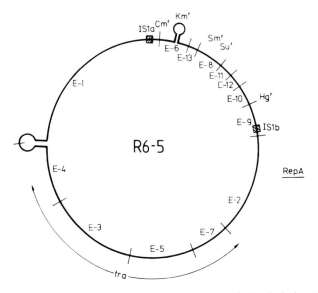

Fig. 1. Restriction endonuclease cleavage map of plasmid R 6-5. The *cross bars* represent cleavage sites for *Eco*RI. Cm^r, Km^r, Sm^r, Su^r, Hg^r are determinants for resistance to chloramphenicol/fusidic acid, kanamycin, streptomycin/spectinomycin, sulphonamide, and mercury salts, respectively; *IS1a*, *IS1b* insertion sequences that bracket the resistance determinant segment of the plasmid; *RepA* region containing determinants for replication; *tra* region containing determinants for conjugal transfer. The *stem* and *loop structures* indicate the locations of two transposable elements, Tn*601*/Tn*903* (Km^r) and a derivative of Tn*10 (large loop)*, containing an insertion within its Tc^r determinant

2. Experiment 1: Cloning of DNA Fragments of Plasmid R6-5 Using the pBR322 Vector

a) Introduction

The majority of cloning experiments presently carried out involve the use of plasmid vectors (mainly pBR322 and its relatives), although cosmids and vectors based on other plasmid replicons, and phage genome vectors are also widely employed. In the following experiment, the pBR322 vector will be used in the cloning of DNA fragments of the R6-5 plasmid.

In procedure 1, the principle of insertional inactivation will be illustrated. The pBR322 vector specifies resistance to ampicillin (Ap^r) and tetracycline (Tc^r). The Ap^r determinant contains the unique *Pst*I site of the plasmid and the Tc^r determinant contains the unique *Eco*RV, *Bam*HI, *Sph*I, *Sal*I, and *Xma*III sites. When pBR322 is used for the cloning of *Pst*I fragments, transformant bacteria are selected on Tc-containing plates, whereas when it is used for cloning, e.g., *Sal*I fragments, transformant bacteria are plated on Ap-containing plates. Transformant clones carrying hybrid molecules will thus be Tc^r Ap^s and Ap^r Tc^s, respectively, whereas those containing recircularized nonhybrid vector molecules will be Ap^r Tc^r in both cases.

The transfer of transformant clones obtained on plates containing one antibiotic to plates containing the other, readily identifies clones carrying hybrid plasmids. In procedure 1, *Pst*I generated fragments of R6-5 will be cloned in pBR322 and hybrid plasmids identified as exhibiting the phenotype $Tc^r\ Ap^s$.

Other useful cloning sites of the vector, such as *Eco*RI, are not located in either of the two genes specifying readily scored phenotypes. The cloning of DNA fragments in these sites, therefore, requires a means of selecting hybrid molecules. In procedure 2, the use of alkaline phosphatase to prevent vector recircularization, and the direct selection of desired hybrid molecules, will be illustrated. Treatment of the vector molecule with alkaline phosphatase results in removal of the terminal 5' phosphate groups. This prevents the formation of phosphodiester linkages between the 3'hydroxyl and the 5' ends of vector molecules, but does not prevent linkage of vector 3' hydroxyl ends to 5' phosphate ends of fragments to be cloned. Hybrid molecules formed in this way are linked in only one strand at each vector terminus; closure of the opposite strand occurs in vivo after transformation. In this experiment, *Eco*RI-cleaved pBR322 molecules will be treated with alkaline phosphatase and used to clone *Eco*RI fragments of plasmid R6-5. Transformant clones will be selected on plates containing Ap, to obtain all possible hybrid molecules, and on plates containing chloramphenicol, kanamycin, and streptomycin, to select directly hybrid molecules containing DNA fragments carrying the antibiotic resistance determinants of R6-5.

b) Procedure 1

Day 1

Label 2 sterile Eppendorf tubes *a* and *b*.
Into tube *a*, pipette

H_2O	28 µl
MRB × 10	6 µl
R6-5 DNA (150 µg/ml)	20 µl
*Pst*I (1 U/µl)	6 µl

Into *b*, pipette

H_2O	43 µl
MRB × 10	6 µl
pBR322 (100 µg/ml)	5 µl
*Pst*I (1 U/µl)	6 µl

Briefly agitate each tube on a vortex mixer, spin a few sec in Eppendorf centrifuge and incubate at 37°C for 60 min.

Inactivate endonuclease by holding tubes at 70°C for 10 min and place on ice.

Remove from each tube two 5µl samples, place in tubes labeled *a1, a2, b1, b2* and to these add 13 µl H_2O plus 2 µl of stop solution; retain for subsequent electrophoresis.

Label a sterile Eppendorf tube *c*, place on ice and add

PstI-cut R6-5 DNA (tube a)   35 μl
PstI-cut pBR322 DNA (tube b) 35 μl
ATP (10 mM)   9 μl
DTT (40 mM)   9 μl
DNA ligase (20 U/ml)   2 μl

Incubate tube c overnight at 14°C.

Day 2

Transfer 15 μl of the ligation mixture to a tube labeled c1 and add to it 3 μl H_2O plus 2 μl of stop solution. Subject the contents of tubes a1, b1, and c1 to electrophoresis on a TBE-buffered 0.8% agarose gel.

Use 5 μl and 50 μl of the ligation mixture (tube c) for transformation of E. coli C600; select transformants on agar plates containing tetracycline (10 μg/ml).

Day 3

Transfer by means of sterile toothpick 200 Tc^r colonies to two series of plates, the first containing Ap (50 μg/ml), the second containing Tc. For ease of comparison of the two series, pick according to a grid placed below the plates and clearly mark a point of reference at the top of each plate.

Incubate both series of plates overnight at 37°C.

Day 4

Compare each pair of plates and identify clones that grow on Tc plates, but not on Ap plates. Streak out 10 Tc^r Ap^s clones on Tc plates for single colony purification and inoculate each one into 10 ml volumes of L-broth containing Tc.

Incubate plates and cultures overnight at 37°C.

Day 5

Make plasmid mini-preparations from the L-broth cultures and analyze with and without digestion with PstI by electrophoresis through a 0.8% agarose gel. As controls include samples a2 and b2 in the gel.

What is the proportion of clones carrying hybrid plasmids in relation to all transformant clones obtained? Do you have the impression that all R6-5 fragments are cloned with equal facility (only possible to answer if sufficient clones are analyzed, e.g., if the results of a class experiment are pooled)? If not, suggest reasons for this.

c) Procedure 2

Day 1

Label two sterile Eppendorf tubes d and e.

To tube d add

H_2O 28 μl
HRB X 10   6 μl
R5-6 DNA (150 μl/ml)   20 μl
EcoRI (1 U/μl)   6 μl

Mix briefly on a vortex, spin a few sec in Eppendorf centrifuge, and incubate at 37°C for 60 min.

Inactivate endonuclease by holding tube at 70°C for 10 min.

To tube e add

H$_2$O	19 μl
HBR × 10	3 μl
pBR322 (100 μg/ml)	5 μl
EcoRI (1 U/μl)	3 μl

Mix briefly on vortex, spin a few sec in Eppendorf centrifuge, and incubate at 37°C for 60 min.

Inactivate endonuclease by holding tubes at 70°C for 10 min, then place on ice.

To tube e, add 60 μl LRB (to dilute NaCl).

Add 2 μl alkaline phosphatase (0.5 mg/ml).

Incubate at 37°C for 60 min.

Extract the contents of tube e with phenol twice and then with ether three times. Precipitate DNA with isopropanol and resuspend in 60 μl LRB.

From tubes d and e, transfer two 5 μl samples to tubes marked d1, d2, e1, e2 and add to each of these latter four tubes 13 μl H$_2$O plus 2 μl of stop solution. Retain for subsequent gel electrophoresis.

Mark three sterile Eppendorf tubes f, g, and h, and place on ice.

To tube f add

EcoRI-cut R6-5 (tube d)	35 μl
EcoRI-cut pBR322 (tube e)	35 μl
ATP (10 mM)	9 μl
DTT (40 mM)	9 μl
DNA ligase (20 U/μl)	2 μl

To tube g add

H$_2$O	5 μl
LRB × 10	2 μl
EcoRI-cut R6-5 (tube d)	8 μl
ATP (10 mM)	2 μl
DTT (40 mM)	2 μl
DNA ligase (20 U/μl)	1 μl

To tube h add

H$_2$O	5 μl
LRB × 10	2 μl
EcoRI-cut pBR322 (tube e)	8 μl
ATP (10 mM)	2 μl
DTT (10 mM)	2 μl
DNA ligase (20 U/μl)	1 μl

Agitate tubes f, g, and h, spin a few sec in Eppendorf centrifuge, and incubate overnight at 14°C.

Day 2

To tubes *g* and *h* add 2 µl stop solution.

Transfer 15 µl from tube *f* to a new tube labeled *f1* and add to this 5 µl H_2O plus 2 µl of stop solution.

Analyze by electrophoresis on a 0.8% agarose gel the contents of tubes *d1, e1, f1, g,* and *h*.

Use 5 µl and 50µl samples from tube *f* for transformation of *E. coli* C600.

Select transformant clones on agar plates containing Ap (50 µg/ml), chloramphenicol (50 µl/ml), kanamycin (50 µg/ml), and streptomycin (20 µg/ml).

Incubate plates overnight at 37°C.

Day 3

Pick 5 colonies from the Ap plates and 2 colonies from each of the plates containing the other antibiotics; streak for single colonies on identical plates and inoculate 10 ml volumes of L-broth containing the appropriate antibiotics.

Day 4

Prepare plasmid mini-preparations from the L-broth cultures and analyze with and without digestion with *Eco*RI by electrophoresis through a 0.8% agarose gel; include as controls, samples *d2* and *e2* in this gel.

What proportion of all clones obtained contain hybrid plasmids? Were transformants obtained on all selection plates? If not, what conclusions can be drawn regarding the disposition of *Eco*RI cleavage sites relative to the antibiotic resistance determinants? Are you able to identify which fragments of R6-5 carry determinants for resistance to Cm, Km, Sm?

d) Results and Discussion

Using the experimental protocol given and using no procedure to prevent vector recircularization, one can expect 10–20% of transformant clones to contain hybrid plasmids. Generally, gene cloning with alkaline phosphatase-treated vector molecules results in 95% of transformant clones containing hybrid molecules, although the absolute number of such clones thereby obtained may be significantly reduced when compared with those obtained using nontreated vectors.

Several examples of pBR322/R6-5 hybrids are shown in Figs. 2 and 3, as are two directly selected Sm^r plasmids and two Km^r plasmids. Note that one Km^r plasmid contains no pBR322 fragment, but instead contains the E-2 fragment of R6-5; it is thus a miniplasmid (note the low amount of DNA on the gel).

The direct cloning of DNA fragments is more generally useful than might perhaps be appreciated, particularly in the cloning and analysis of groups of genes of complex properties. In such cases, the cloning of all essential genes that specify a property may not be possible, either because there are too many genes or because they are organized in blocks that are widely distributed around the chromosome. The cloning of some genes of such a property may be possible, if purified mRNAs or antibodies

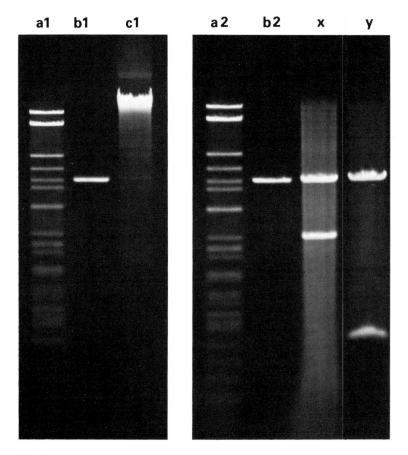

Fig. 2. Agarose gel electrophoresis of *Pst*I-cleaved R6-5, pBR322, and R6-5/pBR322 hybrid plasmids (procedure 1). Tracks *a1*, *b1*, and *c1* contain R6-5/*Pst*I, pBR322/*Pst*I, and R6-5/*Pst*I plus pBR322/*Pst*I ligated together, respectively. Tracks *a2* and *b2* are the same as *a1* and *b1*, whereas *x* and *y* are two *Pst*I-cleaved R6-5/pBR322 hybrid plasmids

against gene products are available. However, a more general approach, at least for bacterial properties, is the inactivation of the property by transposon mutagenesis, the cloning of DNA fragments containing the transposon by direct selection of an antibiotic resistance specified by the transposon, and the use of the cloned transposon-carrying fragment as a DNA hybridization probe to screen gene banks for cloned fragments containing nonmutant genes (e.g., see Lehrbach and Timmis 1983).

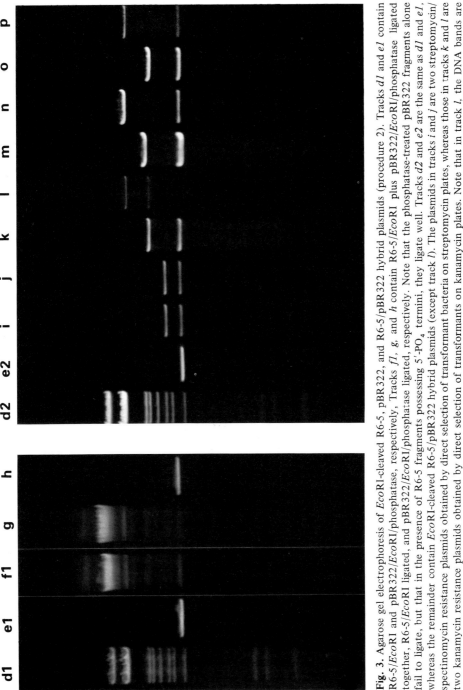

Fig. 3. Agarose gel electrophoresis of *Eco*RI-cleaved R6-5, pBR322, and R6-5/pBR322 hybrid plasmids (procedure 2). Tracks *d1* and *e1* contain R6-5/*Eco*RI and pBR322/*Eco*RI/phosphatase, respectively; Tracks *f1*, *g*, and *h* contain R6-5/*Eco*RI plus pBR322/*Eco*RI/phosphatase ligated together, R6-5/*Eco*RI ligated, and pBR322/*Eco*RI/phosphatase ligated, respectively. Note that the phosphatase-treated pBR322 fragments alone fail to ligate, but that in the presence of R6-5 fragments possessing 5′-PO$_4$ termini, they ligate well. Tracks *d2* and *e2* are the same as *d1* and *e1*, whereas the remainder contain *Eco*RI-cleaved R6-5/pBR322 hybrid plasmids (except track *l*). The plasmids in tracks *i* and *j* are two streptomycin/spectinomycin resistance plasmids obtained by direct selection of transformant bacteria on streptomycin plates, whereas those in tracks *k* and *l* are two kanamycin resistance plasmids obtained by direct selection of transformants on kanamycin plates. Note that in track *l*, the DNA bands are faint and there is no pBR322 fragment. Instead, there is in addition to the R6-5 E-6 fragment carrying the Km^r gene, a large R6-5 fragment, namely fragment E-2, which will be identified in Chap. 3.4 as being a fragment carrying all essential replication determinants of R6-5. The plasmid in track *l* is thus an R6-5 miniplasmid. The plasmids shown in tracks *m–p* are *Eco*RI-cleaved R6-5/pBR322 hybrids that were selected only by their Ap^r phenotype

3. Materials

MRB × 10	(see Chap. 3.2)
HRB × 10	(100 mM Tris-Cl, pH 7.4, 100 mM MgCl$_2$, 1 M NaCl)
LRB	(10 mM Tris-Cl, pH 7.4, 10 mM MgCl$_2$)
TBE	(see Chap. 3.2)
ATP	(see Chap. 3.2)
DTT	(see Chap. 3.2)
pBR322 DNA	(100 µg/ml)
R6-5 DNA	(150 µg/ml)
*Pst*I	(1 U/µl)
*Eco*RI	(1 U/µl)
Alkaline phosphatase	(0.5 mg/ml, calf intestine, Boehringer-Mannheim)
DNA ligase	(20 U/ml)
Stop solution	
Phenol	
Isopropanol	
Agarose	
Ethidium bromide	(5 mg/ml)

Competent cells of *E. coli* K-12 C600
Nutrient agar plates containing
  (a) ampicillin (50 µg/ml)
  (b) tetracycline (10 µg/ml)
  (c) streptomycin (20 µg/ml)
  (d) kanamycin (50 µg/ml)
  (e) chloramphenicol (50 µg/ml)

4. References

Franklin FCH, Bagdasarian M, Bagdasarian MM, Timmis KN (1981) Molecular and functional analysis of the TOL plasmid pWWO from *Pseudomonas putida* and cloning of genes for the entire regulated aromatic ring *meta* cleavage pathway. Proc Natl Acad Sci USA 78:7458–7462

Hohn B, Hinnen A (1980) Cloning with cosmids in *E. coli* and yeast. In: Setlow JK, Hollaender A (eds) Genetic engineering, principles and methods, vol 2. Plenum Press, New York, pp 169–183

Lehrbach PR, Timmis KN (1983) Genetic analysis and manipulation of catabolic pathways in *Pseudomonas*. Biochem Soc Symp 48:191–219

Long SR, Buikema WJ, Ausubel FM (1982) Cloning of *Rhizobium meliloti* nodulation genes by direct complementation of Nod⁻ mutants. Nature 298:485–488

Ozaki LS, Maeda S, Shimada K, Takagi Y (1980) A novel ColE1::Tn*3* plasmid vector that allows direct selection of hybrid clones in *E. coli*. Gene 8:301–314

Roychoudhury R, Jay E, Wu R (1976) Terminal labeling and addition of homopolymer tracts to duplex DNA fragments by terminal deoxynucleotidyl transferase. Nucleic Acids Res 3:101–116

Schumann W (1979) Construction of an *Hpa*I and *Hin*dII plasmid vector allowing direct selection of transformants harbouring recombinant plasmids. Mol Gen Genet 174:221–224

Setlow JK, Hollaender A (eds) (1980) Genetic engineering, principles and methods. Plenum Press, New York

Timmis KN (1981) Gene manipulation in vitro. In: Glover SW, Hopwood DA (eds) Soc Gen Microbiol Symp 31. Genetics a a tool in microbiology. University Press, Cambridge, pp 49–109

Timmis KN, Cabello F, Cohen SN (1974) Utilization of two distinct modes of replication by a hybrid plasmid constructed in vitro from separate replicons. Proc Natl Acad Sci USA 71: 4556–4560

Timmis KN, Cabello F, Cohen SN (1978) Cloning and characterization of *Eco*RI and *Hin*dIII restriction endonuclease-generated fragments of antibiotic resistance plasmids R6-5 and R6. Mol Gen Genet 162:121–137

Ullrich A, Shine J, Chirgwin J, Pictet R, Tischer E, Rutter WJ, Goodman HM (1977) Rat insulin gene: construction of plasmids containing the coding sequences. Science 196:1313–1319

3.4 Mini-Plasmid Formation: The Cloning of Replication Origins

K.N. TIMMIS[1] and J. FREY[2]

Contents

1. General Introduction 171
2. Experiment 1: Generation of Mini-Plasmids of R6-5 172
   a) Introduction 172
   b) Procedure 172
   c) Results and Discussion 173
3. Materials 175
4. References 175

1. General Introduction

Gene cloning is essentially a technique for the deletion of nonrelevant DNA sequences and for the dissection of relevant DNA sequences into desired sub-sets. Although gene cloning is generally used to isolate determinants of phenotypes encoded by replicons, it may also be used to isolate the essential determinants of replicons themselves, namely, their sequences that are required for replication, and thereby generate mini-replicons.

Replication of DNA is a highly complex process involving a variety of *trans*-acting proteins and specific DNA sequences, particularly the site of initiation of replication, the replication origin (see Chap. 8; Kornberg 1980). Extrachromosomes, such as plasmids, contain origin sequences and generally encode some specific replication factors, although they depend largely upon host chromosome-encoded factors (Thomas and Helinski 1979, Timmis et al. 1980). Thus, in order to form a mini-plasmid, it is necessary to generate one or more DNA fragments that contain all essential replication determinants. In contrast, in order to form a mini-chromosome, it is in principle necessary to generate a fragment carrying only origin sequences, since when this is introduced into a host cell, all *trans*-acting factors necessary for replication are provided.

Some extrachromosomes, such as the narrow host range plasmids R6-5 and F, have their essential replication determinants organized together in a gene cluster

---

1 Department of Medical Biochemistry, University of Geneva, CH-1211 Geneva, Switzerland
2 Department of Biochemistry, University of Geneva, CH-1211 Geneva, Switzerland

adjacent to the origin (Timmis et al. 1975, 1978, Lovett and Helinski 1976, Thomas and Helinski 1979), whereas others, such as the broad host range plasmids RK2 and RSF1010, have their replication determinants distributed around the replicon (Thomas and Helinski 1979, Bagdasarian et al. 1982). The generation of mini-plasmids of the former class is obviously easier than that of the latter class.

The strategy for generating mini-plasmids is dictated by the essential phenotype exhibited by replicons, namely, their capacity for replication. The procedure is, in essence, to cleave the replicon and to link the fragments thereby generated to a fragment of DNA that specifies a selectable marker, but that is itself *unable to replicate*. The selection of transformant bacteria expressing this marker results in the isolation of clones containing replicons composed of the DNA fragment carrying the selectable marker linked to one or more fragments from the digested plasmid that specify all essential functions for its autonomous replication. Thus, unlike most cloning experiments, mini-plasmid/mini-chromosome formation does not involve the use of a cloning vector. The technique for miniplasmid/mini-chromosome formation was first developed with bacterial plasmids (Timmis et al. 1975, Lovett and Helinski 1976) and later extended to other replicons including chromosomes (Yasuda and Hirota 1977, Messer et al. 1978).

2. Experiment 1: Generation of Mini-Plasmids of R6-5

a) Introduction

The first experiments to isolate mini-replicons were carried out with the large conjugative plasmids R6-5 and F (Timmis et al. 1975, Lovett and Helinski 1976). In these experiments, nonreplicating *Eco*RI fragments specifying resistance to ampicillin or kanamycin were ligated to the *Eco*RI-digested plasmids in order to clone selectively the DNA fragments capable of autonomous replication. In the case of R6-5, however, a number of selectable functions are encoded by the plasmid itself, namely, resistances to chloramphenicol, fusidic acid, kanamycin, streptomycin/spectinomycin, and mercury salts. Thus, in order to generate mini-plasmids of R6-5, it is not necessary to provide an external DNA fragment specifying a selectable function. In the following experiment, R6-5 DNA will be cleaved with *Eco*RI and the fragments thereby generated ligated together. The ligated molecules will then be used to transform *E. coli* bacteria for resistance to chloramphenicol, kanamycin, and streptomycin.

b) Procedure

Day 1

Cleavage by EcoRI

Into a sterile Eppendorf tube, pipette

H_2O	28 μl
MRB × 10	6 μl
R5-6 DNA (150 μl/ml)	20 μl
*Eco*RI (1 U/ml)	6 μl

Mix briefly on a vortex, spin a few sec in an Eppendorf centrifuge, and incubate at 37°C for 60 min.
Inactivate enzyme by holding tube at 70°C for 10 min, cool on ice 30 min.
Remove two 5 μl samples and to each add 13 μl H$_2$O + 2 μl stop solution (a, c).

Ligation

To the remainder of the digested DNA

ATP (10 mM)	6 μl
DTT (40 mM)	6 μl
T4 ligase (20 U/ml)	2 μl

Mix, spin a few sec in an Eppendorf centrifuge, and incubate overnight at 14°C.

Day 2

Remove a 10 μl sample from the ligation mixture and to this add 8 μl H$_2$O + 2 μl stop solution (b).
Analyze samples a and b by electrophoresis on a horizontal TBE-buffered 0.8% agarose gel.
Use the remainder of the ligation mixture for transformation of *E. coli* C600, 4 μl for one transformation; 40 μl for another.
Plate transformation mixtures on agar containing the single antibiotics kanamycin (50 μg/ml), streptomycin (20 μg/ml), and chloramphenicol (50 μg/ml).

Day 3

Examine plates and count transformant colonies. Were transformants obtained on all plates? If not, how might this be explained?
Pick two well separated colonies from each type of selection plate and (1) use to inoculate 10 ml volumes of L-broth containing the relevant antibiotic and incubate with shaking at 37°C overnight, (2) streak out on fresh antibiotic plates to purify clones by single colony isolation, and (3) pick to alternate antibiotic plates to determine whether the clones express resistance to other antibiotics.

Day 4

Examine plates. Store purified clones.
Make mini-preparations of plasmid DNA from the 10 ml cultures and analyze plasmids by electrophoresis on a 0.8% agarose gel with and without cleavage of DNAs with *Eco*RI. Sample c should be run in parallel.

Note: Because R6-5 has a low copy number, as do also its mini-plasmids, plasmid yields are low; therefore, the largest possible volumes of DNA preparations should be used for electrophoresis.

c) Results and Discussion

The results of the experiment are shown in Fig. 1. Samples a, b, and c are the controls described above. Samples d and e are *Eco*RI-cleaved plasmid DNAs isolated from two

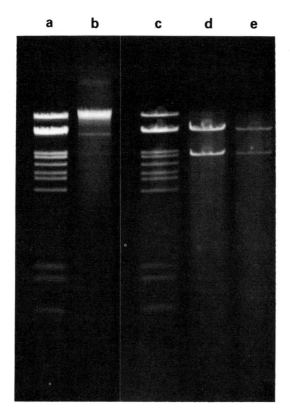

Fig. 1. Agarose gel electrophoresis of R6-5 DNA and two kanamycin resistance mini-plasmids. For details see text

kanamycin-resistant clones and are both seen to be composed of R6-5 *Eco*RI fragments E-6, the kanamycin resistance fragment (see Chap. 3.3), and a fragment belonging to the triplet band E-2/E-3/E-4. This fragment is, in fact, the 13 kb fragment E-2 (Timmis et al. 1975, 1978). Thus, it is shown that a single *Eco*RI fragment of R6-5, E-2, contains all plasmid replication determinants. Indeed, the replication determinants are clustered together on a 2.0–2.5 kb segment of E-2 that has been defined by a second round of mini-plasmid formation using an *Eco*RI mini-plasmid and the *Pst*I endonuclease (Timmis et al. 1978).

If no transformants had been obtained, despite the fact that efficient ligation had been achieved and that good transformation with a control plasmid DNA sample had been obtained, it would have to be concluded that either the essential replication determinants had not been clustered or that all of the antibiotic resistance determinants used for selection had contained *Eco*RI cleavage sites, such that no fragment containing an intact selectable gene had been produced. These two possibilities could be distinguished by attempting to clone with a cloning vector, *Eco*RI fragments containing intact antibiotic resistance determinants.

3. Materials

As Chap. 3.2, plus
- R6-5 DNA   150 μg/ml
- *Eco*RI   1 U/μl

Nutrient agar plates containing (a) kanamycin (50 μg/ml)
(b) streptomycin (20 μg/ml)
(c) chloramphenicol (50 μg/ml)

10 ml volumes of L-broth containing these antibiotics

4. References

Bagdasarian M, Bagdasarian MM, Lurz R, Nordheim A, Frey J, Timmis KN (1982) Molecular and functional analysis of the broad host range plasmid RSF1010 and construction of vectors for gene cloning in Gram negative bacteria. In: Mitsuhashi S (ed) Resistance in bacteria: genetics, biochemistry and molecular biology. Japan Sci Soc Press, Tokyo

Kornberg A (1980) DNA replication. Freeman, San Francisco

Kornberg A (1982) 1982 Supplement to DNA replication. Freeman, San Francisco

Lovett MA, Helinski DR (1976) Method for the isolation of the replication region of a bacterial replicon: construction of a mini-F Km plasmid. J Bacteriol 127:982–987

Messer W, Bergmans HEN, Meijer M, Womack JE, Hansen FG, von Meyenberg K (1978) Minichromosomes: plasmids which carry the *E. coli* replication origin. Mol Gen Genet 162:269–275

Thomas CM, Helinski DR (1979) Plasmid DNA replication. In: Timmis KN, Pühler A (eds) Plasmids of medical, environmental and commercial importance. Elsevier/North-Holland, Amsterdam, pp 29–46

Timmis KN, Cabello F, Cohen SN (1975) Cloning, isolation and characterization of replication regions of complex plasmid genomes. Proc Natl Acad Sci USA 72:2242–2246

Timmis KN, Andres I, Slocombe PM (1978) Plasmid incompatibility: cloning analysis of an IncFII determinant of R6-5. Nature 273:27–32

Timmis KN, Danbara H, Brady G, Lurz R (1980) Inheritance functions of group IncFII transmissible antibiotic resistance plasmids. Plasmid 5:53–75

Yasuda S, Hirota Y (1977) Cloning and mapping of the replication origin of *Escherichia coli*. Proc Natl Acad Sci USA 74:5458–5462

3.5 Gene Cloning with Bacteriophage λ

A. PÜHLER[1]

Contents

1. General Introduction 176
2. Experiment 1: Principle Techniques for the Application of a λ Insertion Vector 178
   a) Introduction 178
   b) Objectives 179
   c) Procedure 179
   d) Results 183
3. Experiment 2: Cloning Part of the *E. coli* Tryptophan Operon in a λ Insertion and a λ Replacement Vector 184
   a) Introduction 184
   b) Objectives 184
   c) Procedure 185
   d) Results 186
4. Materials 188
5. References 189

1. General Introduction

The Genome of Phage λ and its Utilization as a Cloning Vector

After infection of *E. coli,* the temperate phage λ (Hershey 1971) can grow lytically or lysogenize the host cell. As a prophage, the λ-repressor (product of gene c_I) represses all genes necessary for lytic growth. Mutations in gene c_I give rise to mutant phages no longer capable of lysogenization. In contrast to the P1 prophage, the λ-prophage is integrated into the *E. coli* chromosome. The λ-genome is well characterized and a set of genes necessary for lytic or lysogenic growth have been identified and mapped. Figure 1 shows the genetic map of phage λ. Genes *A* to *J* code for head and tail proteins. In the integration region, the phage attachment site *att* and the gene for the integrase *int* are indicated. Next to the recombination region is the regulation region where the genes *N* and c_I are located. The genes *O* and *P* are part of the replication region and genes *S* and *R* are necessary for lysis of the host cell. In addition to this genetic map, detailed restriction maps of the λ-genome for different restriction

---

[1] Lehrstuhl für Genetik, Fakultät für Biologie, Universität Bielefeld, D-4800 Bielefeld 1, Fed. Rep. of Germany

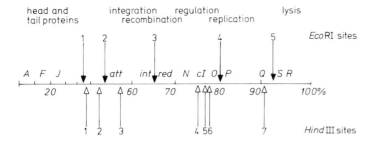

Fig. 1. Organization of the λ-genome. The physical and genetic map of the λ-genome is divided into 100 units. 5 *Eco*RI sites *(arrows above the map)* and 7 *Hind*III sites *(arrows below the map)* are indicated. Some characteristics of the λ-genes *A, F, J, att, int, red, N, c_I, O, P, Q, S,* and *R* are discussed in the text or listed by Hershey (1971)

enzymes have been constructed. In Fig. 1, the five *Eco*RI sites (Thomas and Davis 1975) and the seven *Hind*III sites (Phillipsen et al. 1978) of the λ-genome are shown.

At first sight, the usefulness of phage λ as a cloning vector is not obvious; it contains too many restriction sites. However, λ-genomes with only one or two restriction sites for *Eco*RI and *Hind*III have been constructed (Thomas et al. 1974, Murray et al. 1977, Blattner et al. 1977). Another difficulty lies in the packaging capacity of the λ-head. It packages only DNA between 38–53 kbp according to the size of the λ-genome of 49.4 kbp (Blattner et al. 1977). To circumvent these difficulties, deletion mutants have been isolated which lack up to 30% of the λ-genome. These deletion mutants with reduced numbers of *Eco*RI and *Hind*III restriction sites can be used for gene cloning since they allow the insertion of rather large DNA fragments. For cloning experiments in *E. coli*, two types of λ-vectors are available: the insertion and the replacement vector. Their properties will be discussed in the following sections.

λ Insertion Vector for EcoRI

In Fig. 2, a λ-insertion vector is shown. It carries a deletion in the nonessential *b* region (b^{538}) and the immunity region is replaced by the immunity region of the lambdoid phage 434 (imm^{434}). The vector carries a single target for the restriction endonuclease *Eco*RI, which is located in the c_I-repressor gene of the immunity region, imm^{434}. The deletion of about 18% allows the insertion of up to 9 kb of DNA. Hybrid molecules are recognized by a change from turbid to clear plaque morphology resulting from the inactivation of the c_I-repressor by the inserted DNA. The turbid plaque morphology is caused by losogens which can grow within the plaque.

λ Replacement Vector for EcoRI

This vector has two targets for the restriction enzyme *Eco*RI. These flank a replaceable segment of the phage genome (Fig. 2B). This replaceable segment fills essentially the space between these two restriction sites, since a phage with DNA less than 38 kbp is not viable. Plaques obtained in cloning experiments result from phages that either

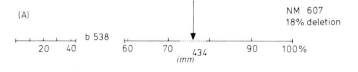

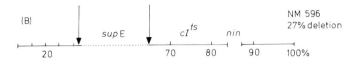

Fig. 2A, B. λ-insertion and a λ replacement vector for EcoRI. **A** The physical map of the λ insertion vector NM607 for EcoRI. It contains a deletion of 18% in the b region (b^{538}) and the immunity region is replaced by the immunity region of phage 434 (imm^{434}). The 5 EcoRI sites of the λ-genome are mutated without changing the viability of the phage. The new imm^{434} region carries a single EcoRI site. The construction of NM607 is described by Murray et al. (1977). **B** The physical map of the λ replacement vector NM596. Of the 5 EcoRI sites of λ, this λ vector contains EcoRI sites number 1 and 3. Between these two sites an EcoRI fragment carrying a suppressor gene (supE) is inserted. A small deletion (nin) exists at the right end of map. The construction of this replacement vector is also described by Murray et al. (1977)

regain their original segment of acquire donor DNA with a molecular weight between 3 and 18 kbp. Thus, the advantage of the replacement vector is its enhanced capacity to accept foreign DNA. The replaceable fragment of the vector carries the readily recognizable phenotype of the suppressor gene E (sup$E$). In a suitable host with an amber mutation in the *lacz* gene, hybrid phages are easily identified by a simple color test on lactose fermentation indicator plates.

2. Experiment 1: Principle Techniques for the Application of a λ Insertion Vector

a) Introduction

In this experiment, a hybrid phage is used whose genome consists of the λ insertion vector L1 and a cloned DNA fragment coding for part of the tryptophan operon of *E. coli*. The λ insertion vector L1 is identical to phage NM607 which is shown in detail in Fig. 2. The inserted DNA fragment carries *trp*DEop, part of the tryptophan operon containing the operator and promotor region, and the genes *trp*E (anthranilate-transferase) and *trp*D (phosphoribosyl-anthranilate-transferase). This DNA fragment has a size of ≈ 7.5 kb and is inserted into the c_I-gene of L1 via EcoRI sites. The scheme of the experiment to be done is outlined in Fig. 3. L1-*trp*DEop DNA is converted by EcoRI digestion into 3 linear DNA fragments, the left arm of L1, the *trp*DEop fragment, and the right arm of L1. Religation of these DNA fragments gives rise to two types of viable phage genomes, the phage vector L1 and the λ-hybrid

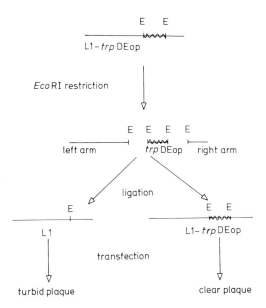

Fig. 3. The cloning experiment. The hybrid genome L1-*trp*DEop consists of the L1 insertion vector which is identical to NM607 of Fig. 2 and the cloned *trp*DEop fragment. After *Eco*RI digestion and religation, 2 types of viable λ-genomes (L1 and L1-*trp*DEop) are present in the ligation mixture. They can be identified after transfection according to their plaque morphology

L1-*trp*DEop. These two types of λ-genomes can be distinguished after transfection of an *E. coli* strain. Phage L1 produces turbid plaques whereas the hybrid phage L1-*trp*DEop is responsible for a clear plaque morphology. Thus, by a very simple experiment, the main features of the λ insertion vector can be demonstrated.

b) Objectives

1. To restrict L1-*trp*DEop DNA by *Eco*RI and to religate the fragments leading to phage particles that produce turbid plaques in the case of L1 and clear plaques in the case of L1-*trp*DEop.
2. To isolate DNA from the hybrid phage L1-*trp*DEop.

c) Procedure

EcoRI Restriction of L1-trpDEop DNA, Religation of the Fragments and Transfection of E. coli Cells with the Ligation Mixture

For this experiment, L1-*trp*DEop DNA is needed at a concentration of 200–300 μg/ml. The isolation of this DNA is described in the second part of the procedures.

Day 1

Digest 5 μg of L1-*trp*DEop DNA with *Eco*RI.
Mix 25 μl of L1-*trp*DEop DNA (200 μg/ml) in TE with 2.5 μl of 10 × *Eco*RI restriction buffer (0.1 M Tris, pH 7.5; 0.1 M MgCl$_2$, 1 M NaCl), and 0.5 μl *Eco*RI enzyme in a sufficient quantity.

Incubate at 37°C for 60 min. To inactivate the enzyme, the mixture is heated at 70°C for 5 min.

Check for completeness of the digestion by EcoRI by agarose gel electrophoresis.

The EcoRI restricted L1-trpDEop DNA solution is now divided: 20 µl is used for ligation, while 5 µl is left for the transfection experiment.

Ligation of EcoRI Restricted L1trpDEop DNA Using T4 Ligase

Mix 20 µl L1-trpDEop DNA, restricted by EcoRI in EcoRI restriction buffer, 2.5 µl 0.1 mM ATP solution, 1 µl T4 ligase in a sufficient quantity, and 2.5 µl 100 mM Dithiothreitol. Incubate overnight at 14°C.

Ligase activity can be checked the next day by agarose gel electrophoresis.

Day 2

Transfection of Competent E. coli Cells with Ligation Mixture

Preparation of Competent Cells

Dilute cells from a fresh overnight culture of E. coli C600 1:50 into L-broth and grow until $OD_{650} = 0.6$.

Chill on ice and spin cells down at 6000 rpm for 5 min. Pour off the supernatant and resuspend the cells in 0.1 M $MgCl_2$ (same volume as original). Leave on ice for 20 min.

Spin again and pour off supernatant. Resuspend the cells in 0.1 M $CaCl_2$ in half the original volume. Leave on ice for 20 min.

Spin and pour off the supernatant. Resuspend the cells in 0.1 M $CaCl_2$ in 1/10 original volume.

Leave the cells on ice for an additional 20 min. The cells are now competent and should be ready for uptake of DNA.

Transfection of E. coli C600 with Different DNA Samples

For introduction of λ-genomes into competent E. coli cells, the following procedure is normally used:

Mix 0.2 ml of competent cells with 0.1 ml DNA solution in SSC and leave on ice for 30 min.

Heat shock for 2 min at 42°C and leave once more on ice for 30 min.

Add 0.1 ml of the suspension to 2.5 ml of top layer agar for plating on LB-plates. If necessary, use serial decimal dilutions of the transfection mixture. For transfection, it is important not only to include the ligation mixture, but also L1-trpDEop DNA and EcoRI restricted L1-trpDEop DNA.

The following transfections are set up:
  A) competent cells + L1trpDEop DNA, untreated
  B) competent cells + L1-trpDEop DNA, EcoRI digested
  C) competent cells + L1-trpDEop DNA, EcoRI digested and religated
  D) competent cells + buffer

The experiments A, B, and C will demonstrate how the transfection rate is changed when the vector DNA is digested (B) or digested and religated (C). Experiment D is a control to show that the competent cells are phage free.

Use the following DNA concentrations:

Experiment	Competent cells of C600	Amount of DNA in 0.1 ml SSC buffer
A)	0.2 ml	0.4 µg
B)	0.2 ml	0.4 µg
C)	0.2 ml	2.0 µg
D)	0.2 ml	0 µg

Day 3

Recording of Results

Count the plaques on the different plates and calculate the transfection frequency for experiments A, B, and C.

Isolation of Phage λ DNA Using the Hybrid Phage L1-trpDEop

The following procedure for the isolation of L1-*trp*DEop DNA can be applied in general for the isolation of phage λ DNA.

Day 1

Purification of a λ-Lysate

For purification of a λ-lysate, single plaques are produced by the following methods:

Preparation of indicator strain:
Spin down an overnight culture of C600 (10,000 rpm, 10 min).
Resuspend to the same volume in 10^{-2} M $MgCl_2$.
Incubate at 37°C for 1 h without aeration.
Store in ice until use.

Plating the phage:
Use phage buffer for serial dilution.
Mix 2.5 ml top agar (45–50°C), 0.1 ml indicator strain, 0.1 ml phage solution and pour on LB plates.
After solidifying, the plates are incubated upside down at 37°C.

Day 2

Production of a Plate Stock from a Single Plaque

Collect a single plaque in 1 ml phage buffer ($\approx 10^6 - 10^7$ pfu) and add 1 drop of chloroform.
Mix 3 ml top agar (45–50°C), 0.1 ml indicator strain, 0.1 ml phage solution and pour on LB plates. After solidfying, the plates are incubated right side up at 37°C.

Note: The indicator strain should be freshly prepared and the LB plates should be freshly poured.

Watch the plates until confluent lysis occurs (5–8 h).

Harvest the phage by one of two methods:

Method A: add 1 ml of LB-broth to each plate, scrape off the top agar layer and collect in centrifuge tube, add 3 drops of chloroform, centrifuge 10,000 rpm for 10 min, collect the supernatant, add 3 drops of chloroform.

Method B: add 3–5 ml of LB-broth to each plate, leave in the refrigerator overnight, pipette off the broth and proceed as in method A.

Titrate the lysate.

Day 3

Large Scale Production of a λ-Lysate

Two methods are used: the liquid stock method and the Blattner method.

Liquid stock method:

Dilute overnight culture of C600 1:20 (or 1:50) into LB (e.g., 250 ml in 2 l flasks) and add MgSO$_4$ to final concentration of 1 mM.

Grow until OD$_{650}$ = 0.3 with vigorous shaking and add phage lysate (no chloroform!) to give final concentration of $2-3 \times 10^8$ pfu/ml.

Follow the OD of the culture until it drops. When the OD reaches a minimum, add chloroform at the rate of 0.5 ml per 250 ml broth and shake for another 15 min.

Clarify the lysate by centrifugation (10,000 rpm for 10 min).

Titrate the lysate.

Blattner method:

Mix a whole plaque ($\approx 6 \times 10^5$ pfu) with 0.3 ml stationary phase culture of indicator and 0.3 ml salt solution (0.01 M MgCl$_2$ and 0.01 M CaCl$_2$).

Pre-adsorb for 10 min at 37°C (without shaking).

Add 250 ml broth and incubate with good aeration.

Follow the OD until it drops. When OD is minimal, add chloroform (0.5 ml per 250 ml).

Clarify the lysate (10,000 rpm for 10 min).

Titrate the lysate.

A phage titer of about 10^{10} pfu/ml with either method is sufficient for the following experiments.

Day 4

Concentration of the Phage Stock

Centrifuge the lysate at 45,000 $\times$ g in a centrifuge at 4°C for 3 h.

Resuspend in 10 ml phage buffer at 4°C with slow shaking overnight.

The next day, clarify the supernatant at 10,000 rpm for 10 min at 4°C.

Day 5

RNase and DNase Treatment

Add RNase and DNase to a final concentration of 10 µg/ml. This treatment is for degradation of ceulluar DNA and RNA still present in the lysate.

Incubate for 1 h at room temperature.

Purification and Concentration of the Lysate by CsCl Centrifugation

Measure the volume of the phage suspension (balance): x ml.
Add CsCl to a final concentration of 41.5% w/w. This means: take x · 0.415/0.585 CsCl
 = x · 0.71 g CsCl and add it to the phage suspension.
Spin for 1 h at 4°C and at 40,000 rpm to clarify the suspension.
Spin for 24 h at 4°C and 40,000 rpm to band the phage.
Remove phage band the next day from the top of the tube using a syringe.

Day 6

Extraction of λ-DNA

Dialyze the CsCl purified λ suspension against TE (10 mM Tris-HCl, 1 mM EDTA, pH 7.5) at 4°C.
Mix the dialyzed solution with an equal volume of freshly prepared phenol and roll gently for about 1 min.
Separate the liquid phases by Eppendorf centrifugation and remove the phenol phase (lower phase!) with a sterile Pasteur pipette.
Repeat the phenolzation 2 ×.
Dialyze the aqueous phase against TE at 4°C, change the buffer 4 × in 20 h.
Measure the absorbance at 235, 259, and 280 nm. Ratios of absorbance 259/235 and 259/280 should be 2.0.

Note: The preparation of the phenol is important. It should be freshly distilled and stored at 4°C under air-free distilled water. Extract immediately before use with an equal volume of 0.5 M Tris-HCl, pH 8.5. Discard the aqueous phase. Do not use more than 1 day old phenol or if it is discolored.

d) Results

In this experiment, the principle techniques for the application of λ insertion vectors are shown. Firstly, the isolation of phage λ DNA is described. It is important to prepare highly purified DNA solutions, which can be restricted completely and ligated. Here, it should be mentioned that λ-lysates produced by the confluent lysis method on agar plates are not suitable for DNA manipulation since these lysates contain a high amount of impure agarose which is difficult to separate from the DNA. DNA solutions containing impure agarose are sometimes also difficult to restrict and show low ligation efficiency.

The amount of λ-DNA that can be isolated from 250 ml phage suspension with a phage titer of 10^{10} pfu/ml can be calculated. After purification and concentration, 2.5×10^{12} phages are available for DNA extraction. The molecular weight of the λ-genome is 3.3×10^7. This means that 3.3×10^7 g of λ-DNA contain 6×10^{23} molecules or 1 μg λ-DNA contains ≈ 2×10^{10} molecules. Thus, 125 μg of DNA can be obtained from 2.5×10^{12} λ-phages. If this amount of DNA is dissolved in 0.5 ml buffer, a DNA concentration of 250 μg/ml can be obtained. It should be noted that λ-DNA concentrations are conveniently measured by UV absorption at 259 nm. An Absorbance of 1.0 (A_{259} = 1.0) represents a DNA concentration of 50 μg double stranded DNA per ml.

The use of λ insertion vectors is demonstrated by restriction and religation of L1-*trp*DEop DNA. The different steps can be checked by agarose gel electrophoresis. After *Eco*RI digestions, 3 DNA bands of 29 kb (left arm of L1), 11 kb (right arm of L1), and 7.5 kb (*trp*DEop fragment) should be visible. It is important that the *Eco*RI digestion is complete, since undigested L1-*trp*DEop genomes give false positives in the ligation experiment. The transfection of the different DNA-samples (untreated, digested, and ligated) show

1. that 1 μg of L1-*trp*DEop DNA gives 10^5 plaques or more
2. that complete *Eco*RI digestion of L1-*trp*DEop DNA reduces this number to nearly zero and
3. that religation of *Eco*RI digested L1-*trp*DEop DNA gives again plaques, this time clearly reduced in number, but with clear and turbid plaque morphology. Usually more turbid plaques than clear can be recognized.

3. Experiment 2: Cloning Part of the *E. coli* Tryptophan Operon in a λ Insertion and a λ Replacement Vector

a) Introduction

For this experiment, the λ insertion vector NM607 and the replacement vector NM596 (see Fig. 2) are used to clone an *Eco*RI-fragment carrying *trp*DEop. This *Eco*RI fragment is part of the plasmid pPK1. This plasmid is composed of the vector plasmid pMB9 (5.3 kb) and the *trp*DEop fragment ($\approx$ 7.5 kb) already introduced in the preceding experiment. As pointed out in the general introduction, L1-hybrids can be recognized by a change in the plaque morphology and L3-hybrids by the lactose fermentation test. But the following difficulty arises. Since we start with plasmid pPK1, the *trp*DEop fragment as well as the vector plasmid pMB9 can be inserted into L1 or L3. Thus, in order to recognize the insertion of the *trp* fragment, a biological assay is used. This assay is based on the expression of the *trp*D or *trp*E gene of the cloned fragment in appropriate *trp* mutants. The details of this experiment are shown in Fig. 4. Hybrid phages are transferred to a minimal plate covered with a lawn of a *trp* deficient mutant strain of *E. coli*. After infection of this lawn by hybrid phages carrying the *trp* fragment, the infected cells can synthesize tryptophan and support the growth of bacterial cells around the plaque. Hybrid phages that carry the pMB9 plasmid do not show such a reaction.

b) Objectives

1. To clone the 2 *Eco*RI-fragments of plasmid pPK1 (a pMB9-*trp*DEop hybrid) into the λ insertion vector L1 and into the λ replacement vector L3.
2. Using a biological test, the cloned *trp*DEop fragment will be recognized in the λ insertion vector L1 and in the λ replacement vector L3.

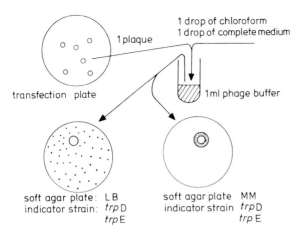

Fig. 4. Biological test for the cloned *trp*DEop fragment in L1 and L3. A whole plaque is transferred with a loop from the transfection plate into a test tube. This plaque suspension is used to infect a *trp*⁻ indicator strain grown on soft agar on two types of plates. One plate contains a rich medium (LB) in which the *trp* deficient strain can grow. The other is a minimal plate where growth of the auxotrophic strain is inhibited

c) Procedure

Before starting the experiments, 1 ml of L1-, L3-, and pPK1-DNA should be prepared to a concentration of 200 μg/ml. As the preceding experiment gives details of the experimental steps required, only a short outline of the experiments will be given here.

Cloning of EcoRI Digested pPK1-DNA into the λ Insertion Vector L1 and into the λ Replacement Vector L3

Day 1

EcoRI Restriction

L1-, L3-, and pPK1-DNA are digested with *Eco*RI. Digestion is checked by agarose gel electrophoresis.

Ligation

The following DNA mixtures are used for the ligase reaction:

1. L1- and pPK1-DNA, both *Eco*RI digested.
2. L3- and pPK1-DNA, both *Eco*RI digested.

Again use gel electrophoresis to check the ligase activity.

Day 2

Transfection

Transfect *E. coli* C600 with the two ligation mixtures. Do not forget to include all the necessary controls, e.g., to transfect *E. coli* with L1-DNA, with L1-DNA digested with *Eco*RI and to plate the competent cells without DNA treatment.

Day 3

Biological Test for the Cloned trpDEop Fragment in L1 and L3

The Trp$^+$ test is described schematically in Fig. 4.

Take a whole plaque from the transfection plate and transfer it into 1 ml of phage buffer. Add 1 drop of chloroform and several drops of complete medium.

Use two types of agar plates:

1. Take a rich (LB) plate and pour a lawn of LB-soft agar containing a *trp*D$^-$ or *trp*E$^-$ strain.
2. Take a minimal (MM) plate and pour a lawn of "water" soft agar containing the *trp*D$^-$ or *trp*E$^-$ strain.

After solidifying, take a loop of the phage suspension and infect the bacterial lawn of the two different plates.

Incubate overnight. L1-*trp* and L3-*trp* phages should support the growth of the bacterial lawn on the minimal agar around the plaque.

Use for this test clear plaques of the L1 experiment. They should carry cloned fragments.

In the case of cloning with λ replacement vector L3, use all plaques for the exchange of the *sup*E fragment against *trp*- or the pMB9-fragment.

d) Results

In order to check *Eco*RI digestion and T4 ligation in the described experiments, agarose gel electrophoresis was recommended. A picture of a typical gel is shown in Fig. 5. *Eco*RI digestion of L1-DNA and pPK1 (identical to pMB9-*trp*DEop) DNA results in each case in two fragments. After mixing of these *Eco*RI fragments and treating the solution with T4 ligase, the typical DNA pattern disappears and a more diffuse DNA band consisting of long DNA molecules can be recognized. This change in the band pattern indicates the successful action of T4 ligase. Whether such a ligation mixture contains hybrid L1 molecules or not is checked by the transfection experiment. Figure 6 shows the result of this experiment. Clear plaques and turbid plaques can be distinguished. It should be mentioned that clear plaques are produced when the *Eco*RI fragment *trp*DEop or the linearized pMB9 is inserted into the L1 vector. The biological test described in the procedures section distinguishes between hybrid phages containing either of these two fragments. Another way to distinguish between the two inserted fragments is based on their different lengths. This can be determined by isolating DNA from the λ-hybrids, restricting it with *Eco*RI, and separing the *Eco*RI fragments on agarose gels.

The goal of this experiment is the successful cloning of *trp*DEop into the λ insertion vector L1 and the λ replacement vector L3. These hybrid phages are used in Chap. DII for studies of gene expression in mini-cells of *E. coli* There it is shown that from the *trp* promotor, the genes *trp*E and *trp*D are transcribed and translated. Gene products of *trp*E and *trp*D will be radioactively labeled in mini-cells of *E. coli* and identified on polyacrylamide gels.

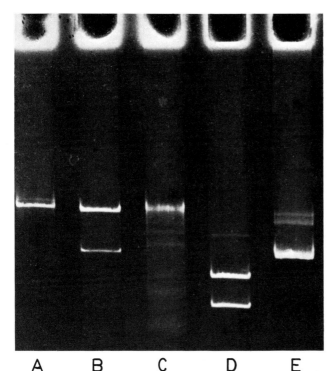

Fig. 5. Cloning of plasmid pPK1 (pMB9-*trp*DEop) into the λ insertion vector L1. The different steps of the cloning experiment are checked by agarose gel electrophoresis. For details see text. Lane *A* L1 DNA; *B* L1 DNA digested with *Eco*RI; *C* mixture of L1 DNA + pPK1 DNA digested with *Eco*RI and ligated with T4 ligase; *D* pPK1 DNA digested with *Eco*RI; and *E* pPK1 DNA

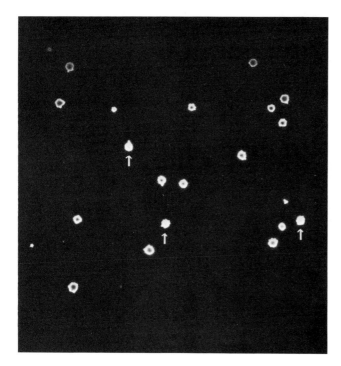

Fig. 6. Turbid and clear plaques produced by L1 phages. Turbid plaques indicate the original L1 vector phage whereas clear plaques *(arrowed)* are produced by L1 hybrid phages carrying foreign DNA inserted into gene c_I

4. Materials

Strains, Phages and Plasmids

E. coli C600: thr^-, leu^-, $B1^-$, gal^-, $lacY^-$, $supE$, $tonA$
E. coli $trpD^-$
E. coli $trpE^-$
L1 ≡ NM607 (see Fig. 2A)
L3 ≡ NM596 (see Fig. 2B)
L1-*trp*DEop (see Fig. 3)
pPK1 ≡ pMB9-*trp*DEop

Media

LB	Luria Broth per liter:
	10 g Bactotryptone
	5 g Yeast Extract
	10 g NaCl
	adjust to pH 7.2
	before use, complete with 10 ml of 10% glucose per liter
LA (LB-plates)	Luria Agar per liter H_2O:
	10 g Bactotryptone Difco
	5 g Yeast Extract Difco
	10 g NaCl
	12 g Bactoagar Difco
	adjust to pH 7.2
	after sterilization, add 10 ml of 10% glucose and 5 ml of 0.5 M $CaCl_2$

Top Agar (also called Soft Agar):
    Same as LA but only 6 g Difco agar

Minimal Medium ≡ Davis synthetic agar
(Minimal Agar plates)
    500 ml agar (2 ×)
    50 ml asparagine 2%
    500 ml Davis base (2 ×)
    10 ml glucose 10%
    8.3 ml 0.1 M $MgSO_4$

Davis Base (2 ×)	7 g K_2HPO_4
	2 g KH_2PO_4
	0.45 g Na_3 citrate $2H_2O$
	1 g $(NH_4)_2SO_4$
	500 ml H_2O, pH 7.2
Agar (2 ×)	14 g Bactogar Difco
	500 ml H_2O

"Water" Soft Agar: per liter H_2O
    8 g Bacto Agar Difco

Buffers

TE	1 mM EDTA, 10 mM Tris/HCl, pH 7.5
SM (phage buffer)	0.1 M NaCl, 0.01 M $MgSO_4$, 0.02 M Tris/HCl, pH 7.5
SSC	0.15 M NaCl, 0.015 M Na citrate

5. References

Blattner FR, Williams BG, Blechl AE, Denniston-Thompson K, Faber HE, Fulong LA, Grunwald DJ, Moore DD, Schumm JW, Sheldon EL, Smithies O (1977) Charon phages: safe derivatives of bacteriophage lambda for DNA cloning. Science 196:161–169

Hershey AD (1971) The bacteriophage lambda. Gen Res Unit, Carnegie Inst, Cold Spring Harbor Lab, Cold Spring Harbor New York

Murray NE, Brammar WJ, Murray K (1977) Lambdoid phages that simplify the recovery of in vitro recombinants. Mol Gen Genet 150:53–61

Phillipsen P, Kramer RA, Davis RW (1978) Cloning of the yeast ribosomal DNA repeat unit in *Sst*I and *Hin*dIII lambda vectors using genetic and physical size selections. J Mol Biol 123: 371–386

Thomas M, Davis RW (1975) Studies on the cleavage of bacteriophage lambda DNA with *Eco*RI restriction endonuclease. J Mol Biol 91:315–328

Thomas M, Cameron JR, Davis RW (1974) Viable molecular hybrids of bacteriophage lambda and eukaryotic DNA. Proc Natl Acad Sci USA 71:4579–4583

3.6 Cloning with Cosmids

U. PRIEFER, R. SIMON, and A. PÜHLER [1]

Contents

1. General Introduction 190
2. Experiment 1: In Vitro Packaging of λ-DNA 191
   a) Introduction 191
   b) Objectives 192
   c) Procedure 192
   d) Results 193
3. Experiment 2: Construction of a Cosmid Bank of an *E. coli* Tn5 Mutant 194
   a) Introduction 194
   b) Objectives 196
   c) Procedure 197
   d) Results 200
4. Materials 200
5. References 201

1. General Introduction

A major disadvantage in constructing a gene bank in plasmid vectors is that the average size of the fragments cloned is relatively small since transformation preferentially selects the smallest hybrid molecules. The number of clones required to ensure the presence of all genomic sequences is therefore fairly high.

Cloning into cosmid vectors followed by in vitro packaging drastically reduces the number of clones necessary to cover an entire genome since this technique allows the selection for long recombinant molecules in the size of 40–50 kb.

The principle of the cosmid cloning and in vitro packaging system is as follows:

1. A cosmid vector, first developed by Collins and Hohn (1978), is a normal plasmid vector, but which contains additionally the cohesive ends of lambda (λ*cos*). For cloning work, it can be used like any other plasmid vector. The presence of the λ*cos* site, however, enables the plasmid to be packaged into λ phage heads. Upon introduction into *E. coli* via infection, the DNA molecule again behaves as a plasmid.

---

[1] Lehrstuhl für Genetik, Fakultät für Biologie, Universität Bielefeld, D-4800 Bielefeld, Fed. Rep. of Germany

2. In addition to λ*cos*, the DNA to be packaged has to be in the size range of 78% (38 kb) to 105% (51 kb) of wild λ-DNA (49 kb) (Feiss et al. 1977, Feiss and Siegele 1979). Since the usual cosmid vectors have a size between 5 and 10 kb, only polycosmids (vector-to-vector ligation products) or cosmids containing an insert DNA in the size of 40–45 kb can be packaged. Provided that the formation of vector concatamers is excluded, only hybrid molecules with large inserts are recovered upon infection.

3. The packaging of the DNA is carried out in vitro. The in vitro system (Sternberg et al. 1977, Hohn 1979) consists of two *E. coli* strains lysogenic for λ phage mutants, each defective in the synthesis of one of the proteins involved in phage assembly. Upon cell lysis, the packaging mix contains all λ specific proteins and appropriate DNA (λ*cos*, size) will be packaged into the phage heads by self-assembly.

In this section, the following experiments will be described: (1) preparation of λ-packaging mixes; (2) test of the efficiency of the mixes by packaging wild λ-DNA; (3) construction of a cosmid library of a Tn5-carrying *E. coli* strain; and (4) characterization of the gene bank.

2. Experiment 1: In Vitro Packaging of λ-DNA

a) Introduction

The in vitro packaging of λ-DNA is based on the morphogenesis of the λ phage head (Fig. 1). At the beginning of the head synthesis, the four proteins pE, pNu3, pB, and pC aggregate to form a starting protein complex, called scaffolded prehead. For further maturation, a bacterial enzyme, pgroE, is needed to remove the pNu3 enzyme (prehead). DNA packaging starts with the activity of two proteins, pA and pNu1, which bind to catenated linear λ-DNA close to the left *cos* site. In the presence of pFI, this DNA/protein complex is attached to the prehead and the head is filled with DNA. Filling the head is followed by the reaction with protein pD. During the packaging process, the left and the right *cos* sites come together and are both cleaved by the function of protein pA to generate the 12 bp cohesive ends. This ensures that one full-size unit of the λ-genome is packaged in each head. After pFII has closed the entrance, the tail, which is assembled separately, is attached to the filled head.

It is important to note that the steps of λ phage synthesis occur by a self-assembly process, which is brought upon by the affinity of λ-proteins to one another. The only exception is the step catalyzed by the bacterial enzyme pgroE.

The in vitro packaging system consists of two *E. coli* strains, BHB 2688 and BHB 2690, which are both lysogenic for different λ-mutants. BHB 2688 contains a prophage deficient for the synthesis of the very early protein pE (*E*am), the mutant in BHB 2690 carries an amber mutation in gene D (*D*am), the protein of which (pD) is involved in a late maturation step. Both prophages additionally are heat inducible (c_{Its}) and lysis deficient (*S*am). They also have the λ attachment site deleted (b2) and an inactivated recombination system (*red*3).

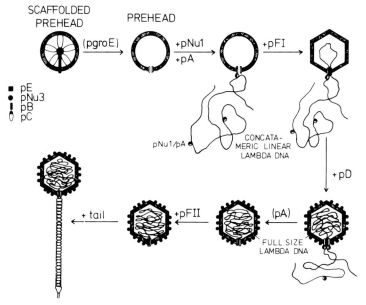

Fig. 1. Morphogenetic pathway of bacteriophage λ. Drawing according to Maniatis et al. (1982). The different steps of the self-assembly of λ-proteins involved in the phage head synthesis are shown. For details see text

The strains are grown separately and the lytic functions are induced by heat treatment. The two cultures are grown for further 2–3 h to allow accumulation of the λ-proteins. They are then combined and lysed by freezing and thawing. The result is a mixture (packaging mix) of a complete set of λ-proteins, which is able to build mature λ-particles by self-assembly in a cell-free in vitro system.

b) Objectives

1. To prepare an in vitro packaging mix.
2. To determine the efficiency of the mix by packaging wild λ-DNA.

c) Procedure

Day 1

Preparation of the Packaging Mix

Inoculate 200 ml prewarmed NZ-amine medium separately with overnight cultures of strains BHB 2688 and BHB 2690 (initial OD should not be higher than 0.05).

Incubate with good aeration at 30°C until an OD_{600} of 0.3–0.4 is reached (both strains should have the same OD).

Dilute the cultures with 200 ml of NZ-amine medium preheated to 65°C.

Incubate the flasks in a 45°C water bath for 25 min to induce lytic functions.
Afterwards, shake the cultures vigorously for 2–3 h at 37°C.
Check chloroform induced lysis. Due to the Sam mutation of the prophages, the induced cells normally do not lyse, but a small amount of chloroform initiates lysis. Take 3 ml samples, add 0.2 ml of $CHCl_3$ and mix carefully.
If the test is positive (clearance within 3 min), the cultures are chilled in an ice water bath, mixed, and centrifuged at 5000 rpm for 10 min at 0°C.
Resuspend the bacteria in 10 ml of cold complementation buffer (CB), transfer the solutions into smaller tubes, and centrifuge again (as above).
Suspend pellet in 5 ml of cold CB + ATP (5 mM) and recover the bacteria by centrifugation as above.
Suspend cells in a minimum of cold CB + ATP (0.5–0.8 ml).
Transfer aliquots of 20–40 μl into sterile, precooled Eppendorf tubes and freeze the extracts in liquid nitrogen.
Store packaging mix at $-70°C$.

Day 2

Testing the Packaging Efficiency with λ-DNA

Thaw a packaging mix in an ice water bath for 1–3 min.
Add 0.1–1 μg of λ-DNA in a small volume of buffer (1–5 μl) to the melting packaging mix. (Alternatively, the melted packaging mix can be added to the λ-DNA.)
Mix thoroughly by stirring with the Eppendorf tip.
Incubate the extract at room temperature for 60 min. After half of the incubation time, another packaging mix which was preincubated 15 min with DNase (0.1 mg/ml) can be added.
Add 1 ml of LB medium and 50 μl of $CHCl_3$; mix carefully until the solution is homogenous.
Centrifuge down cell debris (5 min, Eppendorf centrifuge).
Make serial dilutions (in SM buffer) of the supernatant which contains the in vitro packaged phages.
Mix 0.1 ml of each dilution with 0.1 ml of an indicator strain (e.g., *E. coli* C600) grown in LBMgMal medium to early stationary phase.
Add 2.5 ml LBMgMal soft agar (45°C).
Pour the mixture onto LBMgMal plates and incubate overnight at 37°C.
Controls: Packaging mix, λ-DNA, and indicator strain should be free of contaminating phages.

d) Results

The efficiency of the packaging mix may vary between 10^6 and 10^8 pfu per 1 μg of λ-DNA. Mixes which result in less than 10^6 pfu/μg DNA should not be used in an actual cosmid cloning experiment.

There are several variables which can influence the packaging efficiency, e.g.:

1. The strains BHB 2688 and BHB 2690 should be checked before use, since they are known to revert readily. Single colonies are streaked out on LB plates and incubated overnight at 30°C and 42°C. No colonies should appear at 42°C.

2. It is important to harvest the induced strains in a state when all λ-proteins are present at the highest rate. This is the case when the induced strains can easily be lysed by the addition of chloroform. Therefore, the chloroform test should be performed several times during the incubation (from 1 1/2 to 3 h after heat induction).

3. The packaging efficiency can be low when the packaging mix is not concentrated enough. Therefore, the pellet should be resuspended in the smallest possible volume of complementation buffer.

3. Experiment 2: Construction of a Cosmid Bank of an *E. coli* Tn5 Mutant

a) Introduction

The basic steps of the experiment presented in this section are outlined in Fig. 2 and are summarized as follows:

In order to prepare total cellular DNA of high molecular weight (> 100 kb), it is important to avoid any shearing of the DNA during lysis and isolation to minimize possible DNA breakage (Fig. 2a). The lysis procedure given below usually results in very good DNA preparations. Of course, any other gentle isolation method can also be used, e.g., Sarcosyl lysis followed by CsCl-EtBr gradient centrifugation (see Chap. 1.2).

As indicated in Fig. 1, only DNA flanked by two *cos* sites with an intervening length corresponding to the size of the λ-genome will be packaged. Therefore, the molecular weight of the potential inserts is one of the most important factors. Since the average distance of restriction sites in the DNA is relatively small compared to the size of the desired fragments, only partial digestion will result in large restriction fragments, generated more or less randomly (Fig. 2b). To eliminate larger and smaller digestion products, a size fractionation is recommended. This can either be done by sucrose gradient centrifugation or by preparative agarose gel electrophoresis as described in the experiment (Fig. 2c).

The insert DNA is mixed with completely digested cosmid vector DNA and ligated (Fig. 2d). The absolute DNA concentration in the ligation mixture is of critical importance. Low DNA concentration favors the self-ligation of the molecules, only a high concentration of vector and potential insert DNA will result in concatamers generated by intermolecular ligation. These concatamers can either be constructed by vector-to-vector annealing (polycosmids) or by joining foreign DNA to cosmid vectors (hybrid molecules). The ratio of vector DNA to potential insert DNA is therefore a critical parameter in the cosmid cloning system. High vector DNA concentrations compared to the concentration of target DNA will increase the probability of polycosmid formation. This problem can be minimized by treating linearized vector DNA with alkaline phosphatase prior to ligation.

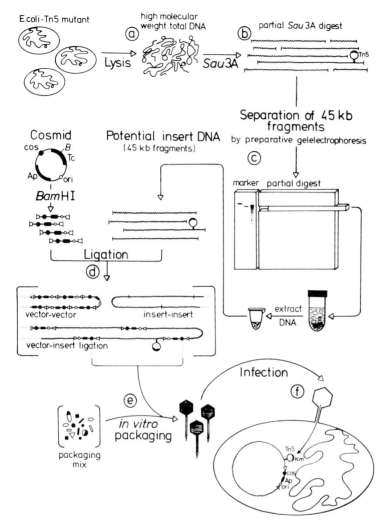

Fig. 2. Cloning total DNA of an *E. coli* Tn5 mutant into cosmids. The basic steps of experiment 2 performed in Sect. 3. For details see text

The ligation mix is subjected to in vitro packaging (Fig. 2e). The *cos* sites flanking the inserted DNA are recognized by the protein pA and the DNA is packaged into phage heads.

The phage particles are used to infect an *E. coli* recipient and recombinant clones are selected (Fig. 2f). To avoid insertion of the cloned DNA into the recipient genome, it is recommended to use a recombination deficient strain for infection. The injected molecules circularize via the λ cohesive end. Since they contain the complete cosmid genome, they can replicate as a plasmid. Recombinant clones are recognized via their drug resistance.

The cosmid pHC79 (Hohn and Collins 1980), which is used in this experiment, is a derivative of pBR322 (Fig. 3). Like pBR322, it offers several restriction sites and allows insertional inactivation. Cloning Sau3A fragments into the BamHI site (as it is done in the experiment) inactivates the Tc^r gene, leaving the Ap^r gene for selection. In addition to pBR322, pHC79 contains the λ cohesive ends (cos) which allow hybrid molecules of λ size to be packaged in vitro. After infection of E. coli, hybrid clones can be recognized via Ap-resistance and Tc-sensitivity.

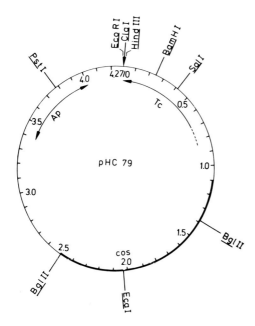

Fig. 3. Map of cosmid pHC79. Cosmid pHC79 consists of plasmid pBR322 carrying antibiotic resistance genes against ampicillin (Ap) and tetracycline (Tc) and a DNA fragment containing the cos site of phage λ. Restriction sites of various endonucleases useful for gene cloning are indicated. The map of pHC79 is modified from Hohn and Collins (1980)

b) Objectives

1. To prepare total cellular DNA of an E. coli Tn5 mutant.
2. To prepare restriction fragments with an average size of 45 kb by partial digestion with the four base specific enzyme Sau3A.
3. To ligate the E. coli Sau3A fragments to BamHI digested pHC79 molecules.
4. To introduce hybrid cosmid molecules into E. coli cells by in vitro packaging and infection.
5. To screen the infected cells for Tn5-containing cosmid clones.

c) Procedure

Day 1

Isolation of Total Cellular DNA of High Molecular Weight

In general, the protocol given below is as described by Meade et al. (1982), but it has been modified in some ways.

Spin down a 5 ml saturated culture of an *E. coli* Tn5 mutant grown in LB medium at 37°C and suspend the pellet in 25 ml of 1.0 M NaCl.
Shake for 1 h at 4°C.
Centrifuge, suspend pellet in 25 ml of cold TES buffer.
Centrifuge, suspend in 5 ml of cold TE buffer.
Add 0.5 ml of lysozyme solution (2 mg/ml TE), mix carefully, incubate 15 min at 37°C.
Add 0.6 ml of Sarcosyl-pronase (10% Sarcosyl and 5 mg/ml pronase in TE), incubate 1 h at 37°C.
Extract 2–3 times with 5 ml of phenol and then with 5 ml of chloroform.
Adjust aqueous phase to 0.3 M NH_4Ac and add 0.54 vol of isopropanol at room temperature.
Instead of removing the precipitated DNA with a glass rod (as originally described by Meade), the DNA can be recovered by normal centrifugation. Wash DNA pellet with 70% EtOH.
Suspend DNA very carefully in 1–2 ml of buffer (if necessary, by heating up to 65°C or by dissolving overnight).
Measure the length of DNA in an agarose gel (0.5%, Tris-acetate, low voltage). Only DNA preparations which migrate more slowly than intact λ-DNA are suitable for the experiment.

Day 2–4

Preparation of Sau3A Fragments in the Size Range of 45 kb and Ligation to BamHI digested pHC79 DNA

1. Partial Sau3A Digestion of Total DNA (Fig. 4)

Establish optimal conditions for partial digestion, e.g., vary the enzyme concentration with fixed DNA amounts and incubation time.
Analyze the size of the fragments by comparison with undigested λ-DNA on an 0.5% agarose gel, preferentially run with low voltage overnight.
Estimate the degree of digestion that produces the maximal amount of DNA in the desired size.
Prepare a large-scale digest (200–500 μg DNA) under the optimized conditions.

2. Isolation of Restriction Fragments of about 45 kb in Size

Separate the partial *Sau*3A fragments by electrophoresis on a low melting agarose gel (0.5%, Tris-acetate, 30 mA, 4°C, 20 h).
Use molecular weight standards in the size of 40–50 kb (e.g., undigested λ-DNA).

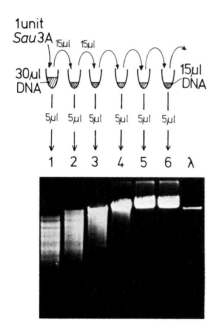

Fig. 4. Partial *Sau*3A digestion of total cellular DNA. The conditions for partial digestion were established by serial dilution of the enzyme. 30 μl of total DNA in restriction buffer was dispensed into Eppendorf tube *1*, and 15 μl into each of tubes *2–6*. To tube *1*, one unit of *Sau*3A was added, mixed and 15 μl of this mixture was transferred to tube *2*. This serial dilution was continued through to tube *6*. Digestion was allowed for 20 min at 37°C, the reaction was stopped by heating the samples for 10 min at 68°C. 5 μl of each digest was loaded onto a 0.5% agarose gel (in Tris-acetate buffer) which was run for 15 h at 25 V. Undigested λ-DNA served as molecular weight standard. For large scale digestion, this enzyme concentration was used which produced the largest amount of fragments slightly smaller than λ-DNA (lane *3*)

Stain the lanes containing the molecular weight marker with EtBr and identify the stretch of the gel that contains DNA fragments of appropriate length.

Cut out the corresponding piece of gel and transfer it into siliconized, screw-capped glass tubes.

Add 5 vol of TE buffer, incubate at 68°C with repeated inversions until the agarose is completely melted.

Extract with phenol several times, until no more interphase is visisble.

Recover the DNA by EtOH precipitation, suspend DNA pellet in 50–100 μl of TE.

Determine the DNA concentration. Make serial 1:1 dilutions of the DNA preparation as well as of a DNA of known concentration (e.g., λ-DNA). Run 5 μl of these dilutions on an agarose gel (1.2%, Tris-borate, 120 V, 10 min). Compare the fluorescence of the sample with the concentration standard and estimate the concentration of the DNA (see Fig. 5).

3. BamHI Digestion of pHC79

Digest pHC79 DNA with *Bam*HI using standard conditions.
Ensure the completeness of the digestion by agarose gel electrophoresis.
Determine the DNA concentration as above.

4. Ligation

Mix 2 μg of insert DNA with *Bam*HI digested vector DNA in a molar ratio of 1:3.
Precipitate with EtOH and redissolve the DNA in ligation buffer in a volume calculated to result in a final DNA concentration of at least 400 μg/ml.
Ligate under standard conditions.

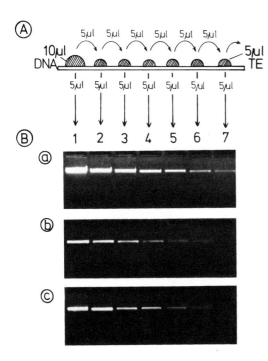

Fig. 5A, B. Rapid estimation of DNA concentration. A Serial 1:1 DNA dilutions were prepared. B 5 μl of each dilution step were loaded onto a 1.2% agarose gel (in Tris-borate buffer) and subjected to electrophoresis at 120 V for 10 min. The concentrations were estimated by comparing the fluorescence of the samples with that of a λ-DNA marker of 20 μg/ml. *a* Standard λ dilution series. The fluorescence in lane *1* corresponds to a DNA concentration of 20 μg/ml, lane *2:* 10 μg/ml, lane *3:* 5 μg/ml etc. *b* Dilution series of the target DNA (*Sau*3A fragments). Estimated concentration: 5 μg/ml. *c* Dilution series of the vector DNA (*Bam*HI digest). Estimated concentration: 10 μg/ml

Day 5

Packaging and Infection

The ligation mix is packaged as for the λ-DNA packaging experiment in the preceding section.

Transfer 0.1 ml of the extract containing the phage particles to 1 ml of the *E. coli* recipient, e.g., *E. coli* HB101, grown in LBMgMal to early stationary phase; store the residual supernatant at 4°C.

Incubate the infected culture at 37°C for 30–60 min.

Pellet the bacteria by centrifugation, wash the pellet, and resuspend again in LB medium.

Plate on Ap-containing medium to select for transfected cells.

Day 6

Screening the Clones

Caculate the number of Ap^r cells per μg of ligated insert DNA.

To determine the percentage of polycosmids (Ap^r Tc^r) replica plate Ap^r colonies to Tc-containing agar.

To identify hybrid molecules with the Tn5-mutated DNA fragment cloned (Ap^r Tc^s Km^r), replica plate also to Km-containing medium.

d) Results

10^4 recombinants per µg ligated target DNA should be obtained. If this is not the case, the section on the preparation of target DNA fragments in the size of 45 kb has to be repeated. Other variables are the molar ratio of vector to insert DNA and the DNA concentration in the ligation mix.

The percentage of clones carrying the double resistance $Ap^r\ Tc^r$ is an important parameter for a cosmid gene bank, since such clones contain at least two ligated cosmid vectors. A low percentage of polycosmids is characteristic for a good cosmid library.

An easy method to ensure the presence of 50 kb sized plasmids in the recombinant clones is to check individual colonies on an Eckhardt gel (see Chap. 1.3) and to compare the length of the plasmids with a molecular weight marker in the size of 40–50 kb. For further characterization, the cosmid bank should be tested to determine whether it is complete. This can be done by transfecting auxotrophic mutants of E. coli and scoring for complementation or by isolating the DNA of several cosmid clones and comparing their restriction pattern.

Restriction analysis is also recommended for the characterization of the identified Tn5-containing clones. Additionally, ^{32}P labeled Tn5 DNA can be used as a hot probe to ensure the presence of Tn5 in the Km^r recombinants. Labeled Tn5 DNA can also be used to screen the gene bank for Tn5-containing clones by colony hybridization.

4. Materials

E. coli Strains

BHB 2688:	N205 recA⁻ ($\lambda imm_{434}, c_{Its}, b2, red3, Eam4, Sam7)/\lambda$)
BHB 2690:	N205 recA⁻ ($\lambda imm_{434}, c_{Its}, b2, red3, Dam15, Sam7)/\lambda$)
C600:	F⁻, thi-1, thr-1, leuB6, lacY1, tonA21, supE44, λ^-
HB101:	F⁻, hsdS20, recA13, ara-14, proA2, lacY1, galK2, rpsL20, xyl-5, mtl-1, supE44, λ^-

Media

NZ-amine:	20 g NZ-amine, 5 g NaCl, 2 g $MgCl_2$, 2 g glucose in 1000 ml, pH 7.5
LB:	10 g Tryptone, 5 g yeast extract, 5 g NaCl, 1 g glucose
LBMgMal:	10 g Tryptone, 5 g yeast extract, 5 g NaCl, 1 g glucose, 4 g maltose, 1 g $MgCl_2$ per 1000 ml

For plates use 15 g, for soft agar 7 g agar per liter medium.

Buffers

For lysis (Meade et al. 1982)

TES:	0.01 M Tris, pH 8.0; 0.025 M EDTA; 0.15 M NaCl
TE:	0.01 M Tris, pH 8.0; 0.025 M EDTA

Complementation buffer (CB):
Prepare 80 mM Tris, 10 mM $MgCl_2$ (pH 8.0) in bidest H_2O; autoclave 50 ml in a 100 ml flask, chill on ice, add 0.1 ml β-mercaptoethanol, 7 ml DMSO and sterile H_2O to a final volume of 100 ml; add 255 mg spermidine (10 mM) and 161 mg putrescine (10 mM), store at $-20°C$; add 3 mg/ml ATP (5 mM) shortly before use.

SM buffer:
100 mM NaCl; 10 mM $MgSO_4$; 20 mM Tris, pH 7.5.

5. References

Collins J, Hohn B (1978) Cosmids: A type of plasmid gene-cloning vector that is packable in vitro in bacteriophage λ heads. Proc Natl Acad Sci USA 75:4242–4246

Feiss M, Siegele DA (1979) Packaging of the bacteriophage λ chromosome: dependence of *cos* cleavage on chromosome length. Virology 92:190–200

Feiss M, Fisher RA, Crayton MA, Egner C (1977) Packaging of the bacteriophage λ chromosome: effect of chromosome length. Virology 77:281–293

Hohn B (1979) In vitro packaging of λ and cosmid DNA. Methods Enzymol 68:299–309

Hohn B, Collins J (1980) A small cosmid for efficient cloning of large DNA fragments. Gene 11:291–298

Maniatis T, Fritsch EF, Sambrook J (1982) Molecular cloning. Cold Spring Harbor Lab, Cold Spring Harbor New York

Meade HM, Long SR, Ruvkun GB, Brown SE, Ausubel FM (1982) Physical and genetic characterization of symbiotic and auxotrophic mutants of *Rhizobium meliloti* induced by transposon Tn5 mutagenesis. J Bacteriol 149:114–122

Sternberg N, Tiemeier D, Enquist L (1977) In vitro packaging of a λ *dam* vector containing *Eco*RI DNA fragments of *Escherichia coli* and phage P1, Gene 1:255–280

Chapter 4 Gene Expression

4.1 Synthesis of Plasmid-Encoded Polypeptides in Maxicells

G.J. BOULNOIS and K.N. TIMMIS [1]

Contents

1. General Introduction 204
2. Experiment 1: Use of Maxicell Gene Expression System to Identify the R6-5 Complement Resistant Protein 205
   a) Introduction 205
   b) Procedure 205
   c) Results and Discussion 208
   d) Final Comments 210
3. Materials 210
4. References 211

1. General Introduction

Bacterial chromosomes specify several thousand polypeptides; the matching of a single gene product with a specific gene is thus difficult or (more usually) impossible. The principal approach that has been developed to circumvent this difficulty is to effect a physical separation of the gene of interest from the chromosome by cloning in a bacterial plasmid. The problem is then resolved by utilization of a procedure for the selective visualization of the products encoded by plasmid genes.

Three systems for selective labeling of plasmid-specified proteins with radioactive amino acids have been developed.

1. Minicells

Minicell mutant bacteria are defective in cell division; septum formation occurs close to one of the cell poles, resulting in the generation of an essentially normal cell plus a minicell. Minicells fail to receive a copy of the chromosome, but do receive copies of plasmids that may be present in the parent cell. Since minicells possess most of the metabolic potential of normal cells (e.g., see Reeve 1978), plasmid genes are expressed and can be labeled with radioactive amino acids.

---

[1] Department of Medical Biochemistry, University of Geneva, CH-1211 Geneva 4, Switzerland

2. Coupled Transcription-Translation in Vitro

Cell extracts, free of chromosomal DNA and capable of transcription-translation of exogenously supplied DNA, also permit selective labeling of plasmid-specified proteins (Zubay 1973, Pratt et al. 1981).

3. Maxicells

Maxicells are obtained from *E. coli* strains highly sensitive to UV light as a consequence of mutations in their *uvr*A, *phr*, and *rec*A genes. The latter mutation causes each DNA molecule damaged during irradiation of the cells to be extensively degraded. Plasmid-containing maxicells present two types of target to UV irradiation, namely, the very large chromosome and the small plasmids. At low UV doses, chromosomes, by virtue of their size, will receive numerous hits, whereas many plasmid molecules will receive none. Thus, after a post-irradiation incubation to permit degradation of UV damaged DNA, the majority of bacteria will contain only intact plasmid molecules that continue to replicate and to direct the synthesis of proteins. The incubation of such bacteria in media containing radioactive amino acids, accomplishes selective labeling of plasmid-encoded proteins (Sancar et al. 1979, 1981; see also Focus [BRL] Vol. 5, No. 4, p. 13, for improvements to the procedure).

2. Experiment 1: Use of Maxicell Gene Expression System to Identify the R6-5 Complement Resistance Protein

a) Introduction

The conjugative plasmid R6-5 confers upon host bacteria resistance to several antibiotics, mercury salts, and the lytic action of complement. A 6 kb *Eco*RI fragment (fragment E-7, Fig. 1), from the region of R6-5 encoding conjugal transfer and containing the complement (serum) resistance determinant, has been cloned in the pACYC184 vector, to produce hybrid plasmid pKT107. This fragment contains three intact genes: *tra*S and *tra*T, which specify plasmid surface exclusion; and *tra*D, which specifies an essential DNA transfer function. Upstream of *tra*S is part of *tra*G, and downstream of *tra*D is part of *tra*I, two genes involved in conjugal transfer. Mutant derivatives of pKT107, containing the ampicillin resistance transposon Tn*3*, have been isolated that fail to mediate complement resistance. In this experiment, the gene products specified by pKT107 and several pKT107 mutant derivatives will be analyzed by means of maxicells to identify the product of R6-5 that mediates resistance to complement.

b) Procedure

Day 1

Inoculate 5 ml of M9 minimal medium containing 1% casamino acids with CSR603 and CSR603 derivatives carrying plasmids pACYC184, pKT107, pKT116, pKT117,

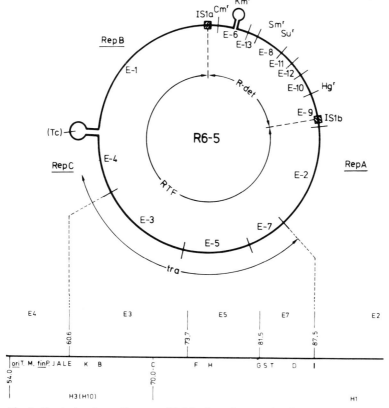

Fig. 1. Physical and genetic map of R6-5. *Cross bars* on the circular map and *vertical bars* above the linear expansion indicate the locations of *Eco*RI cleavage sites, whereas *vertical bars* below the linear expansion indicate *Hind*III sites. Cm^r, Km^r, Sm^r, Su^r, Hg^r resistance to chloramphenicol, kanamycin, streptomycin, sulphonamide, and mercury salts, respectively; Tc inactivated tetracycline resistance genes; *Rep* replication functions; *tra* conjugal transfer functions; *R-det*, *RTF* the resistance determinant and the resistance transfer factors, respectively; *IS1* insertion sequence 1; *oriT* origin of transfer replication; *fin* fertility inhibition

and pKT120. Include tetracycline in the medium (10 µg/ml) used to culture plasmid-carrying strains.

Incubate overnight with aeration at 37°C.

Day 2

Inoculate 10 ml quantities of fresh medium with 0.5 ml of each overnight culture. Incubate at 37°C with shaking until the absorbance at 590 nm is 0.6.

Transfer the cultures into plastic petri dishes and irradiate each culture (remove plate lids!) with UV light (wavelength 260 nm), using a bacteriocidal lamp, such that the survival is reduced to less than 500 bacteria/ml of culture. (The exact dose to accomplish this level of killing should be determined prior to the experiment.)

Plate 0.1 ml samples of each culture on nutrient agar plates to monitor the extent of killing.

Transfer the irradiated bacteria to Erlenmeyer flasks and incubate at 37°C for 1 h.

Add 200 µl of a fresh solution of D-cycloserine (20 mg/ml) and continue incubation overnight. Cycloserine kills bacteria that survive the irradiation and that would otherwise grow during the incubation period and synthesize chromosomal proteins during the subsequent labeling step.

Day 3

During the labeling of the maxicells, prepare a 15% SDS-polyacrylamide gel as follows:

Thoroughly wash the glass plates of a gel electrophoretic apparatus with detergent and wipe with acetone.

Assemble the glass plates and spacers and mount in the electrophoretic chamber. Prepare a 15% separating gel by mixing the following solutions:

Buffer A	16.3 ml
Acrylamide	11.1 ml
H_2O	4.4 ml
Ammonium persulphate (10 mg/ml, freshly prepared)	1.2 ml
TEMED	100 µl

Mix well and pour between the glass plates until the level is about 1 cm from the bottom of the slot-forming comb. Overlay with 0.1% w/v sodium dodecyl sulphate (SDS) from a spray gun and allow 30 min for polymerization to occur.

Pour off the SDS, rinse the top of the gel with water, and blot the glass plates dry with filter paper, taking care not to damage the gel.

Prepare a 7% stacking gel by mixing

Buffer B	7.5 ml
Acrylamide	2.5 ml
H_2O	5.4 ml
Ammonium persulphate	0.38 ml
TEMED	30 µl

Mix well, pour onto the separating gel and place the slot-forming comb into position. Allow 30 min for polymerization and then carefully remove the comb.

Place running buffer into the top and bottom reservoirs of the gel apparatus and, using a Pasteur pipette, rinse the slots well with running buffer.

During the preparation of the SDS-acrylamide gel, label the maxicells as follows:

Collect the irradiated cells by centrifugation at 2500 rpm in a bench top centrifuge.

Resuspend each cell pellet in 5 ml of M9 minimal medium supplemented with threonine, leucine, proline, and arginine and incubate at 37°C for 1 h.

Add 50 µCi of ^{35}S-methionine to each suspension and continue incubation for 1 h.

Collect the cells by centrifugation, discard the supernatant fluids, and resuspend cell pellets in the residual medium in the centrifuge tube.

Add 100 µl of sample buffer and boil for 3 min.

Load 20 μl amounts of each sample into slots of the gel. Load also ^{14}C-labeled polypeptide size standards into one of the slots.

Apply a constant current of 30 mA across the gel for about 4 h. When the blue tracking dye has reached the bottom of the gel, turn off the power, dismantle the gel apparatus, and place the gel in fixer.

Day 4

Process the gel for scintillation autoradiography (Bonner and Laskey 1974) as follows:

Discard the fixer solution, replace with dimethyl sulphoxide (DMSO), and agitate gently for 30 min.

Replace the DMSO with fresh DMSO and continue agitation for 30 min.

Discard the DMSO and replace with a 22% solution (w/v) of 2,5-diphenyloxazole (PPO) in DMSO.

After 1 h, replace this solution with water and over a period of 1 h replace with fresh water several times.

Dry the gel on filter paper and expose to X-ray film overnight at $-70°C$.

Day 5

Develop the X-ray film; if it has been either under- or over-exposed, repeat the autoradiography and adjust the exposure time accordingly.

c) Results and Discussion

The autoradiograph obtained from such an experiment is shown in Fig. 2. No polypeptides are synthesized by plasmid-free CSR603 bacteria (Track B). Polypeptides specified by the vector plasmid pACYC184 can be seen in Track C: the major product has a relative molecular weight (M_r) of 26,000 and is chloramphenicol acetyltransferase (CAT), whereas a minor polypeptide of M_r 35,000 is probably a product of the tetracycline resistance determinant of the plasmid. A fourth polypeptide of M_r 24,500 may be related to CAT. Polypeptides specified by pKT107 can be seen in Track D. Since this plasmid was constructed by insertion of an *Eco*RI fragment of R6-5 into the CAT gene of pACYC184, pKT107 does not direct the synthesis of CAT. The major polypeptide of M_r 25,000 specified by pKT107 is the product of the surface exclusion gene *tra*T. In addition to the tetracycline resistance proteins, other polypeptides specified by pKT107 include one of M_r 75,000 and another of M_r 18,000 that were previously shown to be the products of the *tra*D and *tra*S genes, respectively (Kennedy et al. 1977). The minor polypeptides of M_r 24,000 and 19,000 may be degradation products of the *tra*T protein (Moll et al. 1980). The polypeptide of M_r 45,000 is probably a fusion polypeptide consisting of the C-terminal region of the *tra*I product and the N-terminal moiety of CAT. Tracks E–G contain polypeptides specified by Tn*3* derivatives of pKT107 that fail to mediate resistance to complement. These mutant plasmids direct the synthesis of the *tra*D and *tra*S polypeptides, but not the *tra*T gene product nor its putative degradation products. The presence of Tn*3* in the mutant plasmids reduces the expression of *tra*D and the

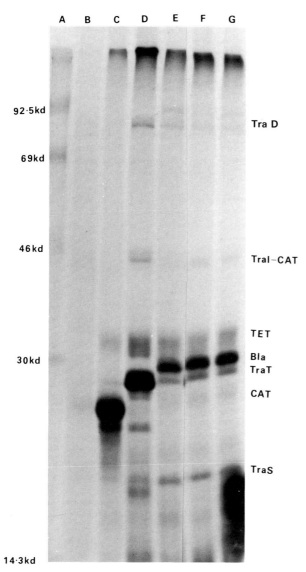

Fig. 2. Identification of the polypeptides responsible for complement/serum resistance. Maxicells were prepared and labeled as described and analyzed by SDS-polyacrylamide gel electrophoresis. The gel was processed for scintillation autoradiography. Track A shows ^{14}C-labeled polypeptides used as size standards. Track B shows polypeptides made by plasmid-free bacteria; Tracks C, D, E, F, and G, those made by bacteria containing the cloning vector pACYC184, pKT107, pKT116, pKT117, and pKT-120, respectively

putative CAT-*traI* fusion, but has no effect on the expression of *traS*. Note that the Tn3 element in pKT116, 117, and 120 is not completely polar on the expression of *traD*, suggesting that *traD* may have its own promoter. As expected, Tn3 insertion mutants show the synthesis of a novel polypeptide of M_r about 32,000 which corresponds to β-lactamase.

This experiment strongly suggests that the *traT* gene product, a major outer membrane protein that mediates plasmid surface-exclusion, is responsible for bacterial resistance to complement.

d) Final Comments

The choice of an expression system to study any given plasmid gene will be determined by consideration of several criteria. Although originally reported to be suitable only for high copy number plasmids, such as derivatives of pMB1, ColE1, and p15A, the maxicell system can also be used successfully with low copy number plasmids like pSC101 (G.J. Boulnois, unpublished data). The maxicell system is technically the most simple and, unlike the minicell system, permits the simultaneous analysis of many plasmids. Furthermore, subcellular fractionation is more readily and satisfactorily accomplished with maxicells than with minicells. However, neither in vivo expression system provides the level of experimental manipulation and hence, sophistication of the in vitro transcription-translation systems.

3. Materials

Day 1

M9 minimal medium (6 g Na_2HPO_4, 3 g KH_2PO_4, 0.5 g NaCl and 1 g NH_4Cl per liter, supplemented with 1 mM $MgSO_4$, 0.1 mM $CaCl_2$, 0.2% glucose, 10 μg/ml B1 and 1% casamino acids).
Tetracycline (10 mg/ml solution in 50% ethanol).

Day 2

M9 minimal medium
Tetracycline (10 mg/ml)
Petri dishes
Bacteriocidal lamp
D-cycloserine (20 mg/ml in M9 minimal medium)

Day 3

Materials for polyacrylamide gel electrophoresis:
Buffer A (0.75 M Tris-HCl, pH 8.8, 0.2% SDS, Bio Rad)
Buffer B (0.25 M Tris-HCl, pH 6.8, 0.2% SDS)
Acrylamide (44% acrylamide, 0.8% bis acrylamide, Bio Rad)
Ammonium persulphate (10 mg/ml, freshly prepared)
TEMED (Bio Rad)
0.1% SDS in a spray gun
Running buffer (0.25 M Tris, 0.191 M glycine, 0.2% SDS, pH 8.5)
Bench top centrifuge and tubes
M9 minimal medium supplemented with threonine (0.1 mg/ml), leucine (0.1 mg/ml), proline (0.2 mg/ml), and arginine (0.2 mg/ml)
^{35}S-methionine (> 100 Ci/mmol)
Sample buffer (0.125 M Tris-HCl, pH 6.8, 4% SDS, 20% glycerol, 1.4 M mercaptoethanol, 0.001% bromophenol blue)

^{14}C-labeled polypeptides (Amersham)
Fixer (30% propan-2-ol, 10% acetic acid)

Day 4

DMSO
22% PPO in DMSO
Gel drier and filter paper (Whatman 3 mm)
X-ray film

Day 5

Developer and fixer for X-ray film

4. References

Bonner WM, Laskey RA (1974) A film detection method for tritium-labelled proteins and nucleic acids in polyacrylamide gels. Eur J Biochem 48:83–88
Kennedy N, Beutin L, Achtman M, Skurray R, Rhamsdorf U, Herrlich P (1977) Conjugation proteins encoded by the F sex factor. Nature 270:580–585
Moll A, Manning PA, Timmis KN (1980) Plasmid-determined resistance to serum bacteriocidal activity: a major outer membrane protein, the *tra*T gene product is responsible for plasmid-specified serum resistance in *Escherichia coli*. Infect Immun 28:359–367
Pratt JM, Boulnois GJ, Darby V, Orr E, Wahle E, Holland IB (1981) Identification of gene products programmed by restriction endonuclease DNA fragments using an *E. coli* in vitro system. Nucleic Acids Res 9:4459–4474
Reeve JN (1978) Selective expression of transduced or cloned DNA in minicells containing plasmid pKB280. Nature 276:728–729
Sancar A, Hack AM, Rupp WD (1979) Simple method for identification of plasmid-coded proteins. J Bacteriol 137:692–693
Sancar A, Wharton RP, Seltzer S, Kacinski BM, Clarke ND, Rupp WD (1981) Identification of the *uvr*A gene product. J Mol Biol 148:45–62
Zubay G (1973) In vitro synthesis of protein in microbial systems. Annu Rev Genet 7:267–287

4.2 Synthesis of Bacteriophage and Plasmid-Encoded Polypeptides in Minicells

J.N. REEVE[1]

Contents

1. General Introduction 212
2. Experiment 1: Minicell Technique for the Identification of Polypeptides Encoded by
   Plasmid or Bacteriophage DNA 215
   a) Introduction 215
   b) Objectives 215
   c) Procedure and Results 215
3. Materials 220
4. References 222

1. General Introduction

Minicells are small, chromosomeless, bacterial cells produced by asymmetric cell division of mutant strains of rod-shaped bacteria (Frazer and Curtiss 1975) (Fig. 1). Minicells produced by plasmid carrying bacteria often contain copies of the plasmid DNA and synthesize mRNAs and polypeptides encoded by the plasmid DNA. As chromosomal DNA is absent, the plasmid-encoded gene products are the only products synthesized in such minicells and can readily be identified by incorporation of radioactive precursors into these macromolecules. The first part of the exercise will involve the isolation of minicells from E. coli DS410 (pKB280) and subsequent identification of polypeptides encoded by the plasmid pKB280. This plasmid was constructed by in vitro recombination (Backman and Ptashne 1978) and was designed to direct the synthesis of large amounts of the bacteriophage λ repressor protein. Incubation of pKB280-containing minicells in the presence of ^{35}S-methionine results in the synthesis of large amounts of radioactively labeled repressor. Radioactively labeled polypeptides synthesized in minicells will be separated by electrophoresis through polyacrylamide gels. Following electrophoresis, gels will be dried and bands produced by radioactive polypeptides identified by fluorography. Bacteriophage λ repressor has a molecular weight of 26,228 (Sauer and Anderegg 1978) and a band of this molecular weight should be the dominant feature on the autoradiograms produced in this experiment. Minicells containing plasmids produced by recombinant DNA techniques

---

[1] Department of Microbiology, Ohio State University, Columbus, OH 43210, USA

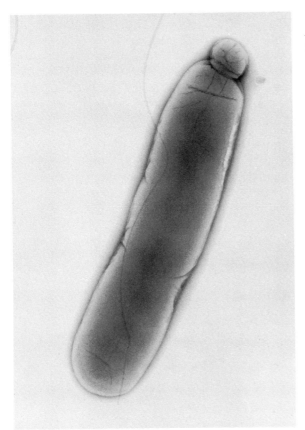

Fig. 1. Electron micrograph of E. coli DS410 showing minicell production

often synthesize anomalous polypeptides which are assumed to result from premature termination of translation (Meagher et al. 1977). Truncated forms of λ repressor may be detected among the polypeptides synthesized in pKB280 containing minicells (Fig. 2) (Reeve 1978).

Minicells are readily infected by bacteriophages and the infecting viral genome is transcribed and translated (Reeve 1977). In the second part of this exercise, polypeptides synthesized in λ infected minicells will be identified. The phages to be used are the λ-*trp*DE recombinant phages constructed as described in Experiment 1. Minicells infected by these phages synthesize polypeptides encoded by both the λ and *trp*DE segments of the viral genomes. Synthesis of λ polypeptides can be inhibited by infection of minicells containing a high concentration of λ repressor protein. A comparison will be made of the polypeptides synthesized in λ-*trp*DE infected minicells in the presence and in the absence of λ repressor.

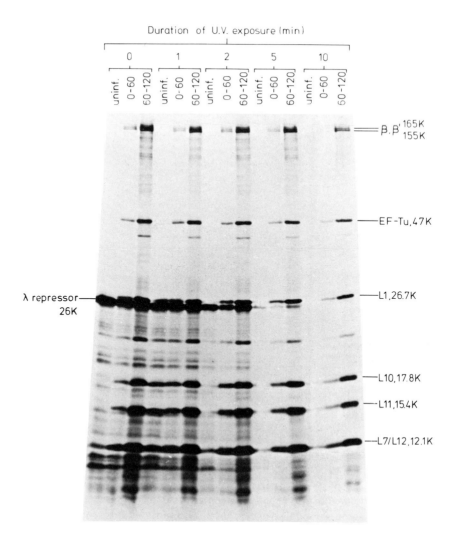

Fig. 2. Autoradiogram of the electrophoretic separation of the polypeptides synthesized in uninfected and λdrifd18-infected minicells of E. coli DS410 (pKB280). Minicells prepared as described and suspended (2×10^{10} per ml) in M9 minimal salts medium containing 20 μg D-cycloserine/ml were exposed to UV irradiation (290 μWcm^{-2}) for the indicated times. Aliquots of the infected minicells were infected (MOI = 5) with λdrifd18 phage. Uninfected and infected minicells were incubated at 37°C in the presence of ^{35}S-methionine assay medium. The uninfected minicells and one aliquot of infected minicells (0–60) were allowed to incorporate the radioactivity for 1 h. A second aliquot of infected minicells was incubated in M9 glucose salts minimal medium for 1 h at 37°C before addition of ^{35}S-methionine assay medium. These samples were then allowed to incorporate radioactivity for 1 h (60–120). Radioactively labeled polypeptides synthesized in the minicells were separated by electrophoresis through a 14–20% polyacrylamide gradient gel. The dried gel was used to expose X-ray film to produce the autoradiogram shown. The figure demonstrates the UV sensitivity of pK280 DNA in minicells. The major product encoded by pKB280 DNA is the λ repressor protein. Infection by λdrifd18 results in synthesis of polypeptides encoded by the E. coli DNA incorporated into the genome of this transducing phage. These products (β and β'-subunits of RNA polymerase, elongation factor EF-Tu, and ribosomal proteins L1, L10, L7/12, and L11) are indicated to the right of the figure. The presence of λ repressor prevents the synthesis of λ gene products encoded by the genome of the transducing phage. (Reprinted with permission from Reeve 1978)

2. Experiment 1: Minicell Technique for the Identification of Polypeptides Encoded by Plasmid or Bacteriophage DNA

a) Introduction

As pointed out in the general introduction, minicells of *E. coli* can be used to identify polypeptides encoded by plasmid or bacteriophage DNA. In this experiment, plasmid pKB280 and hybrid λ phages carrying a *trp*DE fragment are analyzed. In Table 1, important properties of strains used in the experiment as well as references are summarized.

Table 1. Phages and bacterial strains

Number	Important properties	Reference
DS410	Minicell producer; λ sensitive; non-suppressing	Reeve (1979)
DS410 (pKB280)	Plasmid pKB280 confers tetracyclin resistance; λ immunity	Backman and Ptashne (1978)
$\lambda i^{434} trp$DE	λ insertion vector NM607, b_{538}; carring the *trp*DEop region of the *E. coli* chromosome inserted in the *c1* gene. 434 immunity	Murray et al. (1977) Reeve (1978)
λ*trp*DE	λ replacement vector NM596, *nin*5 carrying the *trp*DEop region of the *E. coli* chromosome. λ immunity	Murray et al. (1977)

b) Objectives

1. To isolate minicells from *E. coli* DS410 carrying plasmid pKB280.
2. To identify polypeptides encoded by plasmid pKB280.
3. To identify polypeptides synthesized in *E. coli* minicells infected with λ-*trp*DE hybrid phages in the presence and in the absence of λ repressor.

c) Procedure and Results

The technique described should result in the isolation of an amount of minicells sufficient for several experiments. Smaller preparations can be made by starting with initial culture volumes as low as 50 ml.

Day 1

Preparation

Pipette 30 ml of 23% w/v sucrose dissolved in M9 minimal salts into 6 cellulose nitrate disposable centrifuge tubes. Convenient tubes to use are those designed for use with the Beckman ultracentrifuge rotor SW27.

Place these tubes at $-20°C$ until the solution in each tube is completely frozen.

The tubes are then placed at 4°C overnight during which time the frozen solution gradually thaws. During this process a gradient of sucrose solutions is automatically formed in each tube. (Alternatively 10–30% sucrose gradients can be made by using a sucrose gradient-making device.)

Inoculate 750 ml of L-broth with *E. coli* DS410 and 750 ml of L-broth containing 5 µg tetracyclin/ml with *E. coli* DS410 (pKB280).

Incubate these cultures, with shaking, at 37°C.

Continue incubation for 3 h beyond the time at which the culture reaches the stationary phase of growth (approx. 15 h of incubation).

Day 2

Isolation of Minicells

Chill the cultures of *E. coli* DS410 and *E. coli* DS410 (pKB280) by shaking in an ice water bath.

Remove cells and minicells from suspension by centrifugation at 4°C for 10 min at 10,000 × g (e.g., Sorvall rotor GSA, 8000 rpm). Minicells produced from the two cultures are to be kept separate and as subsequent handling is identical, care should be taken to avoid mixing the two preparations.

Thoroughly resuspend the pellets in a total of 10 ml of supernatant.

Layer the concentrated cell suspensions on top of the sucrose gradients (5 ml of suspension per gradient).

Centrifuge the gradient in a swinging bucket rotor (e.g., Sorvall rotor HB4; 5500 rpm) at 4°C for 20 min at 5000 × g). The majority of nucleated cells form a pellet and the minicells form a broad band within the sucrose gradient. The region between the minicells and the cell pellet contains a mixture of minicells and small nucleated cells. The region immediately above the band of minicells contains lysed cells.

Remove the minicells from the gradient using a 10 ml syringe inserted through the side wall of the disposable centrifuge tube. Remove only the central region of the band of minicells and avoid collecting material from regions immediately above and below the band.

Remove the minicells from the sucrose solution by centrifugation at 4°C for 10 min at 20,000 × g (e.g., Sorvall SS-34 rotor; 13,000 rpm).

Resuspend the minicells in 5 ml of ice cold M9 minimal salts solution.

Layer these suspensions on sucrose gradients and repeat the sucrose gradient centrifugation.

Remove the central region of the resulting band of minicells, pellet, and resuspend the minicells in 5 ml of ice cold M9 minimal salts medium. The minicells should now be suitable for subsequent experimentation.

Check using a phase-contrast light microscope that an adequate separation of cells and minicells has been accomplished. It should not be possible to detect more than one intact rod-shaped cell in a field of minicells at a 500-fold magnification. A small number of lysed cells may be observed. This is permissible as these dead cells do not affect the experiments to be undertaken.

The number of minicells isolated can be estimated by measuring the absorbance of a 1:10 dilution of the minicell suspension. An OD_{660} of 0.1 indicates a minicell density of approx. 10^9 per ml.

Having determined the absorbance of the minicell suspension calculate the volume in which the minicells should be resuspended to obtain an absorbance of 2.

Pellet the minicells (10 min; 8,000 × g) and resuspend to a final absorbance of 2 in M9 minimal salts containing 30% v/v glycerol.

Distribute 1 ml aliquots (equivalent to approx. 2×10^{10} minicells) into small storage tubes and freeze these aliquots by immersion in liquid nitrogen. The frozen stocks may be stored indefinitely in liquid nitrogen or at $-70°C$.

Day 3

UV Sensitivity of pKB280-Directed Polypeptide Biosynthesis

The objectives of this experiment are (1) to analyze the polypeptides synthesized in pKB280 containing minicells and (2) to determine the length of exposure to UV light required to inactivate the template capacity of the plasmid within the minicells.

Remove 1 tube of pKB280 containing minicells from storage, allow the suspension to thaw at room temperature, and dilute to 10 ml with ice cold M9 minimal salts solution containing 0.4% w/v glucose and 20 μg D-cycloserine per ml.

Pellet the minicells by centrifugation at 4°C for 10 min at 8,000 × g and resuspend in 1 ml of M9 minimal salts, glucose, D-cycloserine solution.

Irradiate 150 μl aliquots of the minicell suspension with UV light for increasingly longer periods (1, 2, 5, 10, 15 min). Care should be taken to ensure that the distance between the lamp and minicell suspension is constant, that the suspension is continually shaken during illumination, and that the UV light is not allowed to irradiate unprotected skin or eyes. Keep all minicell suspensions on ice before and after irradiation.

Following irradiation, pipette 100 μl of each irradiated suspension and 100 μl of non-irradiated minicells into 1.5 ml Eppendorf centrifuge tubes and place the tubes at 37°C.

After 10 min incubation, add 5 μl of Difco methionine assay medium containing ^{35}(S)-L-methionine.

Continue incubation at 37°C for 1 h.

Centrifuge the samples for 1.5 min in an Eppendorf centrifuge. *Carefully* remove the supernatants and collect these supernatants as *radioactive waste*.

Mix the pellets on a vortex so that they are no longer firm and form a slurry in the very small volume (2–3 μl) of residual supernatant.

Add 50 μl of electrophoretic sample buffer to each tube, place at 100°C for 3 min and then replace on ice.

The amount of ^{35}S-methionine incorporated must be determined. Mix 5 μl of each sample with 5 μl of lysozyme solution (1 mg/ml) and add 2 ml of 5% w/v TCA.

Collect the resulting precipitates by filtration through glass-fiber filters (e.g., Whatman GF/C), dry the filters, and determine the radioactivity bound to the filter by scintillation counting.

Determine the length of UV exposure required to completely inhibit the incorporation of ^{35}S-methionine. This information will be required in a subsequent experiment.

Load aliquots (20 μl) of the electrophoretic sample buffer containing the dissolved, radioactive polypeptides into the wells of a polyacrylamide gel. Preparation of polyacrylamide gels and the subsequent techniques of electrophoresis, gel drying, and fluorography are presented in this volume. The remaining 25 μl of each sample can be stored at $-20°C$ and used if repetition of the electrophoretic analysis is required.

Day 4

Determination of Optimal Multiplicity of Infection

Remove 1 tube of plasmid-free minicells from storage, allow to thaw, dilute to 10 ml with ice cold M9 minimal salts solution containing 0.4% w/v glucose and 20 μg D-cycloserine per ml.

Pellet the minicells by centrifugation at 4°C for 10 min at 8,000 × g and resuspend in 1 ml of the M9 minimal salts, glucose, D-cycloserine solution.

Pipette increasing volumes of each phage stock to be tested into 1.5 ml Eppendorf centrifuge tubes.

Add 50 μl of minicell suspension to each tube and include a tube without phage to provide an uninfected control (50 μl of minicell suspension contains approx. 1×10^9 minicells and the volumes of phage chosen should provide estimated input m.o.i.s. of between 0.5 and 10, as based on the titer of the phage suspension determined by plaque assay).

Place the tubes at 37°C.

After 30 min incubation add 2 μl of $5^{-3}H$-uridine and 0.5 μl of nutrient broth to each tube. Continue the incubation for 30 min at 37°C.

Add 20 μl of the lysozyme solution, incubate for 2 min at 37°C, add 20 μl of Sarcosyl solution, mix on a vortex, and add 1 ml of ice cold 5% TCA containing 200 μg uridine/ml.

Determine the radioactivity incorporated into TCA precipitable material by collection on glass-fiber filters, drying the filters, and scintillation counting. The optimal ratio of phage to minicells is indicated by the sample in which the maximal incorporation of ^3H-uridine has taken place. Addition of too few or too many phages results in submaximal incorporation. Uninfected minicells should incorporate very little ^3H-uridine and infection must produce a significant (at least 3 ×) stimulation of radioactivity incorporated for subsequent experiments to be successful.

Synthesis of Radioactively Labeled, Phage-Encoded Polypeptides in Infected Minicells

Resuspend frozen stocks of plasmid-free and pKB280-containing minicells in M9 minimal salts, glucose, D-cycloserine as described above.

Irradiate with UV 3 × 150 μl aliquots of the plasmid-containing minicells to completely destroy the template activity of the plasmid.

Pipette the irradiated minicells and non-irradiated plasmid-free minicells into 1.5 ml Eppendorf centrifuge tubes (3 tubes; 100 μl suspension into each tube).

Infect both types of minicells with the optimal number of λtrpDE and $\lambda i^{434} trp$DE phages as determined above. One aliquot of each minicell type remains as an uninfected control.

Incubate the mixtures at 37°C for 45 min and then add 5 μl of the ^{35}S-methionine mixture to each tube. Continue incubation for 1 h at 37°C.

Remove the minicells from suspension by centrifugation for 1.5 min in an Eppendorf centrifuge. *Carefully* remove the supernatants and place them in the *radioactive waste*.

Mix the tubes on a vortex to convert the minicell pellets to slurries.

Add 50 μl of electrophoretic sample buffer to each tube, place at 100°C for 3 min and then replace on ice.

Determine the amount of ^{35}S-methionine incorporated into each sample as described above.

Load aliquots (20 μl) of each sample onto a polyacrylamide gel, carry out electrophoresis, gel drying, and fluorography to determine the number and molecular weights of the polypeptides synthesized in the infected minicells.

Day 7

Conclusion of Experiments

Develop the autoradiograms produced by exposure of X-ray film to the dried electrofluorograms at −70°C. If the autoradiograms are under- or overexposed, replace the electrofluorogram at −70°C in contact with a new X-ray film and develop the second film after an appropriately longer or shorter time. After this type of experiment has been repeated several times, a correlation can be made between the amount of ^{35}S-methionine incorporated into TCA precipitable material and the optimal length of exposure. This correlation can then be used to estimate the required length of exposure in future experiments.

Analysis of Autoradiograms

The effect of UV exposure on synthesis of repressor protein in UV irradiated pKB280 containing minicells should be qualitatively obvious (see Fig. 2). To quantitate the effect, it would be necessary to measure the extent of exposure of the film by the radioactively labeled repressor protein. This measurement can be made with a scanning densitometer. If quantitation is desired care must be taken to choose a length of exposure in which a direct correlation exists between radioactivity present in the repressor protein and extent of exposure of the photographic emulsion.

The polypeptides synthesized in λ*trp*DE and λi^{434}*trp*DE infected minicells will depend on the type of minicell used. In plasmid-free minicells, both λ and the *trp*D and *trp*E polypeptides will be synthesized. In the UV irradiated pKB280-containing minicells, the presence of repressor will inhibit expression of λ genes on the λ*trp*DE genome and only the *trp*D and *trp*E gene products should be synthesized. Expression of the λi^{434}*trp*DE genome is immune to λ repressor and, therefore, both phage and *trp*D and *trp*E genes will be expressed. The expected result is shown in Fig. 3.

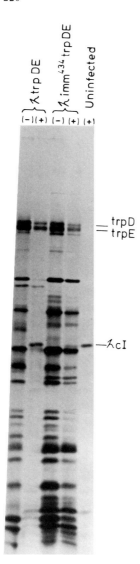

Fig. 3. Comparison of the polypeptides synthesized in infected minicells in the presence and absence of λ repressor. Minicells prepared from *E. coli* DS410 (−) and *E. coli* DS410 (pKB280) (+) were UV irradiated for 8 min (290 μWcm^{-2}) at a concentration of 2×10^{10} minicells per ml. Following irradiation, minicells were infected with λ replacement vector NM596, *nin5*, carrying the *trp*DEop fragment of *E. coli* chromosome between *Eco*RI sites 2 and 3 of λ, and λ insertion vector NM607, *b538*, carrying the *trp*DEop fragment of the *E. coli* chromosome inserted in the *Eco*RI site located in the imm^{434} *cI* gene. Infected and uninfected minicells were allowed to incorporate ^{35}S-methionine for 2 h at 37°C in M9 minimal salts medium. Electrophoresis and other methods were used as described in Fig. 2. In the presence of the λ repressor protein, only the *trp* operon is expressed from the λ*trp*DE phage genome. The presence of λ repressor has no effect on the expression of the hetero-immune λimm^{434}trpDE genome. The *trp*D and *trp*E gene products are indicated. (Reprinted with permission from Reeve 1978)

3. Materials

Day 1

200 ml	23% w/v sucrose in M9 minimal salts solution
6	disposable, cellulose nitrate centrifuge tubes
750 ml	L-broth in 3 liter Erlenmeyer flask
750 ml	L-broth containing 5 μg tetracyclin/ml in 3 liter Erlenmeyer flask

Day 2

6	250 ml centrifuge bottles
6	30 ml plastic centrifuge tubes
4	10 ml disposable syringes with needles
10	1.5 ml plastic storage tubes suitable for immersion in liquid nitrogen
10	10 ml pipettes
1	Refrigerated centrifuge, e.g., Sorvall RC5B; large and small angle-rotors, e.g., GSA and SS34; swinging-bucket rotor, e.g., HB4
3	Cuvettes
1	Spectrophotometer to measure cell absorbance
1	High-resolution phase contrast microscope
2	Microscope slides and cover slips
50 ml	M9 minimal salts solution
20 ml	Glycerol
1	Liquid nitrogen or $-70°C$ storage facility
1	Large ice bucket and large amount of ice for cooling cultures
1	Vortex

Day 3

2	15 ml centrifuge tubes
20 ml	M9 minimal salts solution containing 0.4% w/v glucose and 20 µg D-cycloserine/ml
2	10 ml pipettes
1	Refrigerated centrifuge
1	Rotor, e.g., SS34
1	UV germicidal lamp
1	Very small dish in which to irradiate minicells
1	Stopwatch
1	Ice bucket and ice
6	1.5 ml Eppendorf disposable centrifuge tubes in tube rack
1	37°C incubator or water bath
1	100°C water bath
1	Eppendorf centrifuge
1	Vortex
30 µl	^{35}S-methionine in Difco assay medium (high specific ^{35}S-methionine, $\sim$ 1000 Ci/mmol, is diluted 1:10 into Difco methionine assay medium)
0–20 µl	Variable volume micropipetter
0–200 µl	Variable volume micropipetter
10	Disposable pipette tips
50 µl	Lysozyme solution (1 mg/ml) in 50 mM Tris-HCl; 5 mM EDTA; 100 mM NaCl, pH 7.6
50 ml	5% w/v trichloracetic acid (TCA) containing 200 µg methionine per ml
6	Glass-fiber filters
1	Filter apparatus
6	Scintillation vials; scintillation fluid and a scintillation counter

350 μl   Electrophoretic sample buffer
1   Polyacrylamide gel, electrophoretic apparatus, power pack, etc.

Day 4 (Assuming two phage preparations are to be tested)

4	15 ml centrifuge tubes
50 ml	M9 minimal salts medium containing 0.4% w/v glucose and 20 μg D-cycloserine/ml
1	Refrigerated centrifuge
1	Rotor, e.g., SS34
18	1.5 ml Eppendorf centrifuge tubes plus rack
0–20 μl	Variable volume micropipetter
0–200 μl	Variable volume micropipetter
5	10 ml pipettes
50	Disposable pipette tips
25 μl	^{3}H-uridine (1–30 mCi/mmol; 1 mCi/ml)
10 μl	L-broth
1	37°C incubator or water bath
1	100°C water bath
500 μl	Lysozyme solution (see day 3)
500 μl	Sarcosyl solution (2% w/v in 50 mM Tris-HCl; 5 mM EDTA; 100 mM NaCl, pH 7.6)
75 ml	5% TCA containing 200 μg uridine per ml
18	Glass-fiber filters
1	Filter apparatus
18	Scintillation vials; scintillation fluid and scintillation counter
1	UV germicidal lamp
1	Dish for irradiation of minicells
30 μl	^{35}S-methionine in Difco methionine assay medium
1	Eppendorf centrifuge
1	Vortex
1	Polyacrylamide gel, electrophoretic apparatus, power pack, etc.

For autoradiography

1	X-ray film per gel, e.g., Kodak XR-5
1	Clamp per gel to hold gel and film together
	70°C storage space
	Film developer and fixer

4. References

Backman K, Ptashne M (1978) Maximizing gene expression on a plasmid using recombination in vitro. Cell 13:65–71

Frazer AC, Curtiss R III (1975) Production, properties and utility of bacterial minicells. Curr Top Microbiol 69:1–84

Meagher RB, Tait RC, Betlach M, Boyer HW (1977) Protein expression in *E. coli* minicells by recombinant plasmids. Cell 10:521–536

Murray NE, Brammar WJ, Murray K (1977) Lambdoid phages that simplify the recovery of in vitro recombinants. Mol Gen Genet 150:53–61

Reeve JN (1977) Bacteriophage infection of minicells. A general method for identification of in vivo bacteriophage directed polypeptide biosynthesis. Mol Gen Genet 158:73–79

Reeve JN (1978) Selective expression of transduced or cloned DNA in minicells containing plasmid pKB280. Nature 276:728–729

Reeve JN (1979) Use of minicells for bacteriophage directed polypeptide biosynthesis. Methods Enzymol 68:493–503

Reeve JN (1981) ϕX174-directed DNA and protein syntheses in infected minicells. J Virol 40:396–402

Sauer RT, Anderegg R (1978) Primary structure of the λ repressor. Biochemistry 17:1092–1100

4.3 Determination of Coding Regions on Multicopy Plasmids: Analysis of the Chloramphenicol Acetyltransferase Gene of Plasmid pACYC184

W. KLIPP and A. PÜHLER [1]

Contents

1. General Introduction... 224
2. Experiment 1: Isolation and Characterization of Tn5 Induced Mutants of the Chloramphenicol Acetyltransferase Gene of Plasmid pACYC184................. 226
   a) Introduction... 226
   b) Objectives.. 228
   c) Procedure... 228
   d) Results... 229
3. Experiment 2: Gene Product Analysis of Wild Type and Tn5 Mutagenized pACYC184 Plasmids in Minicells of *E. coli:* Determination of the Chloramphenicol Acetyltransferase Coding Region... 230
   a) Introduction... 230
   b) Objectives.. 230
   c) Procedure... 230
   d) Results... 232
4. Materials... 234
5. References.. 235

1. General Introduction

It is often difficult to identify proteins encoded by cloned DNA fragments, to localize their coding regions, and to determine the direction of transcription. Such problems can be solved for fragments cloned on multicopy plasmids by a method in which the following techniques are used: (1) Tn5 transposon mutagenesis of the cloned fragment and determination of the Tn5 insertion sites; (2) isolation of minicells harboring hybrid plasmids with cloned wild type or Tn5 mutagenized fragments; and (3) radioactive labeling of gene products synthesized in minicells by the hybrid plasmids, separation of the proteins on SDS-polyacrylamide gels, and autoradiography of the gel.

Evaluation of the protein data in combination with the Tn5 insertion sites leads to the construction of the coding region map of such a cloned fragment. This method was first used to determine the coding region map of the Klebsiella nitrogen fixation gene cluster (Pühler and Klipp 1981). The same method was successfully employed

---

[1] Lehrstuhl für Genetik, Fakultät für Biologie, Universität Bielefeld, D-4800 Bielefeld 1, Fed. Rep. of Germany

in the case of a *Rhizobium meliloti Eco*RI fragment coding for component II of nitrogenase (Pühler et al. 1982). An *Eco*RI fragment of the T-DNA of octopine tumors was also analyzed by the method described above resulting in the determination of the coding region for lysopine dehydrogenase (Schröder et al. 1981).

To describe the details of the method, the well-characterized chloramphenicol acetyltransferase gene (Cm gene) of the multicopy plasmid pACYC184 (Chang and Cohen 1978) will be used. The location of the Cm gene on the restriction map of the plasmid is known (Fig. 1) and parts of the gene have already been sequenced (Marcoli et al. 1980). The purpose of the proposed experiments is to localize the gene and to determine its coding region and direction of transcription. The theoretical background to the experiment is outlined in Fig. 1.

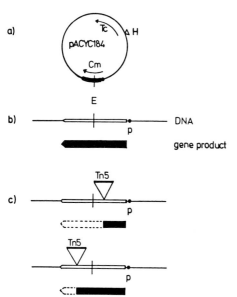

Fig. 1. Determination of the coding region of the chloramphenicol acetyltransferase gene on plasmid pACYC184. a Circular map of vector plasmid pACYC184 with the approximate location of the two antibiotic resistance genes; b a part of pACYC184 with the coding region of the chloramphenicol acetyltransferase and the expression of this gene. The gene product is identified by radioactive labeling in *E. coli* minicells; c determination of the coding region is achieved by analyzing truncated polypeptides caused by different Tn5 insertions. E *Eco*RI site; H *Hin*dIII site; Tc tetracycline resistance; Cm chloramphenicol resistance; p promoter

Figure 1a shows a simplified map of the plasmid pACYC184. It confers chloramphenicol (Cm) and tetracycline resistance (Tc) to its host. The coding region for chloramphenicol resistance carries a unique *Eco*RI site which can be used for *Eco*RI cloning. The gene product of this resistance gene is a chloramphenicol acetyltransferase (CAT) (Fig. 1b). The CAT protein can be identified in *E. coli* minicells as a 22K protein band on polyacrylamide gels after radioactive labeling of polypeptides encoded by pACYC184. Minicells are chromosome-free cells which are segregated by a special filament forming *E. coli* mutant. If such a minicell producing strain harbors a multicopy plasmid, this plasmid segregates into the minicells. After separation of plasmid containing minicells from parental cells, plasmid encoded proteins can be selectively labeled by incorporation of radioactive amino acids.

Using transposon mutagenesis, the coding region of the Cm gene can then be mapped. First, Tn5 carrying pACYC184 plasmids are selected that are no longer chloramphenicol resistant. The Tn5 insertion sites can then be determined by restriction fragment analysis (Fig. 1c). The pACYC184 plasmids carrying Cm::Tn5 mutations are then analyzed in *E. coli* minicells. It is assumed in these experiments that a Tn5 insertion in an operon stops transcription at the insertion site and that a Tn5 insertion in a coding region results in a truncated polypeptide. From the molecular weights of wild type and truncated CAT proteins and the location of the Tn5 insertion sites, the Cm coding region and its direction of transcription can be determined (Fig. 1c).

The proposed experiments include the isolation of Tn5 induced mutants of the chloramphenicol acetyltransferase gene and the determination of the integration site by restriction fragment analysis. In addition, the coding region of the Cm gene will be determined by analyzing proteins encoded by pACYC184 and Cm::Tn5 mutants of this plasmid.

2. Experiment 1: Isolation and Characterization of Tn5 Induced Mutants of the Chloramphenicol Acetyltransferase Gene of Plasmid pCYC184

a) Introduction

As discussed in the preceding section, transposon induced mutations can help in the analysis of the coding region of a selected gene. In the case of the Cm gene of plasmid pACYC184, we were able to develop a simple method for the isolation of Cm::Tn5 mutations. This method starts with a selection step for the isolation of pACYC184 plasmids carrying Tn5. In a second step, such plasmids are then tested for Tn5 insertions into the Cm gene. The selection step makes use of a gene dosage effect. An *E. coli* strain carrying Tn5 integrated in the chromosome (e.g., strain S605) is less resistant to neomycin than a strain that contains Tn5 integrated into the multicopy plasmid pACYC184 (e.g., strain C600 pACYC184::Tn5) The difference in the minimal inhibition concentration is shown in Fig. 2. This gene dosage effect can be used to select transposition of Tn5 from the chromosome onto the multicopy plasmid. The experimental procedure is quite simple: an overnight culture of strain S605 (pACYC184) is plated onto agar plates containing an increasing concentration (150–500 μg/ml) of neomycin. Large colonies growing on these plates usually indicate clones that carry Tn5 inserted into the pACYC184 plasmid. Sometimes, however, a spontaneous high level neomycin resistance mutation, which is chromosomally based, can be mistaken for a Tn5 transposition. Therefore, it is often necessary to confirm the Tn5 transposition onto the plasmid by measuring the plasmid length. pACYC184 plasmids carrying Tn5 show an increase in plasmid length of about 6000 bp. Since pACYC184 is only 4300 bp long, it is very easy to distinguish between pACYC184 and pACYC184::Tn5 plasmids. However, this plasmid analysis step is not necessary since Tn5 insertions into the Cm gene will be isolated and such mutants can be identified by retesting the large colonies described above on agar containing chloramphenicol. Chloramphenicol sensitive colonies indicate a Cm::Tn5 mutation.

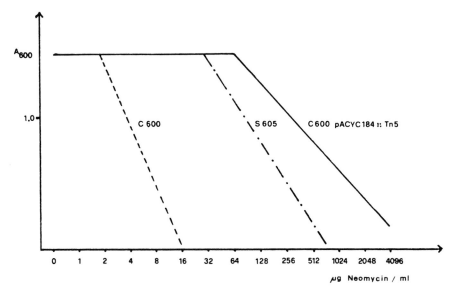

Fig. 2. Determination of minimal inhibitory concentration for neomycin. Liquid medium with the indicated neomycin concentration was inoculated with a bacterial titer of 10^7/ml and incubated for 12 h at 37°C. The optical density (A_{600}) was measured. Isogenic strains carrying no Tn5 (C600), one copy of Tn5 (S605), and about 50 copies of Tn5 (C600 pACYC184::Tn5) are compared

The next step is the exact mapping of the Tn5 insertion into the Cm gene. This can be done by restriction fragment analysis, since transposon Tn5 contains in its inverted repeats a useful set of restriction sites (*Xho*I, *Pst*I, and *Hin*dIII), which are symmetrically located (Fig. 3). To determine the integration sites of Tn5 into the Cm gene of pACYC184, *Eco*RI-*Hin*dIII, and *Eco*RI-*Xho*I, double digests of plasmid

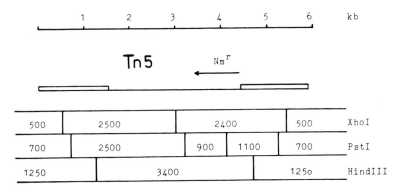

Fig. 3. Restriction map of transposon Tn5. The restriction sites for *Hin*dIII, *Pst*I, and *Xho*I as well as the molecular weights of the resulting fragments (bp) are indicated. The inverted repeats of Tn5 are marked (▭▭). (Simplified, from Jorgensen et al. 1979)

DNA are used. Form an *Eco*RI-*Xho*I digest, it is possible to determine the distance of the Tn5 insertion site from the *Eco*RI site. By the *Eco*RI-*Hin*dIII digest, it is possible to decide whether Tn5 integrated in the smaller or larger *Eco*RI-*Hin*dIII fragment of pACYC184 (see Fig. 1). By combining these two results, the exact integration site of Tn5 in the Cm gene of plasmid pACYC184 can be determined.

b) Objectives

1. To isolate Tn5 insertion mutants of the chloramphenicol acetyltransferase gene of plasmid pACYC184.
2. To map Tn5 integration sites by restriction fragment analysis.

c) Procedures

Day 1

Isolation of Colonies Carrying pACYC184::Tn5 Plasmids

To select for Tn5 transposition from the *E. coli* chromosome onto plasmid pACYC-184, neomycin gradient plates are used.

Preparation of neomycin gradient plates:

0.2 ml of neomycin stock solution (50–75 mg/ml) are dropped in the middle of an agar plate containing the normal antibiotic concentration for plasmid selection (5 µg Tc/ml). By diffusion, a neomycin gradient is developed which inibits growth of strain S605 (pACYC184) in the middle of the plate, but allows the growth of a bacterial lawn at the edge. On these freshly prepared gradient plates, 0.2 ml of an overnight culture is spread and incubated overnight at 37°C. Large colonies growing in the middle of the gradient plate are presumed to carry pACYC184::Tn5 plasmids.

Day 2

Isolation of Tn5 Insertion Mutants of the Chloramphenicol Acetyltransferase Gene of Plasmid pACYC184

Large colonies growing in the middle of the gradient plate are taken and retested for Tc-resistance and Cm-resistance. Tn5 insertion mutants of the Cm gene can be identified by Tc-resistance and Cm-sensitivity.

Day 3

Mapping of the Tn5 Integration Sites in the Cm Gene

To localize the Tn5 insertion, plasmid DNA is prepared by one of the rapid isolation methods described in this volume. Using restriction enzymes, preferably those with target sequences within the inverted repeats, the Tn5 insertion site can be mapped physically. To map the Tn5 insertions in the Cm gene of pACYC184, double digests with *Eco*RI-*Xho*I and *Eco*RI-*Hin*dIII are used.

d) Results

A typical result of this experiment is shown in Fig. 4. By restriction fragment analysis, 8 independently isolated Tn5 insertion mutants of the Cm gene of plasmid pACYC184 were mapped. The Tn5 insertion sites are all located very near to the *Eco*RI site of plasmid pACYC184. This is consistent with the genetic map of plasmid pACYC184 (Chang and Cohen 1978).

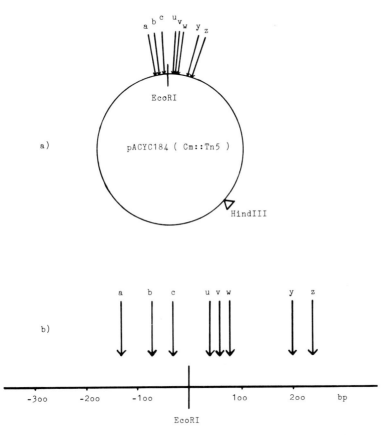

Fig. 4. Localization of Tn5 insertions in the chloramphenicol acetyltransferase gene of plasmid pACYC184. a Circular map of pACYC184 with the restriction sites for *Eco*RI and *Hind*III; *arrows* insertion sites of Tn5 in independently isolated Cm::Tn5 mutants; b a part of the pACYC184 restriction map with the Tn5 insertion sites drawn on a larger scale

3. Experiment 2: Gene Product Analysis of Wild Type and Tn5 Mutagenized pACYC184 Plasmids in Minicells of *E. coli*: Determination of the Chloramphenicol Acetyltransferase Coding Region

a) Introduction

For gene product analysis, the wild type and Tn5 mutagnenized pACYC184 plasmids are introduced into the minicell forming strain DS410. Minicells containing the different plasmids are isolated by sucrose gradient centrifugation and then incubated in a medium containing ^{35}S-methionine. Since minicells harbor the whole transcription and translation machinery of a normal *E. coli* cell plasmid genes are expressed and their gene products are radioactively labeled with ^{35}S-methionine. After separation on SDS-polyacrylamide gels, the protein bands can be recognized by fluorography. In this experiment, it is important to identify the wild type Cm gene product and the Cm::Tn5 gene products on the autoradiogram and to determine the molecular weights of the protein bands. Since Tn5 insertions into a gene terminate protein biosynthesis (Auerswald et al. 1981), a linear correlation between the Tn5 insertion site and the molecular weight of the synthesized gene product exists. In this way, it is possible to determine where the coding region begins. The end of a coding region is given by the molecular weight of the wild type gene product. To do such calculations, one can assume an average molecular weight of 120 for an amino acid. With this assumption, it is possible to calculate the number of amino acids of a protein whose molecular weight was measured. Knowing the number of amino acids, the length of the coding region can then be determined because of the triplet nature of the genetic code.

b) Objectives

1. To determine the molecular weights of the gene products of the Cm and Cm::Tn5 genes of plasmid pACYC184 in minicells of *E. coli*.
2. To locate the chloramphenicol acetyltransferase coding region on plasmid pACYC-184.

c) Procedure

Day 1

1. Minicell Isolation

Plasmids are introduced into the minicell producing strain DS410 by transformation. Use for this purpose the plasmid DNA of pACYC184 and pACYC184::Tn5 isolated for restriction fragment analysis as described in the previous experiment. Competent cells of DS410 are prepared by the $CaCl_2$-technique (this volume) and transformants are selected on Tc-containing medium.

Plasmid containing minicell strains are grown overnight to early stationary phase in 250 ml L-broth with the appropriate antibiotic (in this case, Tc). Minicells are purified from parental cells by three successive sucrose gradient centrifugations as described (this volume).

2. Labeling of Plasmid Containing Minicells

500 μl of a minicell preparation at a concentration of 2×10^{10}/ml in M9 medium containing 4 mM cycloserine are labeled by adding 100 μl of methionine assay medium (Difco) and approx. 10 μCi of very high specific activity ^{35}S-methionine (up to 1000 Ci/mM). Incorporation of label is allowed for 60 min at room temperature.

3. Minicell Lysis

Plasmid specific radioactively labeled proteins are extracted as described for phage specific polypeptides (Reeve, this volume).

4. SDS-Polyacrylamide Gel Electrophoresis (Laemmli 1970)

The labeled proteins are separated on a 14–20% SDS-polyacrylamide gradient gel. The molecular weights of the plasmid encoded proteins are determined by using protein molecular weight standards. To calibrate the 14–20% SDS-polyacrylamide gels, ^{14}C-labeled protein molecular weight standards with a high molecular weight range (from 12,300–200,000) are used.

Preparation of the Gradient Gel

The acrylamide solutions for the separating gel are prepared as described in Sect. 4. Use a gradient making device and pump the gel solutions into a slab gel apparatus at
    a slow rate (2 ml/min).
Overlay the gel carefully with water.
Allow the gradient separating gel to polymerize for 60 min at room temperature.
Fill in the stacking gel and allow polymerization for 6 min.
Load the gel immediately after the stacking gel has polymerized.
Run at 80 V for about 2 h until the front has reached the separating gel and then run
    the gel overnight at 160 V.

Day 2

Preparation of the SDS-Polyacrylamide Gel for Fluorography (Bonner and Laskey 1974)

Shake gel in DMSO for 1 h, but change DMSO after half an hour. Use pre-used
    DMSO the first time and fresh DMSO the second time.
After 1 h remove DMSO, replace with DMSO + PPO (dissolve 30 g PPO in 120 ml
    DMSO) and shake for 3 h.
Replace the PPO solution with water, discard the PPO solution. PPO can be used
    again.
Keep gel in water for 1 h chaning once after half an hour. The gel is dryed with a slab
    gel dryer onto a sheet of Whatman filter paper.

After drying, place gel in contact with X-ray film. Exposure time depends on the amount of radioactivity run on the gel. Approximately 30,000 cpm per track requires 3—5 days contact with the film to produce a correctly eyposed autoradiogram. Films are developed as recommended by the manufacturer.

d) Results

In Fig. 5, a typical autoradiogram demonstrating gene products of the Cm and of the Cm::Tn5 genes is shown. In the case of plasmid pACYC184, only one strong polypeptide with a molecular weight of 22,000 can be seen. This polypeptide represents the Cm gene product since Tn5 mutants of the Cm gene non longer produce a gene product with the same molecular weight. All these mutants synthesize truncated

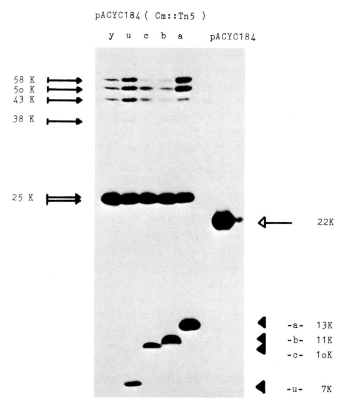

Fig. 5. Autoradiogram of gene products of plasmid pACYC184 and Cm::Tn5 mutants of pACYC-184. The gene products of the plasmid pACYC184 and of the Cm::Tn5 mutants of pACYC184 were synthesized in the E. coli minicell system. a, b, c, u, and y indicate plasmids with specific Tn5 insertions in the Cm gene. The insertion sites are those shown in Fig. 4. The gene products were radioactively labeled (^{35}S), separated on a 14–20% SDS-polyacrylamide gel, and visualized by fluorography. The protein bands are marked as follows: ◁—— chloramphenicol acetyltransferase (CAT); ◀—— truncated polypeptides of CAT; ◀—⊣ Tn5 specific proteins

polypeptides of different molecular weights according to the Tn5 insertion sites in the gene.

It should be noted that gene products of the tetracycline resistance gene of pACYC184 cannot be detected by this method. This can be explained by a weak expression of the gene or by an interaction of the gene product with the bacterial membrane. After insertion of transposon Tn5 into the plasmid Tn5, specific proteins appear on the protein gel. These proteins are specifically marked in Fig. 5 and play a role in kanamycin resistance and Tn5 transposition.

The comparison of the integration sites of Tn5 in the Cm gene of pACYC184 with the corresponding truncated polypeptide is summarized in Fig. 6. Starting with the molecular weights of the truncated polypeptides, the beginning of the coding region of the Cm gene was determined. It is located at a distance of 200 bp from the *Eco*RI site of the Cm gene. By DNA sequencing, Marcoli et al. (1980) found that the Cm coding region starts at 215 bp from the *Eco*RI site, thus confirming the reliability of this procedure. The end of the coding region is determined by the molecular weight of the CAT protein which was found to be 22K corresponding to a coding region of 580 bp. The direction of transcription can also be obtained from the results of this procedure. From Fig. 6, it can be concluded that the Cm gene is transcribed from right to left.

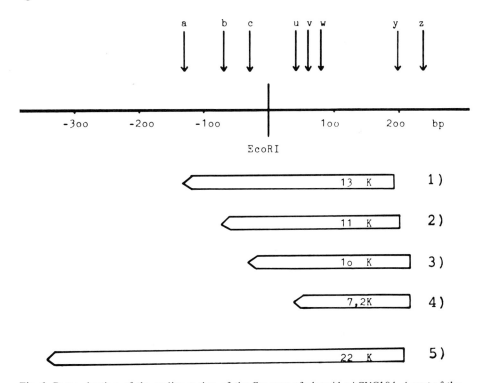

Fig. 6. Determination of the coding region of the Cm gene of plasmid pACYC184. A part of the pACYC184 restriction map with the different Tn5 insertion sites in Cm gene is shown. ⇐⎯⎯⎯⎯ represents coding regions that are expressed in minicells of *E. coli* (see Fig. 5). *1—4* are coding regions for truncated polypeptides, *5* the coding region for the Cm gene. For details see text

4. Materials

Bacterial Strains and Plasmids

E. coli S605:   E. coli C600 met::Tn5
E. coli DS410:   Minicell producer
Plasmid pACYC184:   Vector plasmid carrying chloramphenicol and tetracycline
                         resistance (Chang and Cohen 1978)

SDS Gel Electrophoresis

Buffers:

10 × reservoir buffer   Tris-base 30.2 g
                         Glycine 144.0 g
                         Water to 1000 ml; then filter
Before use: add SDS to 0.1% in 1 × reservoir buffer.

4 × stacking gel buffer Tris-base 61 g
                         HCl; water to pH 6.8
                         Final volume 1000 ml
                         Filter

4 × separating gel buffer Tris-base 182 g
                           HCl; water to pH 8.8
                           Final volume 1000 ml
                           Filter

Sample buffer   4 × stacking gel buffer   0.125 ml
                         10% SDS   0.30 ml
                         50% glycerol   0.2   ml
                         2-mercaptoethanol   0.05 ml
                         H_2O 0.325 ml

Other Solutions

Acrylamide 30% w/v   ⎫
Bis-acrylamide 0.8% w/v ⎬ made as one solution in water
Filter. Store in dark and cold
Glycerol 50% in H_2O, filter
SDS, 10% w/v in H_2O, filter
Ammonoium persulphate, 10% in H_2O,
made fresh immediately for each experiment

SDS-Polyacrylamide Gradient Gel 14–20%

For one separating gel:	14%	20%
50% glycerol	0.33 ml	1.33 ml
4 × separating gel buffer	5.0 ml	5.0 ml
Acrylamide-bis (30%; 0.8%)	9.3 ml	13.3 ml
H_2O	5.0 ml	–

Degas by evacuation	14%	20%
TEMED	1.7 µl	0.6 µl
SDS (10%)	0.2 ml	0.2 ml
$(NH_4)_2S_2O_8$ (10%)	0.15 ml	0.15 ml

Stacking Gel

H_2O	5.8	ml
4 × stacking gel buffer	2.5	ml
Acrylamide-bis (30%; 0.8%)	1.9	ml

Degas by evacuation

TEMED	7.5	µl
SDS (10%)	0.1	ml
$(NH_4)_2S_2O_8$ (10%)	0.08	ml

5. References

Auerswald EA, Ludwig G, Schaller H (1981) Structural analysis of Tn5. In: Movable genetic elements. Cold Spring Harbor Symp Quant Biol, vol XLV, pp 107–113

Bonner WM, Laskey RA (1974) A film detection method for tritium labelled proteins and nucleic acids in polyacrylamid gels. Eur J Biochem 46:83–88

Chang ACY, Cohen SN (1978) Construction and characterization of amplifiable multicopy DNA cloning vehicles derived from the P15A cryptic miniplasmid. J Bacteriol 134:1141–1156

Jorgensen RA, Rothstein SJ, Reznikoff WS (1979) A restriction enzyme cleavage map of Tn5 and location of a region encoding neomycin resistance. Mol Gen Genet 177:65–72

Laemmli UK (1970) Cleavage of structural proteins during the assembly of the head of bacteriophage T4. Nature 227:680–685

Marcoli R, Tiola S, Bickle TA (1980) The DNA sequence of an IS1 flanked transposon coding for resistance to chloramphenicol and fusidic acid. FEBS Lett 110:11–14

Pühler A, Klipp W (1981) Fine structure analysis of the gene region for N_2-fixation *(nif)* of *Klebsiella pneumoniae*. In: Bothe H, Trebst A (eds) Biology of inorganic nitrogen and sulfur. Springer, Berlin Heidelberg New York, pp 276–286

Pühler A, Arnold W, Horn D, Jäckel B, Priefer U, Simon R, Weber G (1983) Genetic analysis of *R. meliloti* plasmids and *R. meliloti nif*-genes. In: Clark KW, Stephens JHG (eds) Proc 8th North American Rhizobium Conf. Printing Services The University of Manitoba, Canada, pp 90–114

Schröder J, Hillebrand A, Klipp W, Pühler A (1981) Expression of plant tumor-specific proteins in minicells of *Escherichia coli:* a fusion protein of lysopine dehydrogenase with chloramphenicol acetyltransferase. Nucleic Acids Res 9:5187–5202

4.4 Identification of Gene Products by Coupled Transcription-Translation of DNA Fragments in Cell-Free Extracts of *E. coli*

J.M. PRATT[1], G.J. BOULNOIS[1,2], V. DARBY[1,3], and I.B. HOLLAND[1]

Contents

1. General Introduction 236
2. Experiment 1: Preparation and Characterization of an S30 Extract 237
   a) Introduction and Objectives 237
   b) Procedure 237
3. Experiment 2: Identification of Polypeptides Coded by Cloned Sequences Similar in Molecular Weight to Those Specified by the Vector 240
   a) Introduction and Objectives 240
   b) Procedure 240
   c) Results and Discussion 242
4. Experiment 3: Mapping Polypeptide Coding Sequences on a Staphylococcal Plasmid . . 242
   a) Introduction and Objectives 242
   b) Procedure 243
   c) Results and Discussion 244
5. Final Comments 245
6. Materials 246
7. References 247

1. General Introduction

A central objective of many gene cloning experiments is the identification of the products of cloned genes. Several systems are currently available, including (a) infection of UV irradiated, lysogenic hosts with λ phage vectors carrying foreign DNA (Ptashne 1967); (b) UV irradiation of UV sensitive bacteria carrying high copy number hybrid plasmids (maxicell system, Sancar et al. 1979; see also Chap. 4.1); and (c) the use of anucleate minicells, either containing hybrid plasmids, or infected with λ phage vectors carrying foreign DNA (Reeve 1978; see also Chap. 4.2). These systems are limited to the use of specialized transducing phages or multicopy plasmids in conjunction with certain mutant strains of *E. coli*. Occasionally, the need to use such bacteria precludes the analysis of some genes and their products. Moreover,

---

1 Department of Genetics, University of Leicester, Leicester, LE1 7RH, Great Britain
2 Present address: Department of Medical Biochemistry, University of Geneva, CH-1211, Geneva 4, Switzerland
3 Department of Genetics, University of Madison, Wisconsin, USA

these systems do not readily permit the investigator to establish whether or not a gene has been cloned with its natural promoter.

As an alternative to these semi in vitro systems, cell-free synthesis of polypeptides (Zubay 1973, Collins 1979) can be exploited as a method for gene product identification. This is a particularly powerful approach since short DNA fragments, e.g., those generated by restriction endonucleases, can be used to program protein synthesis (Yang et al. 1980, Pratt et al. 1981). As a result, sequences coding for polypeptides and their corresponding promoters may be localized to specific regions of cloned DNA segments (Pratt et al. 1981). Furthermore, since vector and cloned DNA fragments can be analyzed separately, fusion polypeptides specified by vector:insert boundary regions can be readily identified.

This chapter describes the preparation of an extract that supports coupled transcription-translation of exogenously supplied linear and circular DNA. The extract will be used to identify polypeptides specified by a cloned sequence carrying the *env*A gene of *E. coli* and to locate sequences coding for several polypeptides specified by the staphylococcal plasmid pC221 with respect to the restriction map of the plasmid.

2. Experiment 1: Preparation and Characterization of an S30 Extract

a) Introduction and Objectives

The first part of the experimental work involves the preparation of an extract of *E. coli* that supports coupled transcription-translation of purified DNA. The activity of the extract in supporting protein synthesis is greatly dependent on the concentration of magnesium ions and the amount of extract used. These parameters, therefore, need to be optimized.

Since RNAase is a potent inhibitor of transcription-translation in the extract, it is essential that all water and apparatus used in the experiment is free from such activity. This is most conveniently achieved by adding 1 ml of diethyl-pyrocarbonate (DEPC; a powerful protein denaturing agent) per liter of water and stirring for 1 h, followed by autoclaving to destroy residual DEPC. All apparatus should be soaked in water containing DEPC and then sterilized. To further minimize the risk of RNAase contamination, a strain (MRE600) defective in RNAase I activity, is normally used as a source of the extract.

b) Procedure

Day 1

Inoculate 100 ml of growth medium with strain MRE600 and culture overnight at 37°C.
Assemble and sterilize the fermenter containing 10 liter basic medium.
After sterilization, add the other ingredients aseptically.

Day 2

Add the overnight culture to the medium in the fermenter until an absorbance at 450 nm of about 0.07 is reached.
Incubate overnight at 37°C.

Day 3

The absorbance of the overnight culture should be 2–3. Harvest the cells by centrifugation at 4°C for 15 min at 5000 rpm using the Sorvall GSA rotor and pre-weighed bottles.
Estimate the weight of the cell pellet and wash the cells 3 × in S30 buffer supplemented with mercaptoethanol (25 ml per gram of cells).
Centrifugations should be performed using the SS34 rotor at 10,000 rpm for 20 min at 4°C.
After the final spin, store the cells as pellets overnight at −70 to −80°C.

Day 4

Allow the cell pellets to thaw on ice and resuspend in S30 buffer containing mercaptoethanol (10 ml per gram of cells).
Harvest the cells by centrifugation using the SS34 rotor and pre-weighed tubes.
Determine the weight of the cell pellet and resuspend in S30 buffer (63.5 ml per 50 g cells) under vacuum. This is best achieved by transferring the cells to a Buchner flask, adding the S30 buffer, and mixing with a sterile spatula accompanied by frequent degassing.
When the cells are thoroughly resuspended disrupt them by passage through a French press at 8,400 psi; add 100 μl of 0.1 M dithiothreitol (DTT) to every 10 ml of "pressate" obtained.
Centrifuge the pressate at 4°C in the SS34 rotor at 15,500 rpm for 30 min.
Remove the upper 4/5 of the supernatant and recentrifuge the supernatant.
Again remove the upper 4/5 and store protected from light in an Erlenmeyer flask at 4°C.
During the centrifugations, prepare the preincubation mix as follows:

Pyruvate kinase (10 mg/ml as $(NH_4)_2SO_4$ solution)	25 μl
Tris-acetate (pH 8.2, 2.2 M)	1.0 ml
Magnesium acetate (3 M)	23 μl
ATP (38 mM, pH 7.0)	2.63 ml
PEP (phosphoenolpyruvate; 0.42 M, pH 7.0)	1.5 ml
DTT (0.55 M)	60 μl
20 amino acid mix (50 mM)	6 μl
H_2O	7.5 ml

To every 25 ml of supernatant obtained, add 7.5 ml of preincubation mix and incubate at 37°C for 80 min. This step results in the "run off" of ribosomes from pre-existing mRNA and degradation of this RNA, which would otherwise contribute to background translation in the system.

Transfer the incubated extract to dialysis tubing and dialyze against four changes of 50 vol of S30 buffer.

Change the buffer every 45 min.

Distribute the extract in 0.5 ml aliquots in Eppendorf tubes and freeze in liquid nitrogen.

Day 5

Prepare the low molecular weight mix (LMM) as indicated:

Stock solution	Volume
Tris-acetate (2.2 M, pH 8.2)	40 µl
DTT (0.55 M)	5 µl
ATP (38 mM, pH 7.0)	50 µl
Triphosphate mix: (CTP, 88 mM; GTP, 88 mM; UTP, 88 mM) pH 7.0	15 µl
PEP (0.42 M, pH 7.0)	100 µl
19 Amino acid mix (55 mM each, methionine-free)	10 µl
Polyethylene glycol – 6000 (40% in H_2O)	75 µl
Folinic acid (2.7 mg/ml)	20 µl
cAMP (50 mM, pH 7.0)	20 µl
tRNA (*E. coli*; 17.4 mg/ml)	15 µl
Inorganic mix (Ammonium acetate, 1.4 M; potassium acetate, 2.8 M; calcium acetate, 0.38 M)	40 µl

Mix the components in the order given above. The mixture may then be kept frozen for a few weeks.

Allow an aliquot of the extract to thaw slowly on ice, transfer to a 1.5 ml reaction tube and centrifuge in an Eppendorf centrifuge for 5 min.

Remove the supernatant and store on ice.

Set up a series of reactions as described in Table 1.

Mix DNA (any purified plasmid DNA available will suffice), ^{35}S-methionine, low molecular weight mix, magnesium acetate, and buffer in a 1.5 ml reaction tube and incubate at 37°C for 5 min.

Add the S30 extract, which should be prewarmed to 37°C for a few minutes before use.

Continue the incubation for a further 30 min and then add 5 µl of 44 mg/ml solution of methionine.

Allow a further 5 min at 37°C before transferring the reaction tubes to ice.

The magnesium concentration and amount of S30 extract at which maximal protein synthesis occurs is determined by measuring incorporation of radioactivity into trichloroacetic acid (TCA)-precipitable material.

Place 2 µl samples from each reaction mixture on numbered squares of filter paper and wash the filters in 10% TCA supplemented with 100 µg/ml methionine at 4°C for 30 min.

Discard the TCA, add fresh 5% TCA with methionine and heat at 90°C for 10 min.

Wash the filters again 3 × at 4°C with fresh TCA supplemented with methionine, then once in acetone, and allow to dry.

Place filters in a vial containing scintillant and count in a scintillation counter.

Table 1. Optimization of the magnesium concentration

Sample	DNA	^{35}S met	LMM	Mg	T/A	S30
1	5 µl	2	7.5	1	9.5	5 = 30 µl
2	5 µl	2	7.5	1.5	9	5
3	5 µl	2	7.5	2.0	8.5	5
4	5 µl	2	7.5	2.5	8	5
5	5 µl	2	7.5	3.0	7.5	5
6	5 µl	2	7.5	3.5	7	5
7	5 µl	2	7.5	4.0	6.5	5
8	5 µl	2	7.5	1.5	6	8
9	5 µl	2	7.5	2.5	5	8
10	5 µl	2	7.5	3.5	4	8
11	–	2	7.5	2.5	13	5
12	–	2	7.5	2.5	10	8

Requirements: DNA (preferably 300–500 µg/ml; ^{35}S-methionine 7 µCi/µl; LMM – low molecular weight mix; Magnesium acetate 0.1 M; T/A buffer, 1 mM Tris-acetate, pH 7.0; S30 extract

Conditions that give optimal incorporation of radioactivity should be adopted for subsequent experiments. One can expect a 15–50-fold stimulation of protein synthesis when DNA is added and the 2 µl sample should contain more than 50,000 cpm.

3. Experiment 2: Identification of Polypeptides Coded by Cloned Sequences Similar in Molecular Weight to Those Specified by the Vector

a) Introduction and Objectives

In this experiment, the S30 extract previously prepared will be used to analyze polypeptides encoded by a plasmid (pLG510) carrying the *env*A gene of *E. coli*. This plasmid was constructed by ligating a 2.5 kb *Eco*RI-generated fragment from the 2 min region of the *E. coli* chromosome into the unique *Eco*RI site of vector plasmid pKN410. This low copy number vector is a miniderivative of the large plasmid R1, specifies resistance to ampicillin, and carries a mutation that causes "runaway replication" of the plasmid in bacteria incubated at 37°C (Uhlin et al. 1979). Bacteria harboring this plasmid and its derivatives are propagated at 30°C, but plasmid DNA is extracted after incubation at 37°C for 6 h. The experiment, which should run concurrently with experiment 3, involves using plasmids pLG510 and pKN410 and the purified 2.5 kb cloned fragment to identify polypeptides encoded by the latter.

b) Procedure

If necessary, the various DNA templates may be prepared on days 1 and 2 while the culture of MRE600 is growing.

Day 6

Cleave 25 μg of pLG510 DNA with *Eco*RI and separate the 2.5 kb fragment from the vector molecule (15.4 kb) on a 0.5% preparative agarose gel.

Purify the 2.5 kb fragment from the gel as follows:

Cut a slice of agarose containing the 2.5 kb fragment from the gel and carefully place in a dialysis bag containing 1 ml of 1/10 concentration electrophoretic buffer supplemented with 1 μg ethidium bromide per ml.

Expell excess buffer, close the bag, and place in an electrophoretic tank containing 1/10 electrophoretic buffer.

Apply a constant current of 80 mA until the DNA has completely migrated out of the gel slice and has accumulated on the inner wall of the dialysis bag.

Reverse the current for 30 s, remove the bag from the tank, and gently agitate to dissolve the DNA in the buffer.

Remove the buffer from the bag without taking any agarose.

Reduce the volume of buffer 10-fold by repeated extraction with butan-2-ol (also effectively extracts ethidium bromide from the DNA).

This procedure yields about 2.5 μg of DNA fragment which should be ethanol precipitated and redissolved in 5 μl of 1 mM Tris-Cl, pH 7.5, 0.1 mM EDTA (1/10 TE). Ethanol precipitate also 2.5 μg of pLG510 and pKN410 plasmid DNA, again resuspending in 5 μl of 1/10 TE.

Day 7

Set up the in vitro protein synthesis reactions with the DNA from day 6 and the procedure adopted on day 5 (see Table 1) using the magnesium concentration and amount of S30 extract that gave maximal incorporation of radioactivity. Include a reaction lacking DNA as control.

Process the samples as described for day 5 and determine TCA precipitable counts.

Add 30 μl of sample buffer to the remainder of the sample and boil for 5 min.

Store the boiled sample at $-20°C$ until ready for analysis by polyacrylamide gel electrophoresis.

Reboil prior to loading on the gel.

While the above reactions are incubating, prepare a polyacrylamide gel as described in Chap. 4.1, adjusting the volumes of acrylamide and water so that the final concentration of acrylamide is 11%.

Load samples into the slots of the gel using as many counts per slot as possible, but ensure (1) that all slots have an equal number of counts and (2) that not more than one half of any sample is used.

For the control sample, load 30 μl. Include in the gel molecular weight standards (^{14}C-labeled polypeptides).

Run the gel and fix overnight as described in Chap. 4.1.

Day 8

Dry down the fixed gel onto thick filter paper and expose to X-ray film.

Day 9

Develop the X-ray film and analyze results. Repeat the autoradiography for a longer period of time if the exposure has been too brief.

c) Results and Discussion

An autoradiograph obtained from such an experiment is shown in Fig. 1. When the products obtained in vitro from plasmid pLG510 (Track 2) are compared with those from the vector (Track 3), no clear differences are evident. Plasmid pKN410 directs the synthesis of two polypeptides of apparent molecular weight M_r 31,500 and 32,500. The smaller polypeptide can be shown to be β-lactamase by immunoprecipitation. If pKN410 is linearized by cleavage with EcoRI and used as a template, the M_r 32,500 polypeptide is not made (data not shown). Its coding sequence presumably, therefore, spans the EcoRI cleavage site. Thus, although insertion of the 2.5 kb fragment into this EcoRI site prevents synthesis of the M_r 32,500 polypeptide of the vector, a similar-sized product encoded by the cloned sequence is synthesized. This conclusion is confirmed by the fact that the purified 2.5 kb fragment directs the synthesis of an M_r 32,500 polypeptide, which additionally indicates that this fragment carries the natural promotor for this polypeptide. A recent study (Lutkenhaus and Wu 1980) has shown the M_r 32,500 polypeptide to be the envA gene product.

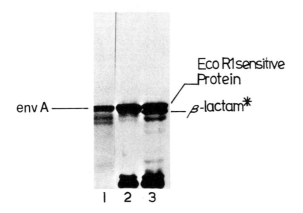

Fig. 1. Identification of the E. coli envA gene product. Polypeptides were labeled in the in vitro system and analyzed by SDS-PAGE. The system was programed with: *1* the 2.5 kb EcoRI fragment from pLG510; *2* PLG510; *3* pKN410. β-lactam* refers to the precursor form of β-lactamase and EcoRI sensitive protein refers to the protein that is lost when pKN410 cleaved with EcoRI is used as a template in vitro

4. Experiment 3: Mapping Polypeptide Coding Sequences on a Staphylococcal Plasmid

a) Introduction and Objectives

This experiment exploits the ability of an E. coli extract to support coupled transcription-translation of heterologous DNA. pC221 is a naturally occurring plasmid of

Identification of Gene Products by Coupled Transcription-Translation of DNA Fragments   243

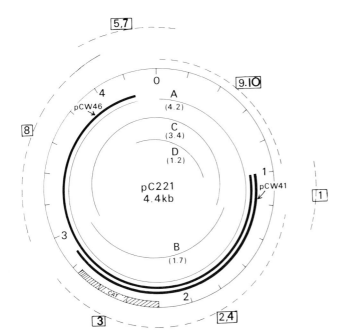

Fig. 2. Map of *S. aureus* plasmid pC221. pCW46 and pCW41, two plasmids derived from pC221, are shown in *thick lines*. *Thin lines* represent 4 restriction enzyme-generated fragments (sizes in kb in parentheses) of pC221. *A* an *Eco*RI-*Hin*dIII fragment; *B Hpa*II fragment; *C Hpa*II fragment; *D Mbo*I fragment. As a result of the analysis of protein synthesis programed by these templates (Fig. 3), the approximate locations *(broken lines)* of 9 polypeptides are shown in *boxes*. Polypeptide 3, which is coded by all three plasmids and fragment B was deduced to be CAT

Staphylococcus aureus that specifies resistance to chloramphenicol. Two pC221 mini-derivatives, pCW41 and pCW46, have been constructed and both code for chloramphenicol resistance (Fig. 2, Wilson et al. 1981). Use of these plasmids and several fragments of pC221 in the in vitro system allows the localization of the sequences coding for several polypeptides on the restriction map of pC221. This experiment should run concurrently with Experiment 2.

b) Procedure

Day 6

If necessary, the various DNA templates may be prepared on days 1–2.

Ethanol precipitate 2.5 µg each of pC221, pCW41, and pCW46 and resuspend the dried DNA pellets in 5 µl of 1/10 TE.

Purify the following DNA fragments from pC221, using the method described previously:

Cleave 5 µg of pC221 with both *Eco*RI and *Hin*dIII, submit to electrophoresis in a 0.7% agarose gel and isolate the 4.2 kb fragment (fragment A).

Cleave 10 µg of pC221 with *Hpa*II and purify the 1.7 kb and 3.4 kb fragments (fragments B and C, respectively).

Cleave 20 µg of pC221 with *Mbo*I and isolate the 1.2 kb fragment (fragment D).

Ethanol precipitate all purified DNA fragments (about 2.5 µg each) and redissolve in 5 µl of 1/10 TE.

Days 7–9

Proceed exactly as described for days 7–9 in Experiment 2.

c) Results and Discussion

The result of such an experiment is shown in Fig. 3. In the *E. coli* in vitro system, plasmid pC221 specifies at least 11 polypeptides that range in size from M_r 38,000 to M_r 11,000. The M_r 26,000 polypeptide has tentatively been identified as chloramphenicol transacetylase, since DNA sequencing studies indicate a molecular weight of 25,900 for this polypeptide (Shaw, W.V., personal communication).

From Fig. 3, it can be seen that:

a) pCW41 encodes polypeptides 1, 2, 3, 4, 6;
b) pCW46 encodes polypeptides 1, 2, 3, 4, 8;
c) Fragment A encodes polypeptides 1, 2, 3, 4, 8, 9, 10;
d) Fragment B encodes polypeptides 2, 3, 4, 6, 11;
e) Fragment C encodes polypeptides 5, 6, 7, 8, 9, 10, 11;
f) Fragment D encodes polypeptides 5, 9, 10.

Given the map positions of the DNA fragments and the DNA segments present in the derivatives of pC221, the regions coding for 9 of the 11 polypeptides specified by pC221 can be localized (Fig. 2). It is not possible, however, to locate the sequences coding for polypeptides 6 and 11 to a unique region of pC221. It is interesting to note that polypeptides 1 and 2 appear to be expressed more efficiently from the small plasmids pCW41 and pCW46 than from pC221.

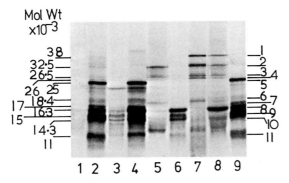

Fig. 3. Polypeptides programed by *S. aureus* DNA in vitro. Equimolar amounts of supercoiled plasmid DNA or fragments were used in the in vitro system. Track *1* no DNA; *2* pC221; *3 Mbo*I fragment D; *4 Hpa*II fragment C; *5 Hpa*II fragment B; *6 Eco*RI-*Hin*dIII fragment A; *7* pCW41; *8* pCW46; *9* pC221. Two polypeptides, 6 and 11 could not be unambiguously assigned to a position on the pC221 map shown in Fig. 2. Polypeptides 1 and 2 appeared to be expressed more efficiently from the small plasmids than from pC221 or fragment A

5. Final Comments

The protein synthesis system described above has a number of advantages over currently used in vivo systems. The preparation of large stocks of both the S30 extract and the low molecular weight mix assures perfect reproducibility of the assay. Identification of polypeptides coded by a few micrograms of a given DNA sequence then becomes a simple matter of incubating the various components of the system and analyzing the products by SDS-polyacrylamide gel electrophoresis and autoradiography. Because the in vitro system does not require a supercoiled template, it is a powerful tool for the localization of sequences coding for polypeptides on a given segment of DNA. Conversely, it may be used to identify intra- and extra-gene restriction enzyme targets.

Extracts that are active in plasmid DNA directed protein synthesis have been prepared from several *E. coli* strains, including some that are RNAase$^+$ (Pratt, J.M., unpublished data). The ability to produce active extracts from a range of *E. coli* K-12 mutants should allow detailed studies on the control of gene expression in vitro. The successful transcription-translation of *S. aureus* DNA indicates that the in vitro system might also be used for the study of a range of heterologous bacterial DNAs. Preliminary results indicate that *B. subtilis* DNA is also expressed in the system (S. Eccles, personal communication).

The efficiency of a linear template may be as low as 25% of that of the equivalent supercoiled template in its ability to direct protein synthesis in the in vitro coupled system. Furthermore, differential effects on the amounts of specific products synthesized from linear and supercoiled templates may be observed. Experiments involving supercoiled and non-supercoiled circular templates and recovery of input linear DNA from the extract show that this reflects exonucleolytic degradation of DNA fragments rather than an absence of supercoiling per se. Attempts to reduce this degradation by using a *recB recC* mutant as a source of extract have proven unsatisfactory; such extracts have a high intrinsic level of protein synthesis directed by contaminating chromosomal DNA fragments. However, if these are removed by purifying the components of the transcription-translation machinery, the decreased exonucleolytic activity of the *recB recC* mutant can be exploited (Yang et al. 1980). Reduced exonucleolytic degradation, and therefore, maximal expression of genes on linear templates, may be achieved more simply by preparing extracts from a *recB*$_{ts}$ mutant grown at 30°C (Jackson, Pratt and Holland, in preparation). The preincubation step is performed at 30°C for 180 min during which time chromosomal DNA fragments are degraded by the *recB*, C-nuclease. Reactions are then carried out at 37°C where this enzyme is inactive.

6. Materials

Day 1

100 ml sterile growth medium contains per liter: 5.6 g KH_2PO_4; 28.9 g K_2HPO_4; 1 g yeast extract; 1.5 mg thiamine, 50 mg each of arginine, methionine, leucine, and histidine. After autoclaving, 40 ml of sterile 25% glucose and 10 ml of 100 mM magnesium acetate is added.

Day 2

Fermenter and materials for making 10 liter growth medium

Day 3

S30 buffer (10 mM Tris-acetate, pH 8.2, 14 mM magnesium acetate, 60 mM potassium acetate, 1 mM DTT) containing 0.5 ml mercaptoethanol/liter

Day 4

S30 buffer supplemented with mercaptoethanol (0.5 ml/liter)
S30 buffer
0.1 M DTT
Preincubation mix
Dialysis tubing

Day 5

Plasmid DNA (e.g., pBR325)
^{35}S-methionine (7 μCi/μl)
Low molecular weight mix

Day 6

Plasmid DNA: pKN410 (2.5 μg), pLG510 (30 μg), pC221 (40 μg), pCW41 (2.5 μg), pCW46 (2.5 μg)
Restriction enzymes: *Eco*RI, *Hin*dIII, *Hpa*II, *Mbo*I
Materials for purification of DNA fragments from agarose gels

Days 7–8

^{35}S-methionine
Low molecular weight mix
0.1 M magnesium acetate
Materials for polyacrylamide gels and autoradiography

7. References

Collins J (1979) Cell-free synthesis of proteins coding for mobilisation functions of ColE1 and transposition functions of Tn3. Gene 6:29–42

Lutkenhaus JF, Wu HC (1980) Determination of transcriptional units and gene products from the *fts*A region of *Escherichia coli*. J Bacteriol 143:1281–1288

Pratt JM, Boulnois GJ, Darby V, Orr E, Wahle E, Holland IB (1981) Identification of gene products programmed by restriction endonuclease DNA fragments using an *E. coli* in vitro system. Nucleic Acids Res 9:4459–4474

Ptashne M (1967) Isolation of the λ phage repressor. Proc Natl Acad Sci USA 57:306–313

Reeve JN (1978) Selective expression of transduced or cloned DNA in minicells containing plasmid pKB280. Nature 276:728–729

Sancar A, Hack AM, Rupp WD (1979) Simple method for identification of plasmid-coded proteins. J Bacteriol 137:692–693

Uhlin BE, Molin S, Gustafsson P, Nordström K (1979) Plasmids with temperature-dependent copy number for amplification of cloned genes and their products. Gene 6:91–106

Wilson CR, Skinner SE, Shaw WV (1981) Analysis of two chloramphenicol resistance plasmids from *Staphylococcus aureus:* Insertional inactivation of Cm resistance, mapping of restriction sites and construction of cloning vehicles. Plasmid 5:245–258

Yang H-Y, Ivashkiv L, Chen HZ, Zubay G, Cashel M (1980) Cell-free coupled transcription-translation system for investigation of linear DNA segments. Proc Natl Acad Sci USA 77:7029–7033

Zubay G (1973) In vitro synthesis of protein in microbial systems. Annu Rev Genet 7:267–287

Chapter 5   DNA Sequencing

G. VOLCKAERT[1], G. WINTER[2], and C. GAILLARD[3]

Contents

1. General Introduction 250
2. Experiment 1: End Labeling with Polynucleotide Kinase 252
   - a) Introduction 252
   - b) Procedure 1: Dephosphorylation 252
   - c) Procedure 2: Forward Kinase Reaction 253
   - d) Procedure 3: Segregation of the Label 253
3. Experiment 2: End Labeling by "Filling In" Single Stranded Ends 255
   - a) Introduction 255
   - b) Procedure 257
4. Experiment 3: Isolation of Labeled DNA Fragments 258
   - a) Introduction 258
   - b) Procedure 1: Polyacrylamide Gel Electrophoresis 259
   - c) Procedure 2: Elution 260
   - d) Procedure 3: Precipitation 260
5. Experiment 4: Sequencing by Chemical Degradation 261
   - a) Introduction 261
   - b) Procedure 1: Degradation 262
   - c) Procedure 2: Thin Sequencing Gels 264
   - d) Procedure 3: Fixing the Gel and Autoradiography 266
   - e) Procedure 4: Sequence Reading 266
6. Convenient Sources of Some Products and Materials 267
7. Experiment 5: Dideoxy Sequencing on M13 Templates 268
   - a) Introduction 268
   - b) Procedure 1: Hybridization 270
   - c) Procedure 2: Preparing the Gel 270
   - d) Procedure 3: Polymerization 270
   - e) Procedure 4: Loading, Running and Fixing the Gel 271
   - f) Procedure 5: Reading a Sequence 272
8. Materials and Products 273
9. Addendum 1: Further Developments in Rapid DNA Sequencing 275
10. References 278

---

1 Rega Instituut, K.U. Leuven, Minderbroedersstraat 10, B-3000 Leuven, Belgium
2 MRC Laboratory of Molecular Biology, Cambridge, England
3 IRBM, Paris, France

1. General Introduction

DNA consists of a continuous chain of deoxyribose units linked via phosphodiester bonds at their 5' and 3' positions and joined to one of the four nucleotide bases, Thymine (T), cytosine (C), the pyrimidines, guanine (G), or adenine (A), the purines, at their 1' positions. It exists in organisms in the form of very long chains (5,000 base-pairs long, in the case of bacteriophage ϕX174, to 240,000,000 bp long, in the case of the largest human chromosome). Thus, in order to determine the sequence of a DNA molecule, the order of nucleotide bases along its sugar-phosphate backbone, it is first necessary to fragment the molecule into segments of a manageable size and to purify each fragment type. This is most readily accomplished by cloning into a plasmid or viral DNA vector. After amplification of the resulting hybrid DNA molecule, the cloned DNA segments (inserts) can be released for sequencing by cleavage with a restriction endonuclease.

The chemical and enzymatic strategies, which have been used to determine the sequence of DNA inserts, are based on three critical procedures:

1. The use of an α^{32}P of γ^{32}P-labeled nucleoside triphosphate to radiolabel the polynucleotide in vitro (at its 5' or 3' end, or internally).
2. The use of a synthetic or degradative technique to generate from the labeled polynucleotides four or more sets of fragments having a fixed common end. Each set consists of a spectrum of fragments with one variable end terminating at or before a particular nucleotide base or set of nucleotide bases. In the simplest case, four sets terminating at T, C, G, or A positions are produced.
3. The use of polyacrylamide gel electrophoresis to separate single stranded DNA fragments differing in length by only one nucleotide.

The sequence of the polynucleotide can then be read by comparison of the different separation patterns (ladders) of the four sets that are revealed by autoradiography of the polyacrylamide gel (Fig. 1).

In the chemical degradation approach (Maxam and Gilbert 1977, 1980), a DNA restriction fragment is first radiolabeled at its 5' or 3' end. Ordinarily, the DNA duplex is thereby labeled in both strands and it is, thus, necessary to segregate the label such that subsequent reactions are carried out with molecules labeled in only one strand. The DNA is then partially degraded in a set of base-specific chemical reactions, which result in excision of the derivatised base and cleavage of the DNA molecule.

In the enzymatic approach (Sanger et al. 1977), an oligonucleotide primer is hybridized to a single strand DNA template adjacent to the target region to be sequenced. The primer is extended across the target by DNA polymerase in the presence of α^{32}P-dATP, to label the newly synthesized strand and dideoxynucleoside triphosphate analogues (ddNTPs) plus deoxynucleoside triphosphates (dNTPs). The ddNTPs compete with their respective dNTPs for incorporation into the growing polynucleotide and, when incorporated, specifically terminate the chain.

The chemical sequencing method starts with the labeling of the restriction fragment at either its 5' end, using polynucleotide kinase and γ^{32}P-rATP (Sec. 2) or at its 3' end, usually by a "3' fill-in" or "3' exchange" reaction with α^{32}P-dNTPs and DNA polymerase (Sec. 3). Labeled fragments are isolated as described in Sec. 4 and chemically degraded as in Sec. 5.

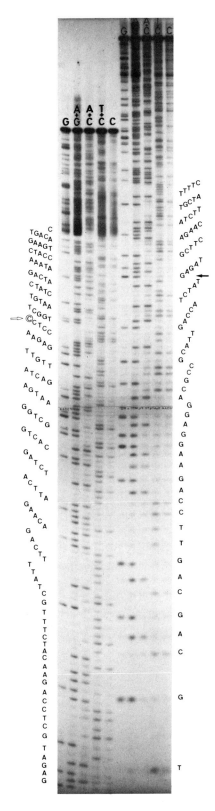

Fig. 1. A 90 cm long 10% sequencing gel with two Maxam and Gilbert loadings. The specificities of the lanes are identified on top of each set. The top of the picture does not correspond to the actual position of the sample wells. The second *(right)* set (of the same degradations) was loaded when the xylene cyanol dye had moved about 60 cm of the gel distance and electrophoresis was continued until bromphenol blue (in this second set) was nearly half-way. One nucleotide has run off. The *filled arrow* shows the postion from where reading can be continued on the first *(left)* sample series. The *encircled C*, indicated by an *open arrow*, represents a gap in the ladder corresponding to a 5-methylated C (see text). Note that this nucleotide is part of an *Eco*RII restriction site

2. Experiment 1: End Labeling with Polynucleotide Kinase

a) Introduction

5' termini are labeled with γ^{32}P-ATP and T4 polynucleotide kinase (Richardson 1965, Lillehaug et al. 1976), usually after dephosphorylation by bacterial alkaline or calf intestinal phosphatase. The kinase exchange reaction described by Berkner and Folk (1977) eliminates the dephosphorylation step, but requires higher amounts of the ATP label and the absolute concentration of ATP is more critical. The kinase exchange reaction can, however, be a useful procedure for the labeling of large amounts of DNA.

Hybrid plasmid purified only by CsCl gradient centrifugation may contain degraded RNA which competes for label in the kinase reaction and which may result in poor labeling of the DNA. Such RNA can be removed from DNA by sucrose gradient centrifugation or by digestion with RNase(s) followed by chromatography on Sephadex G50.

Although kinase efficiently labels single stranded DNA, the reaction is mostly carried out with double stranded restriction fragments, resulting in the introduction of label at both ends. A second restriction digest to segregate the label may be carried out immediately without prior fractionation of the fragments (method 1). Alternatively, the kinased fragments may be fractionated on a polyacrylamide gel (4–10%), extracted, and individually recut with a suitable restriction enzyme (method 2). Strand separation is an alternative procedure to segregate the label if no suitable restriction sites are available. The separation relies, however, on differences in base composition and local secondary structures in the separated complementary strands and is unpredictable. The protocol described later (method 3) is taken from Maxam and Gilbert (1980) and is often successful.

b) Procedure 1: Dephosphorylation

Restrict DNA in a suitable buffer (usually 0.01 M Tris-HCl, pH 7.6, 0.007 M $MgCl_2$, 0.007 M β-mercaptoethanol, but some enzymes require different conditions) with a restriction enzyme in a total volume of 120 μl.

After incubation, add to the reaction mixture 14 μl 1% SDS, 4 μl 1 M Tris-HCl, pH 8.0, and 4 μl bacterial alkaline phosphatase, BAPF (Worthington BAPF in ammonium sulfate suspension, spun down and redissolved in water at a concentration of 1 mg/ml).

Incubate 30–60 min at 56°C.

Add 100 μl distilled phenol (saturated with 50 mM Tris-HCl, pH 8.), mix on a vortex, and spin for 2 min.

Remove the lower phase with a drawn-out Pasteur pipette.

Add 75 μl phenol and repeat the extraction using another drawn-out Pasteur pipette.

Repeat once more with 75 μl phenol.

Extract twice with ether (in excess), discard the upper phases.

Add 15 μl 3.75 M NaCl, mix, add 400 μl ethanol and mix.

Chill at $-70°C$ (dry ice-ethanol bath) for 10 min.
Spin down precipiate, wash pellet with 80% ethanol, and spin again.
Dry in a dessicator.

c) Procedure 2: Forward Kinase Reaction

Dry down γ^{32}P-ATP (in excess, if possible in a threefold molar excess) in a siliconized Eppendorf tube.
Redissolve the ATP in 44 μl H_2O (freshly boiled and cooled) and transfer to the Eppendorf tube containing the dephosphorylated DNA pellet.
Add 5 μl of 0.2 M Tris-HCl, pH 7.6, 0.1 M $MgCl_2$, 0.14 M β-mercaptoethanol and mix.
Add 1 μl kinase (1–3 units).
Mix gently and incubate for 25 min at 37°C.

Alternative 1

With isolated single fragments, the following rapid method can be employed.
Dephosphorylate with BAP (0.5 μl) in 10 μl 50 mM Tris-HCl, pH 8.0 (in the absence of SDS : BAP must be DNase-free!).
Incubate 30 min at 37°C.
Add 0.5 μl 0.1 M NTA (nitrilo-tri-acetic acid).
Mix and heat for 2 min at 100°C.
Quick-chill in ice water.
Add 3 μl 0.2 M Tris-HCl, pH 7.6, 0.1 M $MgCl_2$, 0.15 M β-mercaptoethanol and transfer the dried-down γ^{32}P-ATP to this tube in 16 μl water.
Add 1–3 units kinase and incubate at 37°C for 25 min.

Alternative 2 (Exchange kinase reaction)

Dry down γ^{32}P-ATP in a siliconized Eppendorf tube (200 pmoles or more).
Redissolve in 19 μl 55 mM Imidazole-HCl, pH 6.5, 50 mM KCl, 20 mM $MgCl_2$, 5 mM β-mercaptoethanol.
Transfer to the Eppendorf tube containing the DNA pellet.
Add 2 μl 3 mM ADP and 2 μl kinase (6 units).
Incubate 15 min at 37°C.

d) Procedure 3: Segregation of the Label

Method 1

Inactivate the kinase by heating at 65°C for 10 min.
Add (30–x) μl water and x μl restriction enzyme for the second cleavage (where necessary, adapt the buffer composition accordingly, but note that some changes can disturb the following lanthanum precipitation step) and incubate accordingly.
Add 10 μl 3.75 M NaCl, mix, add 10 μl 0.1 M $La(NO_3)_3$, and mix on a vortex for 20 s.

Keep for 10 min at 0°C (ice water bath).
Spin down the precipitate for 2 min in an Eppendorf centrifuge.
Discard the supernatant and wash the pellet with 100 µl cold 10 mM NaAc, pH 4.0.
Immediately, take up the pellet in 20 µl 0.1 M EDTA.
Add 5 µl running gel buffer (see Sec. 4) and mix vigorously.
Add 5 µl of a solution containing 50% glycerol, 1% Ficoll, 0.05% xylene cyanol, and 0.05% bromophenol blue.
Mix and load onto a polyacrylamide gel (Sec. 4).

Note: Alternatively, the second restriction digest may be carried out after dephosphorylation, but before kination; the newly formed 5' ends are phosphorylated, and hence, will not be labeled if the BAP has been completely removed. The main advantage of this approach is that DNA fragments can be prepared in advance, ready for immediate kinase reaction when the labeled ATP is delivered. It is, however, critical that the sample(s) and the second restriction enzyme are completely phosphatase-free.

Method 2

Add 5 µl 1% SDS and 5 µl of a solution containing 50% glycerol, 1% Ficoll, 0.05% xylene cyanol, and 0.05% bromophenol blue.
Mix and load onto a polyacrylamide gel (Sec. 4).
Run the gel, autoradiograph and extract the fragments according to Sec. 4.
Proceed to a second restriction reaction for each fragment and load the digests again onto a separating, non-denaturating, polyacrylamide gel as before.

Method 3

Load the kination mixture onto a polyacrylamide gel as described in method 2 and isolate the required fragments (Sec. 4).
Dissolve 1 µg labeled DNA fragment in 40 µl 30% dimethylsulfoxide, 1 mM EDTA, 0.05% xylene cyanol, 0.05% bromophenol blue (adapt the volume to the amount of DNA, as appropriate).
Heat at 90°C for 2 min and quick-chill in ice water.
Immediately load into a sample well, having a cross-section of at least 25 mm^2, of a prerun (30 min, 300 V, no warming up!) polyacrylamide gel of large pore size (final composition of 5% acrylamide, 0.1% methylene bis-acrylamide, 50 mM Tris-borate, pH 8.3, 1 mM EDTA; running buffer, 50 mM Tris-borate, pH 8.3, 1 mM EDTA).
Submit to electrophoresis at low voltage to avoid heating of the gel; under such conditions the single stranded fragments move slower than their double stranded counterparts.
Autoradiograph the gel and elute the fragments as described in Sec. 4.

Note: Strand separation is also possible with mixtures of DNA fragments; for this, treat the kination mixture as follows:

Add water to a final volume of 120 µl.
Add 100 µl phenol, mix on a vortex, spin for 2 min. Remove the lower phase with a drawn-out Pasteur pipette.

Extract twice with excess ether, discard the upper phases.
Add 15 µl 3.75 M NaCl, mix, add 400 µl ethanol and mix.
Chill at −70°C for 10 min (dry ice-ethanol bath).
Spin down the precipitate.
Redissolve in 100 µl 0.3 M NaAc, pH 7.5, add 250 µl ethanol, and mix.
Chill again at −70°C for 10 min and spin down precipitate.
Wash the pellet with 80% ethanol and spin again.
Dry in dessicator.
Strand separation is performed as described above. At least 10 µg of DNA fragments may be loaded in a sample well with a cross-section of 25 mm^2.

3. Experiment 2: End Labeling by "Filling In" Single Stranded Ends

a) Introduction

Several methods for the labeling of DNA fragments by extension of their 3′ termini in the presence of α^{32}P-labeled πNTP or dNTP are available and various reaction conditions have been described. For example, calf thymus terminal transferase can be used to elongate single stranded and double stranded DNA, irrespective of the nucleotide sequence at the 3′ end (Kössel and Roychoudhury 19171, Roychoudhury et al. 1976). Excess ribonucleotides may be removed by hydrolysis. Single addition reactions are also possible under certain conditions (Olson and Harvey 1975, Tu and Cohen 1980). The main procedure currently employed involves the use of DNA polymerase to fill in the sticky ends of restriction enzyme generated DNA fragments (Wu 1970, Wu and Taylor 1971, Kleppe et al. 1971, Donelson and Wu 1972). The same enzyme may be used to replace the 3′ nucleotides of flush-ended fragments by their ^{32}P-labeled counterparts (Englund 1972, Donelson and Wu 1972). T4 DNA polymerase, reverse transcriptase, *E. coli* DNA polymerase I, or the corresponding Klenow fragment have all been used for this purpose. End labeling by these procedures, which involves DNA synthesis on a template, is limited to double stranded DNA fragments. The nucleotide sequence at the restriction cleavage site will clearly determine the choice of labeled α^{32}P-dNTP, as illustrated in the following protocols. It is important that the polymerase activity of the enzyme exceeds its 3′ → 5′ exonuclease activity, and therefore, an excess of unlabeled nucleoside triphosphate (protecting nucleotide), corresponding to the 3′ nucleotide, is often added to the reaction. The following protocol is based on that described by Schwarz et al. (1978). The basic protocol (method 1) should be adapted to each type of DNA fragment by selection of the appropriate protecting nucleotide and α^{32}P-dNTP. Labels introduced at both 3′ ends must be segregated as in Sec. 2.

It is often possible to label a single 3′ end by carefully choosing the α^{32}P-dNTP and/or the second restriction digest (methods 2 and 3), so that the second cleavage site cannot accept a labeled nucleotide. Table 1 gives a compilation of some possible combinations of restriction enzymes, labeled dNTP, and protecting dNTP(s). The choice of the protecting nucleotide is arranged so that after introduction of the label, the next addition is blocked. This prevents heterogeneity at the labeled end.

Table 1. Some combinations of restriction digests and dNTPs to allow direct single labeling of DNA fragments by filling in polymerization

Labeled dNTP	Site to be labeled	Second enzyme	Protecting dNTP(s)
dATP	HinfI Sau3A BglII BamHI BclI	SalI AluI PvuII FnuDII HaeI HaeIII	dGTP
		MspI XhoI XmaI SmaI	dGTP + dCTP
	HinfI	TaqI ClaI XbaI	dGTP + dTTP
dCTP	TaqI ClaI SalI	RsaI BclI	dTTP
		BglII HindIII	dTTP + dATP
		BamHI EcoRI HinfI AluI PvuII FnuDII HaeI HaeIII HpaI	dTTP + dGTP
dTTP	SalI	Sau3A BglII BamHI EcoRI AluI PvuII FnuDII HaeI HaeIII HpaI	dGTP
dGTP	BclI	TaqI ClaI XbaI HpaI RsaI	dTTP
dTTP	BstNI	–	dCTP See method 2

DNA Sequencing

Table 1 (cont.)

Labeled dNTP	Site to be labeled	Second enzyme	Protecting dNTP(s)
dTTP/dATP	*Ava*II	–	dGTP (compare with method 2)
dTTP/dATP	*Asu*I	–	dGTP [a]
dTTP/dATP/dGTP	*Fnu*4HI	–	dGTP a

[a] In the examples at the end of the table, separate labeling experiments with different labeled dNTP(s) are run in adjacent lanes on the gel. Since the nucleotide to be filled in can be either one of the four (for *Asu*I and *Fnu*4HI) not all fragments will get labeled

b) Procedure

The protocol starts with suitable fragments prepared by restriction digestion of the DNA, deproteination, and ethanol precipitation to produce a salt-free pellet.

Method 1: *Hin*fI recognition sequence: G–A–N–T–C
*Hin*fI generated fragment ends: 5' pA–N–T–C–

Dry down α^{32}P-dATP (in molar equivalence or excess to 3' ends to be labeled) in a siliconized Eppendorf tube.
Add to the DNA pellet:

2 µl 0.2 M Tris-HCl, pH 7.6, 0.1 M MgCl$_2$
5 µl 0.01 M dithiothreitol
5 µl 1 mM dGTP
5 µl H$_2$O

Mix to dissolve the DNA completely and transfer to the tube containing the dried-down dATP.
Chill to 0°C and add 3 µl Klenow fragment polymerase (1–2 units).
Mix gently and incubate immediately for 5 min at 25°C.
Stop reaction by adding 1 µl 1% SDS.
Add 3 µl of a solution containing 50% glycerol, 1% Ficoll, 0.05% xylene cyanol and 0.05% bromophenol blue.
Fractionate on a polyacrylamide gel and autoradiograph (Sec. 4).

The resulting fragments are labeled at both 3' ends. Label must be segregated by a second restriction digest or strand separation (see Sec. 2, Procedure 3).

Method 2: *Bst*NI recognition sequence C–C–A/T–G–G
*Bst*NI-generated fragment ends: 5' pT–G–G– and 5' pA–G–G–
3' C–C   3' C–C

Dry down α^{32}P-dATP and α^{32}P-dTTP separately in two siliconized Eppendorf tubes (in molar equivalence or excess to ends to be labeled).

Add to the DNA pellet:

    4 μl 0.2 M Tris-HCl, pH 7.6, 0.1 M MgCl$_2$
    10 μl 0.01 M dithiothreitol
    10 μl 1 mM dCTP
    10 μl H$_2$O

Mix to dissolve the DNA completely.

Transfer 17 μl of this solution to each of the tubes containing the dried-down dATP or dTTP.

Proceed further as in method 1 and load the samples in adjacent slots onto a polyacrylamide gel.

The resulting fragments will be either single or double labeled. Single labeled fragments are those visible in both lanes (bearing the label at a different 3′ end of the fragment). Double labeled fragments are only represented in one of the reactions and must be further cleaved or strand separated (Sec. 2, procedure 3).

Note: With complex mixtures, not all co-migrating fragments will necessarily be single end labeled!

Method 3: Filling in of fragments from combined digestions (see Table 1).
    *Sau*3A site (G–A–T–C) to be labeled.
    Generated ends: 5′ pG–A–T–C– and 5′ pC–G–G–
                                3′–   3′ C–

Dry down α^{32}P-dATP in a siliconized Eppendorf tube.

Add to the DNA pellet:

    2 μl 0.2 M Tris-HCl, pH 7.6, 0.1 M MgCl$_2$
    5 μl 0.01 M dithiothreitol
    5 μl 1 mM dCTP
    5 μl 1 mM dGTP

Proceed as in method 1.

To determine which fragments are single labeled, a reference sample of the original DNA digested only with *Sau*3A is prepared and run in a lane adjacent to the labeled DNA. After autoradiography and excision of the labeled fragments, the gel is stained with ethidium bromide to visualize the fragments of the reference digest. Radioactive fragments for which there is no corresponding band in the reference digest must be single labeled.

4. Experiment 3: Isolation of Labeled DNA Fragments

a) Introduction

Labeled DNA fragments are usually fractionated by non-denaturing polyacrylamide gel electrophoresis. Tris-acetate is a convenient buffer system for this purpose. An acrylamide concentration of 5% is appropriate for medium-sized DNA fragments (between 150 and 600 bp), but should be higher for short ones.

Isolation of fractionated DNA fragments from polyacrylamide gels requires a rapid procedure that permits the simultaneous pressing of many fragments. The procedure must also concentrate the DNA, be as quantitative as possible, and not result in degradation of the DNA. The procedure described fulfils these requirements, although the final solution of DNA may still contain so-called soluble acrylamide, which can inhibit further enzyme digestions. To reduce this inhibition, double labeled fragments can be further purified by precipitation with lanthanum nitrate (Molemans et al., in preparation). An alternative way to minimize the problem of soluble acrylamide involves eluting DNA fragments from non-denaturing thin ($>$ 10%) polyacrylamide gels (Sanger and Coulson 1978). The strip of gel is not crushed, but gently soaked overnight at 37°C in 10 vol of 0.5 M ammonium acetate, 0.1% SDS, 0.1 mM EDTA. The eluant is then carefully removed by an Eppendorf pipette.

DNA fragments larger than 600 bp are conveniently separated on low temperature gelling agarose gels (see Chap. 1: Basic Methods).

b) Procedure 1: Polyacrylamide Gel Electrophoresis

Take two 20 cm × 20 cm or 20 cm × 40 cm glass plates.
Insert two 2 mm thick spacers and seal with silicon rubber tubing, 1 mm/3 mm inner/ outer diameter; see also preparation of thin sequencing gel in Sec. 5, procedure 2.
Clamp the mould together with steel clips.
Prepare the polymerization mixture from the following stock solutions according to Table 2:

Acrylamide-Bis: 40% acrylamide
2% N,N'methylene bis-acrylamide dissolved in water and stored at 4°C
Buffer stock: 1 M Tris
0.5 M sodium acetate
0.5 M EDTA
brought to pH 7.8 with glacial acetic acid
Ammonium persulfate: N,N,N',N' tetramethylethylendiamine, stored at 4°C

Table 2. Preparation of non-denaturing gels

	20 × 40 cm (150 ml)		20 × 20 cm (75 ml)	
	8%	5%	8%	5%
Acrylamide-Bis	30 ml	18.75 ml	15 ml	9.4 ml
Buffer stock	6 ml	6 ml	3 ml	3 ml
Adjust to:	148 ml	148 ml	74 ml	74 ml
Ammonium persulfate	2 ml	2 ml	1 ml	1 ml
TEMED	0.2 ml	0.2 ml	0.1 ml	0.1 ml

Pour the gel and insert the comb (2 mm thick, each tooth at least 5 mm wide) immediately.
After polymerization, remove the comb and the tubing and install the gel vertically in the electrophoretic box.

Make contact between gel and the upper buffer trough with a double Whatman 3 MM paper wick, unless a special apparatus to contain notched plates is available.

Running buffer is a 1/25 dilution of the buffer stock.

Add to samples 1/5 to 1/10 vol of a solution containing 50% glycerol, 1% Ficoll, 0.05% xylene cyanol, 0.05% bromophenol blue and load onto gel.

Run at 8 V/cm or less and stop electrophoresis before bromophenol blue runs off.

Remove one glass plate, cover the gel with polyethylene foil, attach two or more radioactive markers to the gel, and autoradiograph at room temperature.

c) Procedure 2: Elution

After exposure, mark the edges of the gel on the X-ray film with a lead pencil using the radioactive markers as a reference; put the film under the gel, place on a light box, and precisely cut out the bands.

Put the gel piece on a disposable weighing tray and mash to a paste with the plunger of a 2 ml or 5 ml disposable syringe.

Place the paste in an Eppendorf tube and add 450 μl 1 M NaCl.

Clamp the tube to a vortex agitator and shake at room temperature for at least 30 min (or overnight for long fragments).

Spin down the gel for 3 min in the Eppendorf centrifuge and remove the supernatant carefully.

Extract twice more with 150 μl 1 M NaCl, applying the vortex manually a few seconds.

Combine the extracts (about 700 μl) and filter the eluate through siliconized glass wool to remove remaining gel fragments. For this purpose, prepare a 1000 μl Gilson pipette tip plugged tightly with long fiber glass wool and siliconize (add 500 μl of Repelcote, spin for a few seconds at low speed in a clinical centrifuge and repeat 3 X with 1 ml H_2O).

Set the tip in a siliconized glass tube (10 cm X 0.8 cm inner diameter), pipette in the eluate, and centrifuge at 3000 rpm for 2 min.

Rinse with 100 μl of 1 M NaCl by centrifugation.

Transfer filtered eluate plus rinse (final volume 750–800 μl) to an Eppendorf tube.

d) Procedure 3: Precipitation

1. Fragments to be Digested with Another Restriction Enzyme

Add to the eluate 10 μl 1 M Imidazole-HCl, pH 7.0, containing 1 mM Tris-HCl and 50 μl 0.1 M La$(NO_3)_3$; mix and keep at 0°C (ice water bath) for 10 min.

Centrifuge (2 min in an Eppendorf centrifuge), remove supernatant fluid, wash pellet with 100 μl cold 10 mM NaAc, pH 4.0, mix on a vortex, centrifuge again and carefully remove supernatant completely.

Resuspend pellet immediately in 10 μl 0.1 M EDTA and add 15 μl 4 M ammonium acetate.

Mix thoroughly, ensure that the pellet dissolves completely.

Add 30 μl water, 150 μl ethanol and mix.

Chill at $-70°C$ for 30 min, spin down the precipitate, wash the pellet with 80% ethanol and centrifuge again.
Dry in a dessicator.

Note: Formation of the final precipitate may be rather slow and must be monitored carefully to avoid loss.

2. Single Labeled Fragments Ready for Chemical Modification Reactions

Add to the eluate 8 μl carrier DNA (sonicated calf thymus DNA, 0.5 mg/ml).
Add 750 μl isopropanol (or fill the tube completely), mix, chill at $-30°C$ for at least 15 min and spin down the precipitate.
Wash the pellet with excess 80% ethanol and centrifuge again.
Dry in a dessicator.

3. Fragments to be Strand Separated

Add carrier DNA to the eluate if necessary.
Add 750 μl isopropanol (or fill the tube completely), mix, chill at $-30°C$ for at least 15 min and spin down the precipiate.
Redissolve in 100 μl 0.3 M NaAc, pH 7.5.
Add 250 μl ethanol.
Chill at $-70°C$ for 10 min.
Spin down precipitate, wash the pellet with excess 80% ethanol and spin again.
Dry in a dessicator.

5. Experiment 4: Sequencing by Chemical Degradation

a) Introduction

Sequencing DNA by the Maxam and Gilbert method (1977, 1980), is based upon the partial degradation of the DNA in a series of different base-specific chemical reactions. Detailed description of the chemistry of these cleavages can be found elsewhere (Kochetkov and Budowskii 1971, Maxam and Gilbert 1980). Briefly, all reactions consist of three consecutive steps: modification of the base, displacement of the modified base, and specific strand scission at this position.

The first reaction, which involves attack of one or two of the four bases, is partial and random, so that on the average, a single nucleotide is modified on each DNA molecule. Modification of the base weakens the glycosidic bond and renders the damaged base vulnerable to excision from its sugar via alkali or amine cleavage. Some fifteen different reaction schemes have already been proposed (Maxam and Gilbert 1980, Friedmann and Brown 1978, Rubin and Schmid 1981, Priess, personal communication) and the five most commonly used are described here. Dimethylsulfate is known to methylate guanine at the N-7 and adenine at the N-3 position (Lawly and Brookes 1963). Subsequent treatment with piperidine specifically leads to displacement of the (ring-opened) methylated G and to β-elimination of both flanking phosphates from the sugar (G lane).

Although under appropriate conditions, it is also possible to release the methylated adenines to produce an A+G lane, a more recent procedure, which relies on protonation of the purines rather than methylation, is now preferred. An A+G lane can also be produced by an initial reaction with diethylpyrocarbonate as described by Peattie (1979) for RNA sequence analysis. With DNA however, piperidine replaces aniline treatment for strand scission and the A and G bands are of comparable intensity. Adenine rings (cytosine to a lesser extent) can also be opened by heating in alkali (Hurst and Kuskis 1958a, b, Jones et al. 1968, Kochetkov and Budowskii 1971a), making them accessible to base catalyzed displacement and strand scission. In molar piperidine, this reaction proceeds to completeness (A+C lane). A useful pyrimidine specific reagent is the nucleophile hydrazine (Kochetkov and Budowskii 1971b), which splits the six-membered rings. Further hydrazinolysis leads to several additional reaction products, all of which can be displaced by piperidine (T+C lane). Finally, reaction of thymine with hydrazine can be suppressed by the presence of molar sodium chloride, a feature that provides for a cytosine specific reaction (C lane). In all cases, piperidine catalyzes β elimination of the phosphates from the free sugar or its derivatives (strand scission), thus leaving two fragments, each carrying one of the flanking phosphates. Only the fragment bearing the ^{32}P-label at the opposite end will produce a band on the sequencing gel. Four or more different reactions are run in adjacent lanes on a polyacrylamide sequencing gel. The consecutive shift of the label from one lane to the other(s) allows deduction of the nucleotide sequence from the labeled end. The precise chemical cleavage reactions used are matters of personal experience and choice. We describe here five different reactions in detail: A+G, G, T+C, C, A+C. This is more than strictly needed to read a sequence, but does provide an additional check should one of the reactions prove less than satisfactory. The thin sequencing gel that we describe is similar to that of Sanger and Coulson (1978): 0.5 mm thick, with 8.3 M urea, 100 mM Tris-borate, pH 8.3, and an acrylamide to methylene bis-acrylamide ratio of 20:1. Sequences from the very first nucleotide up to about 50 can be read from a 20% gel of 40 cm length. For longer sequences, lower acrylamide concentrations (between 10% and 4%), and longer gels (90 cm long) are used which enable up to 400 nucleotides to be read from one end from progressively longer electrophoretic runs. For short reading 20% gels are exposed frozen ($-20°C$ or less). For longer reading on 90 cm gels, the DNA is fixed in the gel with cetyltrimethyl ammonium bromide (Kruykov, personal communication) and exposed at room temperature.

b) Procedure 1: Degradation

Since it is imperative to proceed without stopping from the initial reaction up to the last ethanol wash (see Table 3), it is advisable to collect all materials and solutions needed before beginning the experiment. Put all solutions on ice and prepare a dry ice-ethanol bath ($-70°C$) for the precipitations.

Dimethylsulfate and hydrazine are hazardous reagents and modification reactions and precipitations involving them must be carried out under a well-ventilated fume hood. Gloves should be worn. Solutions containing these products are rendered safe by addition of 5 M NaOH (dimethylsulfate) or 3 M FeCl$_3$ (hydrazine).

Degradation

Redissolve the DNA pellet in 37 µl water (if you wish to save some labeled fragment for further experiments, use a larger volume and adjust the amount of carrier DNA (Sec. 4, Procedure 3, Part 2) accordingly. For each fragment to be sequenced, mark 5 siliconized Eppendorf tubes for the 5 different reactions with a water- and ethanol-resistant felt-tip pen (e.g., Pentel pen, Japan).

For convenience, the reactions are summarized in an easy to follow scheme (Table 3). Start reaction A+G, then proceed with the four others parallel (first heat the A+C sample); in between the manipulations of the latter four reactions, dry down and wash the A+G sample. Carry out piperidine treatment of all 5 samples at the same time.

The modified DNA is relatively stable and whenever necessary, the sequence of reactions may be interrupted after the final ethanol wash and continued later. Samples should be stored at $-30°C$ or $-70°C$. After piperidine treatment, the samples may be stored dry of in loading solution (see below) for several weeks (as long as the radioactivity is sufficient).

c) Procedure 2: Thin Sequencing Gels

Take two glass plates of identical size, either 20 cm × 40 cm, 30 cm × 40 cm or 30 cm × 90 cm, one of them containing a 3 cm notch leaving two ears of about 2.5 cm width. The smaller glass plates should be at least 4 mm thick, 30 cm × 90 cm plates preferably 6 mm. Back and front plates should have the same thickness and be completely flat (ask for mirror glass).

Siliconize one side of the notched plate with Repelcote and wash with ethanol.

Treat the un-notched plate with a saturated solution of KOH in ethanol, rinse extensively with water and wash with ethanol.

For 90 cm gels (adapt dimensions for smaller ones), put a 2.2 m long piece of thin silicon rubber tubing (2 mm/2.6 mm inner/outer diameter) or conformable 0.6 mm rubber thread along the bottom edge of the notched plate; put two 0.5 mm thick 90 cm long plastic spacers (e.g., rigid vinyl copolymer) along the side 0.5 cm in from the edge; place the second plate over the first and clamp together at the bottom with steel clips. Lift the upper plate slightly and carefully pull the tubing or thread inward against the bottom corners of the spacers. Align the tubing or thread along the side edges, stretch slightly, and lower the upper plate (the rubber thread may also be inserted without lifting the upper plate). Clamp the mould together in such a way that the clips press exactly onto the spacers. Make certain that the plates make good contact with the spacers to assure a uniform thickness of the gel, since unequal heating of the gel will otherwise occur during the run and lead to distortion of the lanes. Insert a further spacer at the bottom of the mould.

A mould for smaller gels may also be made up as follows:

Assemble glass plates and side spacers without grease and seal the edges and bottom of the mould with a piece of teflon tape (Scotch, 38.1 mm). Take care to flatten the tape firmly, especially at corners, using a small disposable pipette tip.

Table 3. Systematic description of the chemical sequencing reactions

	A+G	G	C	T+C	A+C
Labeled fragment	10 μl water	190 μl 50 mM sodium-cacodylate, 1 mM EDTA, pH 8.0	20 μl 3.75 M NaCl	15 μl water	40 μl 1.2 M NaOH, 1 mM EDTA
	9 μl	5 μl	5 μl	10 μl	± 8 μl (remainder)
			Chill in ice water		
Reagent	2 μl 4% HCOOH, pH 2.0 with pyridine	5 μl diluted DMS [a] 0°C	25 μl hydrazine 0°C	25 μl hydrazine 0°C	
Final volume	21 μl	200 μl	50 μl	50 μ	48 μl
Reaction	30–45 min 20°C	10 min 20°C	10 min 20°C	10 min 20°C	6 min 90°C
			Chill in ice water		
Stop reaction and isolate modified DNA (insome of the reactions partial base displacement and also strand scission has already occurred)	Lyophilize or dry down; wash with 20 μl water and lyophilize or dry down again	50 μl G-stop solution [b]	200 μl ACT-stop [c]	200 μl ACT-stop [c]	200 μl ACT-stop [c]
		Add 600 μl ethanol (0°C), mix Chill for 5 min at −70°C (dry ice-ethanol bath) Spin down 5 min in an Eppendorf centrifuge Remove supernatant with a Pasteur pipette [d] Rediscolve in 150 μl 0.3 M sodium acetate, pH 7.5 Add 400 μl ethanol and mix Chill at −70°C for 5 min Spin down 5 min in an Eppendorf centrifuge Remove supernatant with a Pasteur pipette [d] Wash the pellet with 90% ethanol (in excess) Spin for 1 min and remove the supernatant Dry in a dessicator (5 min)			
Displacement and strand scission	Redissolve the pellets in 50 μl 1 M piperidine (freshly diluted) Heat for 30 min at 90°C while submerging the tubes in the water bath [e] Lyophilize of dry down in a stream of air Wet the residue with 20 μl water, mix on a vortex, and spin to collect the DNA at the bottom of the tube Lyophilize or dry down again Repeat once more with 10 μl water				

[a] Immediately before starting the reactions, mix in an Eppendorf tube: 30 μl 50 mM sodium cacodylate, pH 8.0, 1 mM EDTA, and 5 μl dimethylsulfate
[b] G-stop solution: 1.5 M sodium acetate, pH 7.0, 1.0 M β-mercaptoethanol, 80 μg/ml carrier RNA (purified by phenol treatment and ethanol precipitation)
[c] ACT-stop solution: 0.3 M sodium acetate, pH 7.5, 20 μg/ml carrier RNA (purified as indicated in footnote b)
[d] Discard supernatants from G-reactions in 5 M NaOH, from C and T+C reactions in 3 M FeCl$_3$
[e] To assure that no water enters the tubes, they are fixed in holes in a 1 cm thick plexi plate and covered with a second plate of the same thickness. In between the plates is a sheet of soft plastic. The plates are then pressed together with high tension steel clips, thus sealing the tubes. This assembly is submerged in a 90°C water bath

Table 4. Preparation of sequencing gels

	30 × 90 cm (150 ml)				30 × 40 cm (60 ml)		20 × 40 cm (40 ml)	
	10%	8%	6%	4%	20%	8%	20%	8%
Urea	75 g	75 g	75 g	75 g	30 g	30 g	20 g	20 g
Acrylamide-Bis	37.5 ml	30 ml	22.5 ml	15 ml	30 ml	12 ml	20 ml	8 ml
Tris-borate	7.5 ml	7.5 ml	7.5 ml	7.5 ml	3 ml	3 ml	2 ml	2 ml
Water	<45 ml	<50 ml	<60 ml	<65 ml	–	<15 ml	–	<10 ml
Dissolve and adjust to:	148 ml	148 ml	148 ml	148 ml	59 ml	59 ml	39 ml	39 ml
Ammonium persulfate	1.5 ml	1.5 ml	0.5 ml	1.5 ml	0.8 ml	0.8 ml	0.6 ml	0.6 ml
TEMED	0.2 ml	0.2 ml	0.2 ml	0.2 ml	0.1 ml	0.1 ml	0.05 ml	0.05 ml

Prepare the polymerizing solution according to Table 4 from stock solutions of:

Acrylamide-Bis:	40% acrylamide
	2% N,N'methylene bis-acrylamide, dissolved in water and stored at 4°C
Tris-borate buffer:	2 M Tris
	2 M boric acid
	0.1 M EDTA, pH 8.3
Ammonium persulfate:	5% solution in water, stored at 4°C in a dark bottle
TEMED:	N,N',N' tetramethylethylenediamin, stored at 4°C

High quality products must be used; whenever necessary, recrystalize, filter, deionize, etc.

The amount of ammonium persulfate is approximate; fresh solutions may be too reactive and cause polymerization before the gel is completely poured. In this case, use less ammonium persulfate (up to 50%), but ensure that polymerization occurs within about 15 min.

After adding persulfate and TEMED, mix the solution and take up in one or two 50 ml syringe(s) without needles (contents about 65 ml each).

Pour the gel at an angle of 45°, applying the solution continuously at one of the ears and keeping continuous contact between the syringe tip and the flowing solution.

Lower the glass plates very slowly to the horizontal as the solution reaches the top of the mould.

Insert the comb immediately (slots of 5–10 mm width) and await polymerization.

When the gel has polymerized, leave it 15 min before removing the comb.

Flush out the slots immediately with water (squeeze bottle) and dry them with 4 mm wide strips of filter paper (Whatman 3 MM). The gel may improve if left overnight, but the slots should be covered with parafilm to prevent their drying out.

Pull out the thin tubing or thread and the bottom spacer from the mould and attach later to a notched buffer compartment, making the contact with the glass plate by means of a short piece of thin tubing or conformable thread.

Install the gel vertically in the bottom buffer compartment and add running buffer: 100 mM Tris-borate, pH 8.3, 5 mM EDTA.

Prepare the samples: redissolve the residue in freshly prepared 80% deionized formamide, 10 mM NaOH, 0.1–0.3% xylene cyanol, 0.1–0.3% bromophenol blue, 10 mM EDTA. The volume to be used depends on the number of gels to be run, the amount of radioactivity present in the samples, and the chain length of the DNA fragment to be sequenced. The residue can be easily recovered in 3 μl loading solution; thus, restrict each loading to a maximum of 3 μl and, whenever possible, 1 μl.

Draw out a thin-walled capillary tube and calibrate it with water to contain 1 μl or the loading volume.

Heat the samples at 90°C for 1 min.

Flush out the gel wells once again and load the samples in adjacent slots in the order: G, A+G, A+C, T+C, C. Rinse the capillary tube with buffer after each application and re-use.

Submit to electrophoresis at 2800–3200 V (90 cm gels) or 1300–1600 (40 cm gels) such that the gel heats up to about 45°C. Adjust the voltage during the run to maintain the gel at this temperature.

For short readings (20% gel), stop electrophoresis before the bromophenol blue dye has moved more than 2/5 of the gel length (a little further for 3' labeled fragments). For longer runs, xylene cyanol is a good reference marker, and on 8% gels, its migration corresponds to that of a 75 base oligonucleotide.

d) Procedure 3: Fixing the Gel and Autoradiography

After electrophoresis, remove the notched (siliconized) glass plate. Do not force the plates apart at the ears as these readily break off.

Spray the gel extensively with a solution of 0.3% cetyltrimethyl ammonium bromide in 0.2 M potassium phosphate, pH 6.0 (wear gloves!), and incubate for at least 10 min.

To remove excess solution, press several tissue papers onto the gel by means of another glass plate.

After 5 min, peel off these papers carefully, cover the gel with a polyethylene foil, and autoradiograph at ambient temperature using Kodak X-omat AR films.

Place a glass plate and some lead weights on the X-ray film to ensure good contact.

e) Procedure 4: Sequence Reading

The five reactions described here provide an abundance of information which enables easy reading of the nucleotide sequence. Although four reactions (G, A+G, C+T, C) are sufficient to deduce a sequence, only the G and C lanes allow a direct deduction. The fifth reaction A+C, provides an independent check.

	(1)	(2)	(3)	(4)	(5)
	G	A + G	A + C	T + C	C
G:	———	———			
A:		———	———		
C:			———	———	———
T:				———	

In the diagram bove, lanes (2) and (4) show the purines and pyrimidines, respectively, the flanking lanes discriminate between the individual nucleotides. The G residues are recognized by their presence in (1) and absence in (3). Conversely A is present in (3), but absent in (1). In case of the pyrimidines, two bands in (3) and (5) flank the C residues of lane (4), both being absent whenever a T occurs. Generally, however, C bands in the A+C lane (3) are much weaker than A and are sometimes scarcely detectable for the shorter oligonucleotides (up to n = 20). Note that the central three lanes constitute the whole sequence ladder which allows us to check the spacing between the bands. Variable gaps between bands (compressions) can indicate the presence of local secondary structure in some of the DNA fragments involved; the sequence deduced should therefore be regarded as only tentative until confirmed. (Do not, however, confuse this effect with the general observation that spacings after G and T are slightly larger than after A and C.) One particular example of unequal spacing is seen whenever a cytosine is 5-methylated (see Fig. 1). Since 5-methyl-C reacts only very slowly with hydrazine (Kochetkov and Budowskii 1971b) and apparently not at all with NaOH, it results in a gap on the sequencing gel.

The cause of an unsatifactory sequencing gel is not easy to identify in a multi-stage procedure such as this. A systematic diagnosis of aberrations in sequencing gels, together with possible causes and suggested solutions, can however be found in the extended protocol of Maxam and Gilbert (1980).

Finally, you are reminded that whenever a band appears at a certain position on the sequencing gel, the corresponding nucleoside has been destroyed. This means that special care must be taken when such sequencing ladders are compared with endonuclease digest, as in S1 nuclease mapping. Also remember that 5' labeled fragments will carry two terminal phosphate residues (one at either end), whereas 3' labeled fragments (obtained by a filling in reaction) only one, i.e., at the chemical breakage point.

6. Convenient Sources of Some Products and Materials (with Catalog Number)

Bacterial alkaline phosphatase: Worthington Biochemical Corp., BAPF
Klenow fragment DNA polymerase I: Boehringer, 104523
T_4-induced polynucleotide kinase: P-L Biochemicals, 0734
                       Boehringer, 174645

Cetyltrimethyl ammonium bromide: Aldrich Chemical Company, 85582-0
Dimethylsulfate: Aldrich Chemical Company, D18630-9
Dimethylsulfoxide: Merck, 2950+
Hydrazine: Eastman Organic Chemicals, 902
Lanthanum nitrate: Merck, 5326
Piperidine: Merck 9724
Repelcote: Hopkin & Williams 99270
Plastic spacer (0.5 mm): rigid vinyl copolymer, yellow, polished both sides, 0.02 in. thick: Shamrock Plastics, 837 South Mikel, Indianapolis, Indiana, U.S.A.
Teflon tape: Scotch, 3M Company, tape no. 5490

Radiochemical firms, e.g., NEN and Amersham not only sell ^{32}P-labeled ribo- and deoxynucleoside triphosphates, but also kits containing enzymes and other reagents.

7. Experiment 5: Dideoxy Sequencing on M13 Templates

a) Introduction

A primer annealed to a single stranded template can be covalently extended in the $5' \rightarrow 3'$ direction by DNA polymerase I (Klenow subfragment) in the presence of deoxynucleoside triphosphates (dNTPs). Klenow polymerase will also specifically incorporate dideoxynucleoside triphosphates (ddNTPs) in place of their dNTP counterparts, but the lack of a 3' hydroxyl group on the ribose moiety of the ddNTP prevents further polymerization. Thus, the ddNTPs act as specific chain terminators and can be used to determine the nucleotide sequence beyond the 3' end of the primer (Sanger et al. 1977). In each of our separate sequencing reactions, the primer is extended in the presence of a different ddNTP (ddTTP, ddCTP, ddGTP, or ddATP) such that there is partial incorporation of the terminator. Clearly the degree of primer extension in each reaction depends on the ratio of each ddNTP to its dNTP counterpart. Since efficient termination requires high ratio (about 100/1) of ddNTP/dNTP, the competing dNTP is conveniently used at a much lower concentration than the noncompeting dNTPs. The four sequencing reactions yield four nests of products, each with a fixed 5' end (the primer) and terminating at either T, C, G, or A. If the primer extension is radioactively labeled by including α^{32}P-dATP in the polymerization, the nucleotide sequence of the template beyond the 3' end of the primer may be deduced after fractionation of the products on a denaturing polyacrylamide gel (Sanger and Coulson 1978) and autoradiography of the gel. The radioactivity of the primer extension is usually maximized by using high specific activity α^{32}P-dATP (at either 400 Ci/mmol or > 3000 Ci/mmol): however, this can result in chain termination due to lack of dATP. After an initial synthesis with carrier free α^{32}P-dATP, the prematurely terminated chains are, therefore, extended by chasing with cold dATP.

The dideoxy technique has been used with double or single strand primers on single stranded templates (Barrell et al. 1979) and with single strand synthetic primers on double stranded templates (Hong 1981). An early strategy devised for sequencing φX174 used the single stranded virion DNA as a template and suitable restriction fragments derived from the double stranded replicative form (RF) as primers. The strategy was subsequently extended to double stranded plasmid templates by first converting the double strands to single strands with exonuclease III (Smith 1979). For example, the ends of foreign DNA inserted into the *Bam*HI site of pBR322 were sequenced by preparing primers of pBR322 DNA from the two regions of pBR322 flanking the inserted DNA and preparing single stranded template from exonuclease III treated *Sal*I or *Hin*dIII cut recombinant plasmid. Internal regions of the foreign DNA were sequenced with restriction fragment primers derived from the recombinant plasmid (Barrell et al. 1979).

The strategies described above were based on the use of a variety of restriction fragment primers on a single template. The alternative of using a variety of recombinant templates with a single universal primer is provided by the M13 shotgun strategy (Sanger et al. 1980). Here, the DNA to be sequenced is digested with restriction enzymes or endonucleases that cleave at random locations. The fragments of DNA generated are cloned into the unique restriction site of the double stranded replicative form (RF) of M13mp2 or M13mp7 (Messing et al. 1980), genetically engineered versions of the single stranded bacteriophage M13. The RF is then used to transform a male (sex factor-carrying) host and the transformed cells are mixed with fresh cells and spread on an agar plate. The transformed cells excrete phage and serve as foci for phage infection, thus, leaving small areas of retarded growth (plaques) in the lawn of cells. In order to distinguish plaques of recombinant phage from the original phage, a complementation assay is used. The vector M13mp2 contains a portion of the *lac* gene of *E. coli* including operator, promotor, and the N-terminal 145 amino acid residues of β-galactosidase (the so-called alpha peptide): this part of the enzyme, whose synthesis is induced by IPTG (isopropyl-β-D-thio-galactopyranoside), is not a functional β-galactosidase. The host strain *E. coli* (JM101), which is deleted of its chromosomal β-galactosidase gene, contains on the sex factor a defective β-galactosidase gene lacking the codons for amino acid residues 11-41. When M13mp2 infects the host the two defective β-galactosidase polypeptides associate and complement each other to produce a functional enzyme. Plaques of infected cells, therefore, produce active β-galactosidase, which is readily identified by including a lactose analogue BCIG (5-bromo-4-chloro-3-indolyl-β-galactoside) in the plate; this colorless compound is hydrolyzed by β-galactosidase to a blue dye, bromo-chloroindole. This confers a blue perimeter on plaques of M13mp2. By contrast recombinant phage, in which the foreign DNA has been inserted at the unique *Eco*RI site within the α-peptide of β-galactosidase, usually yield white plaques; the inserted DNA interrupts the α-peptide gene and prevents complementation of the defective sex factor-specified β-galactosidase. To harvest the recombinant phage, each white plaque is toothpicked into fresh medium and grown as a small culture. The phages are excreted into the medium and can be obtained from the supernatant after spinning down the cells. They are treated with phenol to remove protein coats and the single stranded recombinant DNA is used as template for sequencing. Several universal primers complementary to the region of M13mp2 at the 3' end of the inserted DNA are available either as synthetic primers (Duckworth et al. 1981) (P-L Biochemicals, Collaborative Research or BRL) or as a short restriction fragment (Anderson et al. 1980).

In the experimental protocol detailed below, restriction fragments ($\sim$ 250 bp) from influenza virus cDNA have been ligated into the unique *Eco*RI site of the bacteriophage M13mp2 (Messing et al. 1977, Gronenborn and Messing 1978) and template prepared as described by Sanger et al. (1980) and Winter and Fields (1980). A 30 bp universal primer (Schreier and Cortese 1979) flanking the site of insertion of influenza cDNA was prepared as described in Anderson et al. (1980).

b) Procedure 1: Hybridization

The single stranded M13 template and primer are denatured by boiling and are annealed. The procedure below is described for two clones, but it is readily extended to more.

Draw out a capillary by making the glass pliable in a small Bunsen flame, remove from flame, and pull gently to give a neck of about 3 cm.
Lightly mark the neck with a diamond pencil and break it to give flush ends.
Use a calibrated micropipette to blow into the drawn out tip of the capillary:
  5 μl M13 clone
  1 μl 10 × Hin buffer
  1 μl primer
  3 μl H_2O
Mix by blowing the sample up and down onto a piece of parafilm.
Seal the wide end of the capillary in the Bunsen flame, remove it from the flame, and as the glass cools, the sample should draw back from the capillary tip. Seal the tip.
Leave the capillary for 3 min in a test tube of water placed in a boiling water bath.
Remove test tube from water bath and let it cool to room temperature (15–30 min).

c) Procedure 2: Preparing the Gel

Mix 21 g urea with 7.5 ml 40% acrylamide stock (see below), 5 ml 10 × TBE and water to 50 ml (6% gel). Leave stirring to dissolve the urea. It is not necessary to heat the solution.
Siliconize a set of glass plates and swab with ethanol. Insert spacers, tape gel plates together with yellow tape and clamp plates with two gel clips.
Add 40 μl TEMED and 400 μl 10% ammonium persulphate to the acrylamide solution and mix
Pour the gel from a 50 ml disposable syringe with the glass plates at about 30° from the horizontal.
Insert the comb and leave the gel to set, with the plates raised slightly from the horizontal.

d) Procedure 3: Polymerization

Dry down four 2 μl aliquots of α^{32}P-deoxyadenosine triphosphate (specific activity 400 Ci/mmole) in siliconized glass tubes on a vacuum line.

Draw out eight capillaries and rest each in a siliconized glass tube. It is convenient to arrange the tubes in a 4 × 2 matrix.

Distribute each hybridized clone (2 µl aliquots) along each row of the four drawn out capillaries:

Row 1	.	.	.	. Clone 1
Row 2	.	.	.	. Clone 2

Draw out four capillaries and rest in the four tubes containing dried-down α^{32}P-deoxyadenosine triphosphate.

To the first capillary, add 2 µl dTTP^o mix and 2 µl ddTTP.
To the second capillary, add 2 µl dCTP^o mix and 2 µl ddCTP.
To the third capillary, add 2 µl dGTP^o mix and 2 µl ddGTP.
To the fourth capillary, add 2 µl dATP^o mix and 2 µl ddATP.

It is convenient to arrange these capillaries as a third row of the matrix of clones:

Row 1	.	.	.	. Clone 1
Row 2	.	.	.	. Clone 2
Row 3	.	.	.	. dNTP^o mix/ddNTP/dried-down α^{32}P-dATP
	T	C	G	A

Blow out the dTTP^o mix, ddTTP into the glass tube and take up the radioactive label into the capillary; check this with a Geiger counter.

Add 0.4 µl Klenow polymerase to the capillary and blow out again into the glass tube. To ensure thorough mixing of the enzymes with the triphosphates, roll the drop of liquid around the bottom of the tube.

Take back into the capillary and add 2 µl aliquots to the first capillary of each row of clones (the T column).

Start a stopwatch (t = 0) and thoroughly mix the primer/template and triphosphates/enzymes by blowing up and down the capillary.

Now repeat the procedure except making additions in turn to the second, third, and fourth capillaries of each row of clones with the dCTP^o mix, ddCTP, then with dGTP^o mix, ddGTP, and finally with dATP^o mix, ddATP. This should take about 5 min.

After t = 15 min, add 1 µl 0.5 mM dATP chase to each of the two clones in the T column. Blow out each capillary into its glass tube and discard the capillary.

Now repeat the chase with the C column, G column, and A column; before adding the chase allow each column a full 15 min incubation.

After t = 30 min, add 4 µl formamide dyes to the T column. Similarly repeat with the C column, G column, and A column; before adding the formamide dyes allow each column a full 15 min chase.

Boil the open tubes in a water bath for 10 min to reduce the volume of the samples.

e) Procedure 4: Loading, Running, and Fixing the Gel

Remove the comb from the gel, which should have set, and wash around the slots with distilled water. This removes unpolymerized acrylamide which otherwise sets in the wells. Slash the tape at the bottom of the gel, clamp into the gel apparatus and add 1 liter 1 × TBE to the top and bottom reservoirs.

Wash around the slots with 1 × TBE and, if necessary, straighten them out with a small piece of plasticard (see list of materials).

Remove the samples from the boiling water bath and allow to cool. *Do not plunge in ice.*

Flush out the urea from the slots directly before loading.

Draw up 1.5 μl portions of the samples into a drawn out capillary and load at the bottom of the slots. Do not attempt to load the last 0.5 μl of sample as it is easy to blow air into the slots.

Put the remainder of the sample in the freezer.

Run the gel at 25 mA (about 1.1 kV) for 1.5 h, when the bromophenol blue should have reached the bottom. The gel should be warm to the touch, but not painful.

Remove the plates from the apparatus, cut off the gel tape, and prize the plates apart. The gel should adhere to only one plate; if it adheres to both, prize the plates apart under 10% acetic acid and carefully lift off the upper plate.

Place a piece of netting over the gel and immerse in a bath of 10% acetic acid; the netting prevents the gel from floating away.

Fix the gel for 10 min; to dry the gel, first remove all wrinkles by rolling with a 25 ml pipette. Place several layers of tissue paper on top and roller again the squeeze out surplus liquid. Carefully peel back the tissue paper and cover the gel with Saranwrap.

Place gel under film. To ensure a good contact, place under a heavy steel plate. Overnight exposure is often adequate to read a sequence.

f) Procedure 5: Reading a Sequence

Primed sequencing usually yields bands across all four tracks due to primer fill-in (sticky and restriction fragment primers with a 5' overhang can fill in with α^{32}P-dATP at their 3' ends) and primer loop-back (one strand of the primer may be self-complementary and prime on itself). For the 30 bp primer used, these bands appear at the bottom of the gel in the position of the bromophenol blue (see Anderson et al. 1980).

Bands may also appear in all four tracks unrelated to primer fill-in or primer loop-back and for reasons which are not always clear.

1. A run of Gs (> 5) on the template will often result in a massive band across the gel with poor copying beyond. This effect is thought to result from polymerase stopping on the template, possibly due to stacking of adjacent Gs. The problem may be overcome by sequencing recombinant M13 template in which the other strand has been cloned (see clone turn-around, Winter and Fields 1980, Winter et al. 1981).

2. Local secondary structure of the copied DNA may remain undenatured on the urea-polyacrylamide gel. These structures result in compression of the spacing and are usually followed by a slight expansion of spacing above. At the limit, these compressions appear as bands in more than one track. One remedy is to sequence the other strand when the compression should be displaced a few residues.

Bands in each sequencing track are usually not of equal intensity. For doublets, the following rules apply:

Upper C is always *more* intense than lower C.
Upper G is always *more* intense than lower G, if the G's follow a T.
Upper A is often *less* intense than lower A.

When reading a gel, always read the spaces as well as the band. Ensure that the proposed sequence fits exactly into the space available.

8. Materials and Products

Primer, template, and DNA polymerase (Klenow)

DNA polymerase (Klenow): 0.5 U/µl (Boehringer).
M13 recombinant template: prepared from 1 ml of culture (Winter and Fields, 1980) and made up in 50 µl water.
pSP14 primer (Anderson et al. 1980): derived from an *Eco*RI and *Bam* HI digest of 100 µg pSP14 plasmid. Primer trace labeled at 3' end with α^{32}P-dATP and DNA Klenow polymerase (see Sec. 3), fractionated on 10% thin native polyacrylamide gel and detected by autoradiography. Eluted by soaking gel in 10 vol of 0.5 M ammonium acetate, 10 mM magnesium acetate, 0.1 mM EDTA, 0.5% SDS at 37°C overnight (do not macerate gel). Gel fragments removed by gently spinning eluant through fine holes punched in an Eppendorf microfuge tube. After ethanol precipitation (2.5 vol ethanol, overnight at −20°C) primer DNA spun down in Beckman SW60 rotor at 50,000 rpm, washed with ethanol, and made up to 100 µl with H_2O. Stored frozen in a siliconized glass tube at −20°C.

Buffers and mixes

Convenient stock solutions:	Tris-HCl, pH 7.5	1 M
	$MgCl_2$	1 M
	NaCl	1 M
	EDTA, pH 7.5	100 mM
	Dithiotreitol (DTT)	100 mM
	ddNTP	10 mM
	dNTP	20 mM
	DTT, ddNTP and dNTP are stored frozen, the nucleoside triphosphates in 5 mM Tris-HCl, 0.1 mM EDTA	
10 × Hin buffer (minus DTT):	Tris-HCl, pH 7.5 (1 M)	100 µl
	$MgCl_2$ (1 M)	100 µl
	NaCl (1 M)	500 µl
	H_2O	300 µl
	Store frozen and just before use mix 9 µl with 1 µl 0.1 M DTT	

ddNTP working stocks: The appropriate level of ddNTP is determined by trial and error. As a first approximation try:

ddTTP	0.5 mM
ddCTP	0.1 mM
ddGTP	0.2 mM
ddATP	0.1 mM

Keep frozen in 5 mM Tris-HCl, 0.1 mM EDTA.

dNTP working stocks: 0.5 mM working stocks are made freshly by dilution into 0.1 mM EDTA. Make 80 µl each of dTTP, dCTP, and dGTP and 200 µl of dATP.

dNTP^O mixes:

	dNTP^O mix			
	dTTP^O	dCTP^O	dGTP^O	dATP^O
0.5 mM dTTP	1 µl	20 µl	20 µl	20 µl
0.5 mM dCTP	20 µl	1 µl	20 µl	20 µl
0.5 mM dGTP	20 µl	20 µl	1 µl	20 µl
50 mM Tris-HCl, pH 8.0 1 mM EDTA	5 µl	5 µl	5 µl	5 µl

The dNTP^O mixes keep indefinitely when frozen.

α^{32}P-dATP, 400 Ci/mmole: from the Radiochemical Centre, Ltd., Amersham, Bucks, England, or equivalent.

Formamide dyes: Gently stir 100 ml formamide with 5 g Amberlite MB1 (mixed bed resin) for 30 min. Remove resin by filtration. Add 0.3 g xylene cyanol FF, 0.3 g bromophenol blue, and Na$_2$ EDTA to 20 mM. Store at room temperature.

Repelcote: for silicone treatment (Hopkin & Williams)

40% acrylamide stock:
acrylamide	190 g
bis-acrylamide	10 g
H$_2$O to	500 ml

Stir gently with 5 g Amberlite MB1 for 30 min. Remove resin by filtration and keep stock at 4°C.

10 × TBE buffer:
Tris	108 g
H$_3$BO$_3$	55 g
Na$_2$ EDTA 2H$_2$O	9.5 g
H$_2$O to	1 liter

10% ammonium persulphate:
ammonium persulphate	10 g
H$_2$O	100 ml

can be stored at 4°C for several months.

TEMED: N,N,N',N'-tetramethylethylenediamine

Urea: Ultra-pure

Convenient materials

Melting point capillaries, 10 cm × 1 mm internal diameter (Gallenkamp-Griffin, MFB-210-538L).

1–5 µl calibrated micropets (Clay-Adams). Check that the micropets do not contain alkaline deposits and wash if necessary.

Siliconized glass tubes (1 cm × 5 cm). Siliconize with Repelcote and wash well with distilled water before baking at 150°C for 4 h.

40 cm slab gel apparatus (Raven Scientific Ltd.).

20 cm × 40 cm × 4 mm mirror glass plates, back plate with ears (Raven Scientific, Ltd.).

Side spacers and comb with 18 teeth (each tooth 5 mm wide separated from adjacent tooth by 2 mm gape). Spacers and combs are conveniently cut from 0.3 mm Plasticard available from Raven Scientific, Ltd.

Yellow tape: 'Clipper' Canadian Technical Tape, Ltd. or Tuck Tape.

Dow Saranwrap plastic film (other makes are not usually satisfactory).

Kimwipes (43 cm × 25 cm).

Steel plate (43 cm × 34 cm × 1/8 inch) (Raven Scientific, Ltd.) for weighting down X-ray film.

9. Addendum 1: Further Developments in Rapid DNA Sequencing

An important advantage of M13-dideoxy sequencing compared with the chemical degradation method is that the preparation of DNA templates for sequencing reactions is simple and rapid because only a few manipulations are involved. The Maxam and Gilbert approach requires one or more gel separation steps and fragment-specific enzymatic reactions to introduce and segregate radioactive label. The latter strategy, therefore, usually depends upon a preexisting detailed restriction cleavage map in order to minimize the number of sequencing reactions. In the M13-dideoxy approach, individual routine experiments produce sequences of random stretches of about 300–400 nucleotides that eventually are assembled by virtue of their overlaps to a complete nucleotide sequence.

Similar subcloning strategies for chemical degradation sequencing have now been elaborated in several laboratories. They are based on the use of plasmid vectors to construct a set of subclones of the DNA to be sequenced. These subclones are produced in either a random or ordered fashion such that two (or one, in method 3) specific restriction cleavage sites are located close to the insert. The site closest to the insert serves for the labeling reaction, whereas the other allows subsequent segregation of the label. The small fragment thus created is not removed and after chemical degradation, produces an unreadable region in the lower part of the sequencing gel. This loss of sequence information (30 nucleotides or less) is compensated by a substantial shortening of the overall sequencing procedure.

The pertinent labeling and degradation reactions are essentially as described in the previous sections, but no gel purification step is involved. Moreover, the recombinant plasmids may be prepared by one of the numerous "cleared lysate" procedures now available and are sufficiently pure for sequencing, if 3' end labeled by filling in polymerization (see Sec. 3).

Method 1: Frischauf, A.M., Garoff, H., Lehrach, H. (1981) A subcloning strategy for DNA sequence analysis. Nucleic Acids Res. 8:5541–5549

The approach of Frischauf et al. is applicable to DNA cloned in the *Pst*I site of pBR322 and may be extended to other restriction sites and other vectors with comparable features (see below). The hybrid plasmid is fragmented randomly with DNase I in the presence of Mn^{2+} and *Eco*RI linkers are attached to the linearized molecules. Digestion with *Eco*RI cleaves the oligonucleotide linkers as well as the natural (unique) *Eco*RI site of pBR322. The pool of random length fragments is fractionated by electrophoresis through a low melting point agarose gel and separate fractions differing in length by approximately 100 bp are eluted. The fragments thus obtained are circularized with ligase and used to transform competent cells. Tetracycline resistant transformant clones contain subclone derivatives of the original hybrid plasmid that have lost a DNA segment counterclockwise to the natural *Eco*RI site of pBR322. By limiting the size selection to fragments that are at least 3600 bp long (the distance between the *Eco*RI and *Pst*I sites in pBR322 in the clockwise direction), and at least 700 bp shorter than the original clone (the length of the smaller *Eco*RI-*Pst*I fragment), all sequencing clones will contain a portion of the original insert fused to the *Eco*RI site of pBR322. Plasmid DNA from the subclones is isolated, cleaved with *Eco*RI, labeled at either the 5' or 3' ends, and restricted with *Hin*dIII. The small 29 bp fragment thus produced is not removed from the sample; its degradation products will superimpose on the sequencing ladder of the larger fragment, thus, causing a loss of information of the first 29 nucleotides.

In principle, this method can be extended to any pair of unique restriction sites in a vector if:
1. the order of relevant features of the hybrid plasmid is: insert DNA – site to be labeled (and contained in the oligonucleotide linker) – second restriction cleavage site – selection marker – origin of replication – insert DNA;
2. the restriction site to be labeled is absent in the insert DNA;
3. the vector is small enough to allow size fractionation on an agarose gel prior to subcloning.

This sequencing strategy is target directed rather than shotgun, because the size-fractionated samples provide a systematic series of overlapping sequences of the original insert.

Method 2: Rüther, U., Kaenen, M., Otto, K., Müller-Hill, B. (1981). pUR222: a vector for cloning and rapid chemical sequencing of DNA. Nucleic Acids Res. 9:4087–4098

pUR222 has been constructed by a series of in vitro engineering steps to combine part of the pBR322 sequence with a portion of the *E. coli lac* operon. The ampicillin resistance gene of pBR322 provides a primary transformation selection, whereas the *lac* complementation assay is used to discriminate between recombinant plasmids and the original vector. The latter selection functions exactly as described with M13mp2 (see Exp. 5). Six unique cloning sites (*Pst*I, *Sal*I, *Acc*I, *Hinc*II, *Bam*HI, *Eco*RI) are clustered in the α-peptide DNA fragment of the β-galactosidase gene. This vector segment can complement the defective β-galactosidase gene of the bacterial host (on an F' plasmid) and causes colonies to turn blue on appropriate indicator plates. Recombinant subclones in one of the above cloning sites will have disrupted the expression of the α-peptide fragment and are generally white.

```
------·CTGCAG GTCGAC GGATCC GGG GAATTC·------
        PstI    SalI   BamHI      EcoRI
                AccI
```

Three cleavage sites of the cluster are used in the subcloning sequencing procedure in the following order of sequence: cloning site, labeling site, second cleavage site. For example, one can subclone random segments of the DNA to be sequenced in the *Sal*I site. Small cultures of white colonies are grown, plasmid DNA is isolated and cleaved with *Bam*HI. Filling in polymerization with α^{32}P-dGTP will label the linear DNA at its 3' termini. Subsequent *Eco*RI digestion produces a 9 bp fragment and the remaining vector-insert fragment, bearing the label close to the insert. Again, as in method 1, chemical degradation can be carried out without prior fractionation. The information loss due to the presence of the smaller fragment is only 9 nucleotides.

This strategy is basically a shotgun approach, although the use of double digests could, at least in principle, allow an oriented subcloning of DNA fragments cleaved with the same enzymes. The selection of recombinant plasmids by the color test is not absolute, since about 3% of the recombinant clones are blue (see Rüther et al. 1981). Conversely, low quality restriction enzyme and/or ligase preparations may lead to white colonies by deleting, rather than inserting DNA at the cloning site.

Method 3: Volckaert, G., De Vleeschouwer, E., Blöcker, H., Frank, R. (1983) (In preparation)

pCSV03 and related chemical sequencing vectors have been developed recently for rapid DNA sequencing in either a random or target directed fashion. Only a single restriction digest of the recombinant plasmids is necessary and reading from the very first nucleotide is possible. Moreover, with some of these vectors, inserts can be sequenced in both directions.

The design of these vectors is based on the phenomenon described in Sec. 3, namely the single labeling of asymmetric restriction sites with α^{32}P-dNTP and Klenow polymerase (see Table 1). pCSV03 has the replicon and ampicillin resistance gene of pBR327. It is about 1900 bp long and its pertinent feature is the presence of an 11 bp segment containing a *Sma*I and *Bst*EII cleavage site 23 bp from the *Eco*RI site (*Bst*EII is an isoschizomer of *Eca*I).

```
5' ----- GAATTC ----------- CCCGGGTCACC ----- 3'
3' ----- CTTAAG ----------- GGGCCCAGTGG ----- 5'
         EcoRI               SmaI
                            ---------
                             BstEII
```

Cleavage of pCSV03 with *Bst*EII produces a linearized vector molecule with two asymmetric 5' protruding ends.

```
5'—— GAATTC ——————— CCCGG3'          5'pGTCACC ——— 3'
3'—— CTTAAG ——————— GGGCCCAGTGp5'        3'G ——— 5'
```

Only the left hand side overhang can be labeled with α^{32}P-dCTP and Klenow polymerase in the presence of unlabeled dGTP and dTTP (the absence of dATP in the reaction mix prevents further polymerization). If intracellular dATP remains in the DNA sample after plasmid isolation, e.g., if cleared lysate is used, it must be destroyed by a short incubation with pyrophosphatase at pH 6.5 prior to labeling. The labeled DNA can be used immediately for chemical degradation.

Shotgun sequencing involves insertion of random or restriction digests of the DNA to be sequenced into the *Sma*I site by blunt end ligation. Since no biological selection of recombinant transformants is possible, the *Sma*I cleaved vector DNA is pretreated with phosphatase and purified on a 2% low melting point agarose gel. This procedure results in essentially no background clones ($< 1\%$).

For target directed sequencing, the vector DNA is digested with *Sma*I and *Eco*RI. *Eco*RI linkers are attached to the DNA fragment(s) to be sequenced, which are cleaved randomly with DNase I in the presence of Mn^{2+} and made blunt ended by filling in polymerization with Klenow polymerase and dNTPs. After digestion with *Eco*RI, the DNA mixture is fractionated on a low melting point agarose gel and slices of fragments with increasing length are eluted. Only fragments having an original terminus can be ligated into the vector as only they have an *Eco*RI sticky end at this position. Ligation is in an oriented manner, with fusion of the blunt end terminus to the *Bst*EII, i.e., the labeling site. Nucleotide sequences of recombinant subclones from different gel fractions will thus constitute an array of overlapping sequences.

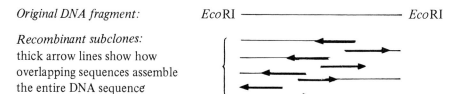

Original DNA fragment: *Eco*RI ————————————— *Eco*RI

Recombinant subclones:
thick arrow lines show how
overlapping sequences assemble
the entire DNA sequence

In principle, only a few subclones of each gel fraction need to be analyzed. The only prerequisite of this method is the absence of *Bst*EII sites in the DNA to be sequences; the *Bst*EII site is, however, rather uncommon.

Chemical sequencing vectors that allow sequencing in both directions have been constructed along the same lines and contain two different *Tth*111I restriction sites as labeling sites, which flank an *Hinc*II blunt end cloning site.

9. References

Anderson S, Gait MJ, Mayol L, Young I (1980) A short primer for sequencing DNA cloned in the single stranded phage vector M13mp2. Nucleic Acids Res 8:1731–1743

Barrell BG, Bankier AT, Drouin J (1979) A different genetic code in human mitochondria. Nature 282:189–194

Berkner KL, Folk WR (1977) Polynucleotide kinase exchange reaction. Quantitative assay for restriction endonuclease-generated 5'-phosphoryl termini in DNAs. J Biol Chem 252:3176–3184

Donelson J, Wu R (1972) Nucleotide sequence analysis of deoxyribonucleic acid. VI. Determination of 3′-terminal dinucleotide sequences of several species of duplex deoxyribonucleic acid using *Escherichia coli* deoxyribonucleic acid polymerase I. J Biol Chem 247:4654–4660

Duckworth ML, Gait MJ, Goelet P, Hong GF, Singh M, Titmas RC (1981) Rapid synthesis of oligodeoxynucleotides. VI. Efficient mechanised synthesis of heptadecadeoxynucleotides by an improved solid phase phosphotriester route. Nucleic Acids Res 9:1691–1706

Englund PT (1972) The 3′-terminal nucleotide sequences of T7 DNA. J Mol Biol 66:209–224

Friedmann T, Brown DM (1978) Base-specific reactions useful for DNA sequencing: methylene blue-sensitized photooxidation of guanine and osmium tetraoxide modification of thymine. Nucleic Acids Res 5:615–622

Gronenborn B, Messing J (1978) Methylation of single-stranded DNA in vitro introduces new restriction endonuclease cleavage sites. Nature 272:375–377

Hong GF (1981) A method for sequencing single-stranded cloned DNA in both directions. Biosci Rep 1:243–252

Hurst RO, Kuksis A (1958a) Degradation of deoxyribonucleic acid by hot alkali. Can J Biochem Physiol 36:919–929

Hurst RO, Kuksis A (1958b) Degradation of some purine and pyrimidine derivatives by hot alkali. Can J Biochem Physiol 36:931–936

Jones AS, Mian AM, Walker RT (1968) The alkaline degradation of deoxyribonucleic acid derivatives. J Chem Soc C:2042–2044

Kleppe K, Ohtsuka E, Kleppe R, Molineux I, Khorana HG (1971) Studies on polynucleotides. XCVI. Repair replication of short synthetic DNAs as catalysed by DNA polymerases. J Mol Biol 56:341–361

Kochetkov NK, Budowskii EI (1971a) Organic chemistry of nucleic acids, part B. Plenum Press, New York, pp 381–400

Kochetkov NK, Budowskii EI (1971b) Organic chemistry of nucleic acids, part B. Plenum Press, New York, pp 401–408

Kössel H, Roychoudhury R (1971) Synthetic polynucleotides. The terminal additon of riboadenylic acid to deoxyoligonucleotides by terminal deoxynucleotidyl transferase as a tool for specific labelling of deoxyoligonucleotides at the 3′ ends. Eur J Biochem 22:271–276

Lawley PD, Brookes P (1963) Further studies on the alkylation of nucleic acids and their constituent nucleotides. Biochem J 89:127–138

Lillehaug JR, Kleppe RK, Kleppe K (1976) Phosphorylation of double-stranded DNAs by T4 polynucleotide kinase. Biochemistry 15:1858–1865

Maxam AM, Gilbert W (1977) A new method for sequencing DNA. Proc Natl Acad Sci USA 74:560–564

Maxam AM, Gilbert W (1980) Sequencing end-labelled DNA with base-specific chemical cleavages. In: Grossman L, Moldave K (eds) Methods in enzymology, vol 65. Academic Press, New York London, pp 499–560

Messing J, Gronenborn B, Müller-Hill B, Hofschneider PH (1977) Filamentous coliphage M13 as a cloning vehicle: insertion of a *Hin*dIII fragment of the *lac* regulatory region in M13 replicative form in vitro. Proc Natl Acad Sci USA 74:3642–3646

Messing J, Crea R, Seeburg PH (1980) A system for shotgun DNA sequencing. Nucleic Acids Res 9:309–321

Olson K, Harvey C (1975) Determination of the 3′ terminal nucleotide of DNA fragments. Nucleic Acids Res 2:319–325

Peattie DA (1979) Direct chemical method for sequencing RNA. Proc Natl Acad Sci USA 76:1760–1764

Richardson CC (1965) Phosphorylation of nucleic acid by an enzyme from T4 bacteriophage-infected *Escherichia coli*. Proc Natl Acad Sci USA 54:158–165

Roychoudhury R, Jay E, Wu R (1976) Terminal labelling and addition of homopolymer tracts to duplex DNA fragments by terminal deoxynucleotidyl transferase. Nucleic Acids Res 3:863–877

Rubin CM, Schmid CW (1981) Pyrimidine-specific chemical reactions useful for DNA sequencing. Nucleic Acids Res 8:4613–4619

Sanger F, Coulson AR (1978) The use of thin acrylamide gels for DNA sequencing. FEBS Lett 87:107–110

Sanger F, Nicklen S, Coulson AR (1977) DNA sequencing with chain terminating inhibitors. Proc Natl Acad Sci USA 74:5463–5467

Sanger F, Coulson AR, Barrell BG, Smith AJH, Roe BA (1980) Cloning in single-stranded bacteriophage as an aid to rapid DNA sequencing. J Mol Biol 143:161–178

Schreier PH, Cortese R (1979) A fast and simple method for sequencing DNA cloned in the single-stranded bacteriophage M13. J Mol Biol 129:169–172

Schwarz E, Scherer G, Hobom G, Kössel H (1978) Nucleotide sequence of *cro, c*II part of the *O* gene in phage λ DNA. Nature 272:410–414

Smith AJN (1979) The use of exonuclease III for preparing single-stranded DNA for use as a template in the chain terminator sequencing method. Nucleic Acids Res 6:831–848

Tu C-PD, Cohen S (1980) 3' End labelling of DNA with $\alpha^{32}P$ cordycepin 5' triphosphate. Gene 10:177–183

Winter G, Fields S (1980) Cloning of influenza cDNA into M13: the sequence at the RNA segment encoding the A/PR/8/34 matrix protein. Nucleic Acids Res 8:1965–1974

Winter G, Fields S, Ratti G (1981) The structure of two subgenomic RNAs from human influenza virus A/PR/8/34. Nucleic Acids Res 9:6907–6915

Wu R (1970) Nucleotide sequence analysis of DNA. I. Partial sequence of the cohesive ends of bacteriophage λ and 186 DNA. J Mol Biol 51:501–521

Wu R, Taylor E (1971) Nucleotide sequence analysis of DNA. II. Complete nucleotide sequence of the cohesive ends of bacteriophage λ DNA. J Mol Biol 57:491–511

Chapter 6 Electron Microscopy

H. BURKARDT[1] and R. LURZ[2]

Contents

1. General Introduction 282
2. Support Films for Electron Microscopy of Nucleic Acids 282
   - a) Introduction 282
   - b) Procedure 1: Parlodion Films 283
   - c) Procedure 2: Carbon Films 283
   - d) References 283
3. Cytochrome C Spreading Techniques 284
   - a) Introduction 284
   - b) Procedure 1: Droplet Method 284
   - c) Procedure 2: Spreading 286
   - d) Procedure 3: DNA Spreading in Carbonate Buffer on a Water Hypophase 287
   - e) Procedure 4: Spreading of Single Stranded Nucleic Acids with T4 Gene 32 Protein . 288
   - f) References 289
4. Length Measurement of DNA 289
   - a) Introduction 289
   - b) References 290
5. Homoduplex Preparation: Visualization of Repeated Sequences with Inverse Orientation (Inverted Repeats) 291
   - a) Introduction 291
   - b) Procedure 293
   - c) References 293
6. Heteroduplex Procedure 294
   - a) Introduction 294
   - b) Procedure 1 295
   - c) Procedure 2 296
   - d) References 296
7. Partial Denaturation 296
   - a) Introduction 296
   - b) Procedure 298
   - c) References 299
8. Visualization of Protein-Nucleic Acid Complexes 299
   - a) Introduction 299
   - b) Procedure 1: BAC Spreading 300
   - c) Procedure 2: Preparation of DNA or DNA-Protein Complexes by Adsorption to Mica 300
   - d) References 302

---

1 Institut für Mikrobiologie und Biochemie, Lehrstuhl für Mikrobiologie der Universität Erlangen-Nürnberg, Egerlandstraße 7, D-8520 Erlangen, Fed. Rep. of Germany

2 Max-Planck-Institut für Molekulare Genetik, Ihnestraße 63–73, D-1000 Berlin 33, Fed. Rep. of Germany

1. General Introduction

The electron microscopy of nucleic acids is a relatively new procedure compared with classical electron microscopic methods that involve, e.g., thin-sectioning or shadowing. It developed, however, into a standard molecular biological technique with the introduction of the basic protein monolayer method by Kleinschmidt and Zahn (1959). Electron microscopy of DNA is generally simple and rapid and enables a large amount of information on molecular structure to be obtained. This chapter covers the first five of the following six basic procedures:

1. Standard preparation (Sec. 3), which provides information on DNA configuration (linear, circular, supertwisted), length (from which molecular weight can be calculated), and concentration;
2. Homoduplex formation (Sec. 5), which demonstrates and enables localization of inverted repeat sequences;
3. Heteroduplex formation (Sec. 6), which shows homologous and heterologous regions of related DNA molecules;
4. Partial denaturation (Sec. 7), which determines the relative A+T content of DNA segments;
5. Visualization of protein-nucleic acid complexes (Sec. 8 and Chap. 7.1), which demonstrates such interactions and localizes the interaction sites;
6. Visualization of DNA:RNA hybrids (R-loops; transcription complexes, see Chap. 7.2) provides information on the organization and localization of genes.

2. Support Films for Electron Microscopy of Nucleic Acids

a) Introduction

Support films, usually mounted on copper grids, are required for the electron microscopy of nucleic acids. Such films must fulfill the general conditions demanded for use in electron microscopical techniques, namely,

1. they must be stable in the electron beam;
2. they should not exhibit any background structure;
3. they must be thin in order to minimize any reduction in specimen contrast.

Parlodion and carbon films are commonly used for DNA preparations. The advantages of parlodion films are (1) that they are easy and rapid to prepare and (2) that films of suitable thickness are easily recognizable by their shininess and color. Their disadvantages are that they (1) are relatively unstable in the electron beam, (2) cannot be used if too old (depends upon storage conditions, such as humidity, etc.), (3) cause heavy contamination of the electron microscope column, if a cooling trap is not used, and (4) sometimes contain particles and holes, particularly if the parlodion solvent is not free of water.

The advantages of carbon films are that they (1) are more stable in the electron beam, (2) result in less contamination of the electron microscope, (3) can be stored for longer periods of time prior to use (but their properties may change), and (4) provide a fine background. Their disadvantages are that (1) they are more difficult to prepare, and (2) their surface properties are less reproducible, and hence, the appearance of DNA on such films is more variable.

b) Procedure 1: Parlodion Films

Parlodion is sold as solid chips and is used as a 3% solution in isoamyl acetate. It is essential that the solution is water-free.

To prepare, dry the chips overnight at 80°C and dissolve in isoamyl acetate (takes about 1 week). Clean, water-free solutions of parlodion may be used for years, if handled carefully.

Arrange copper grids, shiny side up, on filter paper (70 mm diameter) situated at the bottom of a water filled Büchner funnel.

Clean the water surface of contaminating dust particles, etc. by a prior application of the parlodion film. To do this, place one drop of parlodion solution on the middle of the water surface. After evaporation of the solvent, which will take about 5 min, the film will form a thin membrane. Remove the film and trapped dust particles with a needle.

Prepare a second film in the same way. Permit the water to run out of the Büchner funnel so that the film sinks onto the grids.

Dry the filter paper and adhering grids in air.

Procedure 2: Carbon Films

Arrange copper grids on a glass slide and cover them with one drop of a 0.3% solution of neoprene solution in toluene (neoprene is a sticky material that improves the adhesion of the carbon film to grids).

Remove excess neoprene with filter paper and allow grids to dry in air.

Evaporate pure carbon onto a mica sheet in a vacuum evaporator; the thickness of the carbon layer may be estimated by the color of a simultaneously shadowed blank control paper. It should be light gray-brown.

Float the carbon film onto the surface of a water-filled dish.

Coat copper grids with the carbon film by submersing them in the water and lifting them through the carbon film.

Dry grids in air.

d) References

Baumeister W, Hahn M (1978) Specimen supports. In: Principles and techniques of electron microscopy. In: Hayat MA (ed) Biological application, vol 8, pp 1–112

Brack Ch (1981) DNA electron microscopy. CRC Crit Rev Biochem 10:113–169

Ferguson J, Davis RW (1978) Quantitative electron microscopy of nucleic acids. In: Koehler JK (ed) Advanced techniques in biological electron microscopy, vol II. Springer, Berlin Heidelberg New York, p 123

3. Cytochrome C Spreading Techniques

a) Introduction

The analysis of DNA by electron microscopy differs from that of most other biological materials because DNA has certain properties that require special consideration, i.e., DNA molecules are rather long and thin and have no fixed tertiary structure in solution, i.e., they form random coils. The preparation of DNA for electron microscopy must therefore stabilize its delicate structure and spread it out in two dimensions to provide evenly stretched, measurable filaments on the grid. The most commonly used method is the basic protein monolayer technique of Kleinschmidt and Zahn (1959), which involves the use of a monolayer support film of the basic protein cytochrome c (cyt c) to adsorb, unfold, and stabilize DNA. The monolayer is formed on the solution-surface as a result of protein denaturation at the air: solution interface. Two principal variations of the technique, the droplet and spreading methods, are in use.

In the first method, all components (DNA, buffer, and cyt c) are mixed together and deposited in a droplet on a hydrophobic surface, where cyt c denaturation and adsorption of DNA occur simultaneously. This method is very simple, but not very suitable for the examination of single stranded DNA; therefore, its use is restricted to routine preparations of double stranded DNA. For the examination of single stranded DNA, a formamide spreading technique is more suitable. Unfolding of the more flexible, sticky single stranded DNA is achieved by addition of formamide to the solutions to reduce intrastrand hydrogen bond formation.

A two phase system, consisting of a hypophase and a hyperphase (spreading solution), is used for the spreading of DNA. The hyperphase formamide spreading solution contains DNA, buffer, cyt c, and formamide. The hypophase contains buffer at a concentration lower, generally 10% or less, than that of the spreading solution, and formamide. The chosen formamide concentration depends upon the GC content of the DNA and is usually between 30% (for DNA with a GC content of 50%, like that of *E. coli*) and 50% (for GC-rich DNA, like that of many soil bacteria) for the hyperphase. The formamide concentration of the hypophase is usually $\leqslant 0.75$ that of the hyperphase. The greater the formamide concentration difference of the two phases, the larger the area of hypophase covered by the spot of hyperphase, and hence the separation of DNA molecules. After DNA is transferred to the grid, it must be stained to enable visualization in the microscope, if dark field techniques are not used. This is accomplished by staining with ethanolic uranyl acetate, by shadowing with metal vapor, or by a combination of both.

b) Procedure 1: Droplet Method (Fig. 1)

Nucleic acid preparation techniques for electron microscopy are sensitive to the quality of the water available. Water is usually prepared by passage through an ion exchanger and double distillation in a quartz vessel. Rarely, water obtained by this procedure does not permit good preparations of nucleic acids for microscopy, due to peculiar components of the local water supply or to substances released from the

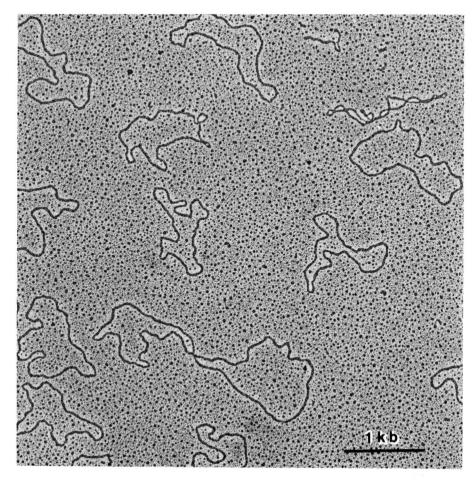

Fig. 1. Electron micrograph of plasmid pBR 322 DNA prepared by the droplet method (see Sec. 3,b). The preparation was rotary shadowed with platinum-iridium (80:20)

ion exchange resin. Such impurities that remain cannot be removed by additional distillations. Therefore, if one of the following procedures fails, it may be worthwhile to obtain water from another source.

Droplet solution:

Mix 2.70 ml H_2O, 0,30 ml 2 M ammonium acetate, pH 7.0, 0.01 M EDTA, 8 μl cyt c, 1 mg/ml

Mix an aliquot of the droplet solution with the DNA sample. The volume ratio of DNA sample and droplet solution may vary between 1:50 and < 1:1000, depending on the DNA concentration of the sample. For samples with DNA concentrations of about 10 μg/ml, mix 0.5 μl DNA with 0.25 ml stock solution.

Place several droplets (40–50 μl) of the mixture on a hydrophobic surface, e.g., a clean plastic petri dish, and leave for 30–60 min to allow cyt c denaturation and DNA adsorption; the longer the time, the more DNA adsorbed.

Pick up parts of the cyt c film and adhering DNA by briefly touching the droplet surface with a coated copper grid (film down).

Dip grid in a tube containing staining solution for 30 s, then rinse briefly in 90–95% ethanol; allow to dry in air. The staining solution is a 1/100 dilution in 90% ethanol of a a stock solution of 5×10^{-3} M uranyl acetate in 0.05 M HCl (ethanol concentrations greater than 95% may damage the Parlodin film). Prior to usage, clean staining solution from small contaminating particles and uranyl acetate precipitates by passage through a Millipore filter.

Shadow grids at an angle of 6.5° and a distance of 4.5 cm with 3.5 cm of a platinum-iridium (80/20) wire (ϕ 0.1 mm) (if the specifications of the shadowing machine available make another distance between wire and grids necessary, the amount of noble metal must be changed accordingly. Other shadowing materials suitable for DNA preparations are pure platinum or platinum-palladium alloy 80/20).

To obtain an even contrast of DNA it is necessary to rotate the specimen during shadowing. If no facility for rotating the specimen is available, shadow twice, changing the shadowing direction by 90° after the first. The latter shadowing method requires more time, but provides better resolution of crossed DNA strands enabling discrimination between the upper and lower strand.

c) Procedure 2: Spreading

1. Hyperphase (Spreading Solution)

We routinely use a hyperphase containing 30% formamide for visualization of both single and double stranded DNA. The quality of spreading depends on the quality of formamide used. We use Fluka formamide (purissimum, Fluka AG, Buchs, Switzerland), which we routinely purify by twofold recrystallization in the cold. The purity is checked by its absorbance at OD_{270}, which should be less than 0.2. In addition, the spreading mixture has buffer and EDTA as follows:

30% formamide
0.1 M Tris-Cl, pH 8.5
10 mM EDTA

This mixture is prepared as a 2 × stock solution and kept at $-20°C$ in small portions. For spreading, mix 10 μl of the 2 × stock solution with 10 μl of DNA solution (2 μg/ml).

2. Hypophase

Pour 10 ml formamide into a 100 ml graduated Erlenmeyer flask (this formamide may be used without purification, if its pH is not less than 6.0).

Add 2 ml of 0.5 M Tris base and 0.5 ml of 0.2 M EDTA.

Add water to 100 ml and adjust pH to 8.5–8.7 by the dropwise addition of 1 N HCl. Use within 1 h.

3. Spreading

Complete the spreading solution by adding to a 20 µl sample 0.4 µl cyt c (5 mg/ml).

Pour enough of the hypophase solution into a clean plastic petri dish (or teflon trough) such that it forms a convex surface.

Clean the surface with a thoroughly cleansed PVC (or teflon) bar.

Insert a glass ramp into the hypophase solution and support it with a second PVC bar as shown in Fig. 2.

Sprinkle a little talc or graphite powder in front of the ramp tip; it should disperse evenly in a thin layer if the surface is clean.

Apply with a tubing-tipped syringe, 3–5 µl of hyperphase to the ramp about 1 cm from the hypophase surface; it will stream down and spread out on the hypophase, displacing the talc as it does so.

On coated grids pick up parts of the hyperphase from an evenly spread area and stain and shadow as described above. Repeat the spreading several times, since the first may not be satisfactory.

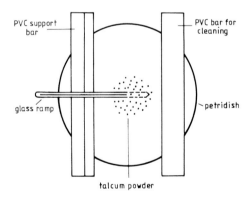

Fig. 2. Schematic representation of DNA spreading on a formamide hypophase

d) Procedure 3: DNA Spreading in Carbonate Buffer on a Water Hypophase

This is an alternative procedure for spreading nucleic acids for electron microscopy and may be used if problems are encountered with the former method. The hyperphase consists of DNA, alkaline Na-carbonate buffer, formamide, and cyt c and is spread on a hypophase of double distilled water. The procedure has a similar spectrum of applications to the previous one, e.g., double and single stranded nucleic acids in heteroduplexes, R-loops, and partially denatured molecules. With this method of preparation, the DNA has a thinner appearance and it is sometimes easier to distinguish between double and single stranded regions. The lengths of double stranded and single stranded DNAs are almost identical. The nucleic acid concentrations needed for this procedure are higher than for a hypophase containing formamide because DNA spreading is more extensive.

Carbonate Buffer

Prepare a fresh solution immediately before spreading by mixing in a plastic tube 400 μl H_2O, 40 μl 1 M Na_2CO_3, and 20 μl 0.2 M EDTA. Store on ice. The pH of this buffer is about 10.5.

Hypophase

Fill a clean plastic petri dish or teflon trough with double-distilled water and clean the surface as described in Sec. 3.

Hyperphase

Mix on ice, 2 μl carbonate buffer, 5 μl formamide, 2 μl DNA (about 10 μg/ml), and 1 μl cyt c (1 mg/ml).

Spreading

Sprinkle some talc or graphite powder on the clean water surface.
Spread 5 μl portions directly on the water surface by touching with the hanging drop at the tip of an Eppendorf pipette.
Pick up the spread film with parlodion-coated grids (carbon-coated grids will give poorly contrasted DNA that has been spread by this method).
Rinse 10 s in 95% ethanol (positive staining is also possible).
Air dry on filter paper.
Shadow the grids with platinum alloy as described previously.

e) Procedure 4: Spreading of Single Stranded Nucleic Acids with T4 Gene 32 Protein

The spreading of single stranded nucleic acids is complicated by their tendency to undergo inter- and intrastrand annealing, which reduces the even spreading and separation of individual strands. This can be prevented by DNA denaturing agents such as formamide (see Sec. 3, b), DMSO (see Sec. 7), and urea. However, even strong denaturing conditions, which are necessary for the spreading of single stranded RNA, do not always give satisfactory results. To improve the visualization of single stranded nucleic acids, Delius and co-workers developed an elegant method based on the use of bacteriophage T4 gene 32 protein (unwinding protein) to bind, extend, and unfold single stranded DNA and RNA.

Gene 32 Protein-Sprading of φX174 Phage DNA

1. Binding of Gene 32 Protein

Mix 16 µl DNA solution (10 µg/ml), 2 µl gene 32 protein (150 µg/ml), and 2 µl 0.1 M potassium phosphate buffer, pH 7.0, and incubate at 37°C for 5 min.

2. Fixation of DNA:Gene 32 Protein Complex

Add 1 µl of 2% glutaraldehyde and incubate at 37°C for 10 min.

3. Spreading

Mix 0.5 µl of sample with 0.25 ml droplet solution and continue preparation as described in Sec. 3, b. Gene 32 protein bound DNA may also be spread using the classical technique (Sec. 3.c and d).

f) References

Davis RW, Simon M, Davidson N (1971) Electron microscope heteroduplex methods for mapping regions of base sequence homology in nucleic acids. Methods Enzymol 21:413–428

Delius HJ, Westphal H, Axelrod N (1973) Length measurements of RNA synthesized in vitro by *Escherichia coli* RNA polymerase. J Mol Biol 74:677–687

Evenson DP (1977) Electron microscopy of viral nucleic acids. Methods Virol 6:219–264

Inman RB, Schnös M (1970) Partial denaturation of thymine- and 5-bromo-uracil-containing DNA in alkali. J Mol Biol 49:93–98

Kleinschmidt AK (1968) Monolyer techniques in electron microscopy of nucleic acid molecules. Methods Enzymol 12:361–377

Kleinschmidt AK, Zahn RK (1959) Über Desoxyribonucleinsäure-Moleküle in Protein-Mischfilmen. Z Naturforsch 14b:770–779

Meyer J (1981) Electron microscopy of viral RNA. Current Topics in Microbiology and Immunology, vol 94/95, pp 210–241

Morris CF, Sinha NK, Alberts BM (1975) Reconstruction of bacteriophage T4 DNA replication apparatus from purified components: Rolling circle replication following de novo chain initiation on a single-stranded circular DNA template. Proc Natl Acad Sci USA 72:4800–4804

Vollenweider HJ (1981) Visual biochemistry: new insight into structure and function of the genome. Methods Biochem Anal (in press)

Westmoreland BC, Szybalski W, Ris H (1969) Mapping of deletions and substitutions in heteroduplex DNA molecules of bacteriophage Lambda by electron microscope. Science 163:1343–1348

Younghusband HB, Inman RB (1974) The electron microscopy of DNA. Annu Rev Biochem 43:605–619

4. Length Measurement of DNA

a) Introduction

Visualization of nucleic acids permits the determination of molecule length by measurement of molecules on an enlarged drawing or print with a map measurer or a digitizer. More recently, partly and fully automated procedures for the measurement of molecules have been described (Littlewood and Inman 1982).

The measurement of a number of molecules of each type (typically 20–200) and a statistical evaluation of the results are essential. Length deviations of identical nucleic acid molecules are caused by errors in measurement and in magnification. In addition to grid-to-grid variation, lengths of molecules on the same grid may vary due to local distortion of the support film, because the length of DNA is very sensitive to torsional forces. If some DNA molecules are visibly stretched, i.e., show preferentially oriented regions, they should not be used for length measurements.

The most precise magnification calibration is obtained by addition to the sample of DNA to be measured, an internal standard, a DNA species of known molecular weight, such as plasmids pBR322 (2.89×10^6 = 4.36 kb) or ColE1 (4.21×10^6 = 6.34 kb), or DNA from phage PM2 (6.64×10^6 = 10.02 kb), as double strand standards, or DNA from phage ϕX174 (1.7×10^6 = 5375 bases), as a single stranded standard. All of these molecules are circular, which enables ready recognition of full length molecules and permits measurements to be restricted to such molecules. Standards should be present on micrographs in excess, in order to reduce errors from poor spreading. Determination of the molecular weight of an unknown DNA is achieved by measurement of a sufficient number of molecules of the unknown and the standard, calculation of the length of the unknown relative to the known, and conversion of the relative value to a molecular weight by multiplication by the molecular weight of the standard DNA. Because plasmids such as pBR322 tend to form dimers, it is important to use a preparation that contains mostly monomers and to be able to recognize dimers on the grid.

The magnification of the microscope can also be calibrated by means of an external standard, although values obtained are less accurate. In the magnification range mostly used for electron microscopy of DNA (about 10,000 X), the most suitable standard is a carbon replica grating (available, e.g., from Balzers AG; distance between lines: 462.9 nm), which is photographed at the same magnification as the sample. The molecular weight of double stranded DNA can be calculated from the measured length of the DNA using the following factor:

1 μm DNA length = 2.07×10^6 (= 3 kb)

b) References

Danbara H, Timmis JK, Lurz R, Timmis KN (1980) Plasmid replication functions: Two distinct segments of plasmid R1, RepA and RepD, express incompatibility and are capable of autonomous replication. J Bacteriol 44:1126–1138

Lang D (1970) Molecular weights of coliphages and coliphage DNA. Contour lengths and molecular weight of DNA from bacteriophages T4, T5 and T7, and from bovine papilloma virus. J Mol Biol 54:557–565

Littlewood RK, Inman RB (1982) Computer-assisted DNA length measurements from electron micrographs with special reference to partial denaturation mapping. Nucleic Acids Res 10:1691–1706

Sanger F, Air GM, Barrell BG, Brown NL, Coulson AR, Fiddes JC, Hutchison CA, Slocombe PM, Smith M (1977) Nucleotide sequence of bacteriophage ϕX174 DNA. Nature 265:687–695

Stüber D, Bujard H (1977) Electron microscopy of DNA: determination of absolute molecular weights and linear density. Mol Gen Genet 154:299–303

Sutcliffe JG (1979) Complete nucleotide sequence of the *Escherichia coli* plasmid pBR322. Cold Spring Harbor Symp Quant Biol 43:77–90

---

3 kb: kilobases for single stranded DNA and kilobase pairs for double stranded DNA

5. Homoduplex Preparation: Visualization of Repeated Sequences with Inverse Orientation (Inverted Repeats)

a) Introduction

Many plasmids contain inverted repeats, some of which have known functional roles in recombination, e.g., IS sequences in transposons. They can be visualized in electron microscopical homoduplex experiments, where they exhibit typical structures in single and double stranded DNA molecules. The preparation of homoduplexes is identical to that of heteroduplexes (see below), except that a single species of DNA is used, and consists of three steps (Fig. 3): (1) denaturation of DNA, i.e., separation into single strands (Fig. 4a); (2) renaturation of DNA, leading to the reformation

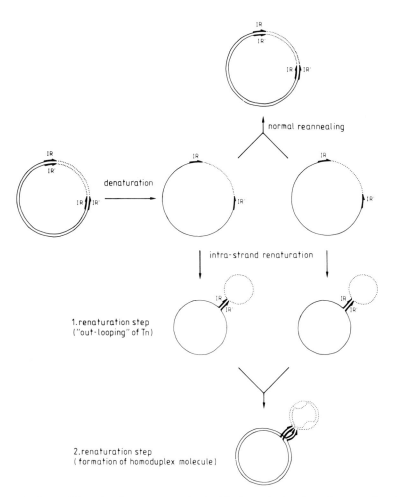

Fig. 3. Possible molecular structures obtained by homoduplex formation experiment

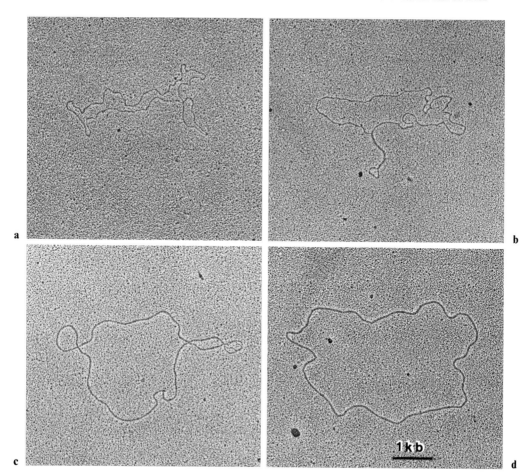

Fig. 4a–d. Electron micrographs of a homoduplex preparation of plasmid pKT710. Plasmid pKT-710 consists of ColE1 plus the *Eco*RI-generated E-6 fragment of plasmid R6-5. Fragment E-6 contains the kanamycin resistance transposon Tn*601/903*, which has terminal inverted repeats of about 1 kb. The DNA was in carbonate buffer and was spread on a water hypophase. Grids were shadowed as in Fig. 1. **a** shows a single stranded pKT710 molecule, whereas **b** shows a molecule in which the inverted repeat sequence of the transposon has annealed – first renaturation step of Fig. 3; **c** the second renaturation step can be seen in which the two complementary strands of pKT710 have reannealed; note that the two single stranded loops did not form a complete double strand, because of the topological constraints imposed upon the annealing of two circular strands (absence of free ends that can rotate); **d** an example of a renaturation that formed the parental molecule without any annealing of the inverted repeats

of double stranded parental DNA (Fig. 4d) and to intrastrand annealing of inverted repeats and the formation of homoduplexes (Fig. 4b, c); and (3) spreading of DNA. A dilution step which improves the quality of single stranded DNA preparations may be introduced prior to spreading. It is generally advisable to carry out renaturation for 2 periods of time, each of which will favor a different structure. Short renaturation times permit intrastrand annealing to occur (1st renaturation step, Figs. 3 and 4b), whereas longer renaturation times also permit interstrand annealing and the formation of homoduplexes (2nd renaturation step; Figs. 3 and 4c), which are unambiguous indicators of inverted repeats. A DNA segment bracketed by inverted repeats (the single strand loop) is best measured on single stranded DNA molecules because the loops of homoduplexes are generally not relaxed and are consequently poorly spread. This is, however, more difficult with short inverted repeats (≤ 100 bp) because the annealed repeats are not readily discerned from a single stranded crossover. In such cases, the existence of inverted repeats can be clearly demonstrated only in homoduplexes.

b) Procedure

Mix in a small Eppendorf tube 5 μl purified formamide (lowers melting point, which is important for renaturation), 1 μl 1 M phosphate buffer, pH 7.0 and 3 μl DNA, and boil for 1.5 min in a water bath (denaturation step).

Add 1 μl 2 M $NaClO_4$ (favors renaturation) and incubate at 40°C in a water bath (renaturation step).

Take aliquots of 2 μl after incubation for 5 min and 30 min (the shorter time allows the renaturation step 1, the longer time, step 2; Fig. 3) and dilute with 10 μl purified formamide and 8 μl Na_2 EDTA (1 mM). This steps dilutes the sodium perchlorate and improves the spreading of single stranded DNA.

Add 0.4 μl cyt c (5 mg/ml) to each sample and spread over 30% formamide in Tris and EDTA, as described in Sec. 3, c).

Pick up DNA, stain and shadow in the usual way.

c) References

Kleckner N, Chan RK, Tye BK, Botstein D (1975) Mutagenesis by insertion of a drug resistance element carrying an inverted repetition. J Mol Biol 97:561–575

Burkardt HJ, Riess G, Pühler A (1979) Relationship of group P1 plasmids revealed by heteroduplex experiments: RP1, RP4, R68 and RK2 are identical. J Gen Microbiol 114:341–348

6. Heteroduplex Procedure

a) Introduction

This extremely useful technique, developed by Davis and Davidson (1968) and Westmoreland et al. (1969), is employed for the analysis of homology between two different DNA species. When these are mixed, denatured, and renatured together, not only homoduplexes, but also hybrids will be formed, if the two types of DNA molecule share homology. After formamide spreading, the unpaired heterologous DNA hybrids will be recognizable as single stranded DNA segments, whereas the homologous DNA will appear as double stranded DNA segments (Fig. 6). The various structures seen in heteroduplexes are shown in Fig. 5.

The quality of the DNA used in heteroduplex experiments is very important for the yield of heteroduplex molecules obtained which are suitable for evaluation. The following points should be observed:

1. The DNA should be relatively pure (plasmid DNA that has been purified through two cycles of dye-CsCl equilibrium centrifugation is usually suitable) and should be in a buffer that does not contain any substances which disturb spreading.

2. The DNA should be fresh. During storage, DNA molecules accumulate single strand breaks that do not influence the length of duplex molecules, but generate single stranded molecule fragments after denaturation. Generally speaking, only full length molecules provide reliable measurements. For circular DNA, the ratio between open (OC) and covalently closed (CCC) DNA should be about 1:1. Linear DNA fragments generated by a restriction enzyme are very convenient for heteroduplexing and their ends provide precise reference points. Restriction enzyme preparations do, however, frequently contain enzymes that nick DNA. Therefore, it is important to use enzyme-cleaved DNA samples as soon as possible for heteroduplexing.

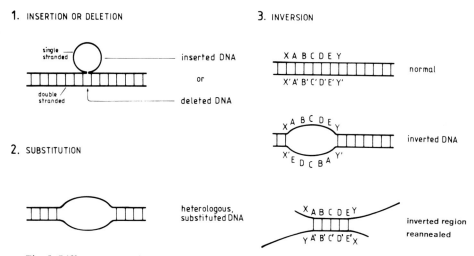

Fig. 5. Different types of hybrid forms of DNA which may be obtained in heteroduplex experiments

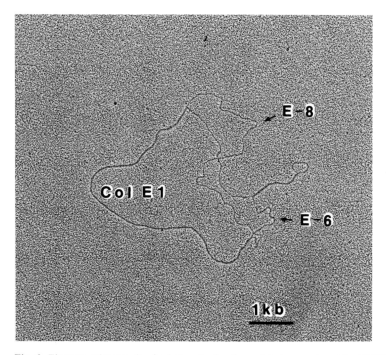

Fig. 6. Electron micrograph of a heteroduplex formed between plasmids pKT710 and pKT670. Plasmid pKT710 is a hybrid between ColE1 and the E-6 fragment of R6-5 (Fig. 4), whereas pKT670 is a hybrid between ColE1 and the E-8 fragment of R6-5. Thus, only ColE1 is common to the two plasmids: this is seen as the duplex region of homology. The E-6 fragment of pKT710 is recognized as the single stranded segment containing the stem and loop structure of Tn*601/903*, whereas the E-8 fragment of the pKT670 is the other segment, which is seen to contain no inverted repeat

3. The DNA concentration should be at least 10 µg/ml; the renaturation time for DNA at lower concentrations must be greatly extended.

4. To obtain the maximal yield of heteroduplexes, the two DNA species to be hybridized should be present at a molar ratio of 1:1.

The proportion of measurable molecules obtained is reduced by the formation of nonhybrid parental DNA, network formation by incomplete heteroduplexes derived from subunit length fragments which aggregate because of sticky, single stranded regions, and by poor spreading.

b) Procedure 1

Mix 5 µl purified formamide with 1.5 µl DNA 1, 1.5 µl DNA 2, and 1 µl 1 M phosphate buffer, pH 7.0. Denature, renature, dilute, spread, stain, and shadow DNA as described for homoduplex preparation.

Modification: heteroduplex formation with bacteriophage DNAs.

Heteroduplex formation between bacteriophage DNA genomes does not require prior isolation of DNA from the phage particles, because DNA release and denaturation may be carried out in a single step, by alkali treatment. Heteroduplex analysis of phage DNA has two advantages over that of plasmid DNA:

1. high DNA concentrations are readily obtained because phage particles can be pelleted and
2. the quality of phage DNA is generally very high because it is protected in the phage head from nucleases and shearing forces.

c) Procedure 2

Mix 3 μl H$_2$O, 2 μl 0.02 M EDTA (pH 8.5), 3.6 μl 1 mM EDTA (pH 7.0), 1.4 μl 1 N NaOH, 0.2 μl phage 1, and 0.2 μl phage 2. The ratio of phage 1 and phage 2 particles should be 1:1. This recipe is appropriate for phage preparations containing 10^{11} pfu/ml. For less concentrated preparations take more phage solution and correspondingly less 1 mM EDTA.

Allow the mixture to stand at 20°C for 10 min (denaturation step).

Neutralize by addition of 1 μl 2 M Tris (pH 7.2), add 11 μl formamide and allow to renature for 20 min at 40°C.

Add 0.5 μl cyt c (5 mg/ml) and spread over 30% formamide with Tris and EDTA as described in Sec 3, c.

d) References

Davis RW, Davidson N (1968) Electron-microscopic visualization of deletion mutations. Proc Natl Acad Sci USA 60:243–250

Davis RW, Simon M, Davidson N (1971) Electron microscope heteroduplex methods for mapping regions of base sequence homology in nucleic acids. Methods Enzymol 21:413–428

Westmoreland BC, Szybalski W, Ris H (1969) Mapping of deletions and substitutions in heteroduplex DNA molecules of bacteriophage lambda by electron microscopy. Science 163:1343–1348

7. Partial Denaturation

a) Introduction

The technique of partial denaturation analysis of duplex DNA was developed by R.B. Inman (1966), who used the method to characterize bacteriophage λ DNA. The conditions chosen for denaturation are such that only partial strand separation takes place (Fig. 7) This separation occurs at nonrandomly located sites along the molecule that are characterized by their high content of A-T base pairs (A-T base pairing, which involves two hydrogen bonds, is less stable than G-C pairing, which involves three). Denatured regions that exceed 50–60 bp (in the case of normal cyt c preparations and bright field microscopy) appear as "bubbles" in the duplex, whose sizes and

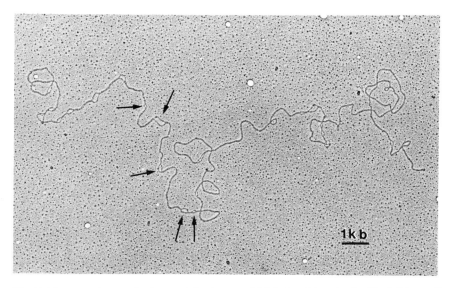

Fig. 7. Electron micrograph of a partially denatured RP4 plasmid molecule. The DNA was linearized by cleavage with *Eco*RI endonuclease, denatured, and spread on a formamide-containing hypophase as described in Sec. 3, c, positivley stained with uranyl acetate, and rotary shadowed with platinum-iridium. *Arrows* indicate some of the denaturation bubbles

positions can be measured. Strand separation in DNA segments with similar A-T content is statistical. Furthermore, different molecules in a single DNA preparation often exhibit a very broad range of degrees of denaturation, i.e., from native to totally denatured molecules. It is therefore essential to measure a number of molecules in order to construct a partial denaturation (A-T) map (Fig. 8).

A typical denaturation histogram consists of an x-axis, which represents the total molecule length, subdivided into fractions for convenient delineation of denatured and undenatured sites, and a y-axis, which shows the number of molecules that are denatured in a particular region. The height of each denaturation peak indicates the probability that such a site will denature under the given reaction conditions and enables its relative A-T content to be deduced. The construction of an A-T map requires alignment of all molecules included in the analysis; this is usually straightforward because the distribution of A-T-rich sites on most DNA molecules is not symmetric and molecule orientation is readily determined. A minority of DNA species, particularly circular molecules may not, however, exhibit an obvious orientation and computerized evaluation (Burkardt et al. 1978, 1979) may be helpful.

Denaturation maps can be used:

1. to characterize DNA molecules;
2. to examine the relationships of different DNAs;
3. to determine the relative A-T-richness of specific DNA segments, e.g., of particular genes or promotor sites.

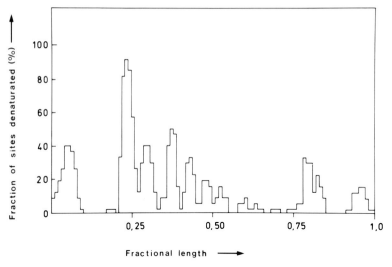

Fig. 8. Partial denaturation histogram (A-T-map) of plasmid RP4

b) Procedure

Denaturation

Mix 0.25 ml DMSO (dimethyl sulfoxide, which lowers temperature required for partial DNA melting), 0.21 ml 0.2 M phosphate buffer, pH 7.0, 0.14 ml formaldehyde (37%, which prevents single strands from reannealing), and add 0.03 ml of this mixture to 0.02 ml DNA solution (concentration about 1–10 μg/ml).

Heat denature to about 55°C for 16–20 min (denaturation temperature and time can be varied according to the G-C content of the DNA to be tested and on the desired extent of denaturation).

Removal of Formaldehyde

The residual free formaldehyde in the denaturation solution interferes with cyt c during the DNA spreading. It must therefore be removed by passage of the solution through Sephadex G 50 as follows:

Equilibrate a small Sephadex G 50 column (5 cm Sephadex in a Pasteur pipette) with chilled 1 mM EDTA.

Apply 20 μl of the denaturation mix to the column, elute with chilled 1 mM EDTA and collect 20 μl volume fractions.

Pool DNA-containing fractions (determine which these will be by use of radioactively labelled DNA in a prior experiment; should be 500–540 μl of the elution volume).

Spreading

Spread 10 μl of the eluate mixed with 10 μl hyperphase (2 × stock solution) over 10% formamide as described in Sec. 3, c).

c) References

Inman RB (1966) A denaturation map of the λ phage DNA molecule determined by electron microscopy. J Mol Biol 18:464–476
Burkardt HJ, Mattes R, Pühler A, Heumann W (1978) Electron microscopy and computerized evaluation of some partial denatured group P resistance plasmids. J Gen Microbiol 105:51–62
Burkardt HJ, Pühler A, Wohlleben W (1979) Adenine + thymine content of different genes located on the broad host range plasmid RP4. J Gen Microbiol 117:135–140

8. Visualization of Protein-Nucleic Acid Complexes

a) Introduction

Use of cyt c for DNA spreading results in a severalfold increase in the diameter of the DNA which can reach 100 Å and more after staining and shadowing. In general, this thickening of the DNA strands is desirable and improves their visualization in the microscope. It is undesirable, however, for examination of DNA:protein complexes because the protein components are obscured by the protein used for spreading. Many attempts have therefore been made to develop procedures for DNA visualization that result in less enlargement of DNA strand diameter.

One way to accomplish this is to replace cyt c with a compound that has similar properties, but a lower molecular weight. Vollenweider et al. (1975) demonstrated that the quaternary ammonium salt, benzyldimethyl-alkyl-ammonium chloride (BAC), which has a molecular weight of about 350, forms a surface film to which DNA can adsorb. DNA prepared with this method has a strand diameter of about 40–60 Å after shadowing, which is low enough for the visualization of all but small DNA-bound proteins. As with cyt c, BAC can be used for spreading as well as for the droplet method. BAC-based methods are the only protein-free procedures which give good results with both double stranded and single stranded nucleic acids. However, they are very sensitive to contaminants and good reproducibility is sometimes difficult to achieve.

Other methods developed for visualization of protein:DNA complexes involve direct adsorption of DNA to a special surface, without use of mono-film. No enlargement of the diameter of nucleic acids other than that produced by the staining process occurs with such procedures (for review, see Fisher and Williams 1979). The properties of the surface to which the nucleic acids adsorb are critical; an ideal surface should be hydrophilic with an evenly distributed positive charge to adsorb negatively charged nucleic acids.

Nucleic acids adsorb to an untreated carbon film (Nanninga et al. 1972), but molecules are often excessively stretched and their appearance is poor. The method of preparing the carbon film and the time between film preparation and use greatly influence the quality of nucleic acids observed; for example, freshly prepared carbon films adsorb DNA relatively well, but older ones do not. Several methods have been developed to improve the properties and reproducibility of carbon films for the adsorption of nucleic acids. These include the treatment of carbon films with amylamine vapor by glow discharge (Dubochet et al. 1971), cyt c or polylysine, to obtain a more evenly distributed positive charge. Treatment with BAC, ethidium bromide, or alcian blue also improves the quality of carbon films for adsorption of nucleic acids.

The method which in our experience gives the best and most reproducible results involves the use of freshly cleaved mica for adsorption of DNA in the presence of ethidium bromide or Mg^{++} ions. The presence of ethidium bromide, which increases the rigidity of double stranded DNA by intercalation between the base pairs, reduces DNA coiling and improves the visualization of long molecules. After adsorption of DNA, the mica pieces are washed extensively with water, dehydrated in ethanol, and dried. They are subsequently rotary shadowed with heavy metal and afterwards shadowed with carbon. The carbon replica is then floated onto a water surface and picked up on a neoprene grid (Koller et al. 1974, Portmann and Koller 1976). Although the mica adsorption procedure is highly sensitive to traces of detergents and other contaminants, in our experience it provides DNA molecules that exhibit the smallest standard deviation in their lengths compared with molecules prepared by other methods. The procedure is thus not only suitable for analysis of DNA:protein complexes, but also for double stranded DNA. However, like other direct adsorption methods, the mica procedure is not useful for examination of single stranded DNA or RNA.

b) Procedure 1: BAC Spreading

Pretreatment of Carbon Grids

Prepare carbon-coated 400 mesh grids as described in Sec. 2, c.
Float the grids for 10–25 min on a filtered (Millipore) solution of ethidium bromide prepared by 100-fold dilution in water of a 10 mg/ml stock solution.
Blot on filter paper and use immediately.

For hyperphase mix:
19 µl formamide, 1 µl BAC (0.2% in formamide), and 5 µl DNA (10 µg/ml).
Use water for hypophase and make spreading of 10 µl portions as described in Sec. 3, c and d) for spreading with carbonate buffer.
Wash grids on water for 15 min, positive stain for 30 s with uranyl acetate, and blot on filter paper.
Examine grids using dark field conditions or rotary shadow with heavy metal.

c) Procedure 2: Preparation of DNA or DNA:Protein Complexes by Adsorption to Mica

The DNA, in either 10 mM Tris-Cl, pH 7.4, 10 mM $MgCl_2$, 6 mM KCl, 6 mM DTT or similar buffer should have a concentration of 0.5–1 µg/ml.
Cut pieces (about 1 cm × 1 cm) of mica (Balzers Union, Liechtenstein), mark the outside surfaces with black felt marker and cleave carefully.
Put droplets (about 20 µl) of the DNA solution on a piece of parafilm and cover with a piece of the freshly cleaved mica (the droplet should thus be spread over the whole area of the mica sheet).

After DNA adsorption for 5 min, wash the mica by floating on a water surface (filled petri dish) for 1 h, and transfer onto an aqueous solution of 2% uranyl acetate for 2 min positive staining.

Dehydrate 10 s in 95% ethanol and rotary shadow as described in Sec. 3, b.

Cover the shadowed site of the mica with a carbon film as described in Sec. 2.

Cut small pieces (2.5 mm × 2.5 mm) from the carbon-coated mica and float the carbon replica on a water surface.

Pick up the carbon replica with neoprene-coated 400 mesh copper grids and examine in the electron microscope (Fig. 9).

Some carbon replicas are difficult to separate from the mica sheet. This problem is usually resolved by dipping the tips of the carbon rods used for evaporation into a solution of 1 M NaCl and drying them before use (Stettler et al. 1979). Separation of carbon films from mica is also improved by storage of the prepared specimen in a wet chamber (petri dish containing wet filter paper) for several hours or overnight.

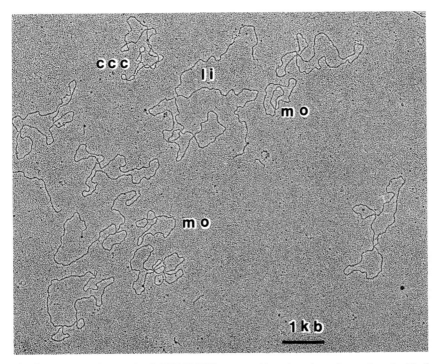

Fig. 9. Electron micrograph of pBR322 DNA prepared by mica adsorption as described in Sec. 8. The pBR322 preparation is the same one used for the droplet technique in Fig. 1. Most of the molecules are open circular dimers, one is supertwisted *(ccc)*, two are linear dimers *(li)* and two open circular monomers *(mo)*

d) References

Abermann R, Salpeter MM (1974) Visualization of deoxyribonucleic acid molecules by protein film adsorption and tantalum-tungsten shadowing. J Histochem Cytochem 22:845–855

Arcidiacono A, Stasiak A, Koller Th (1980) Protein free specimen preparation of nucleic acids and nucleic acid-protein complexes. Proc 7th Eur Cong Electr Micr August 24–29, 1980. The Hague 2:516–523

Chattoraj DK, Gosule LC, Schellmann JA (1978) DNA condensation with polyamines. II Electron microscopic studies. J Mol Biol 121:327–337

Dubochet J, Ducommun M, Zollinger M, Kellenberger (1971) A new preparation method for dark-field microscopy of biomacromolecules. J Ultrastruct Res 35:147–167

Fisher HW, Williams RC (1979) Electron microscopic visualization of nucleic acids and of their complexes with proteins. Annu Rev Biochem 48:649–679

Griffith JD, Christiansen G (1978) Electron microscopic visualization of chromatin and other DNA-protein complexes. Ann Rev Biophys Bioeng 7:19–35

Koller T, Sogo JM, Bujard H (1974) An electron microscopic method for studying nucleic acid-protein complexes. Visualization of RNA polymerase bound to the DNA of bacteriophage T7 and T3. Biopolymers 13:995–1009

Nanninga N, Meyer M, Sloof P, Reijnders L (1972) Electron microscopy of *Escherichia coli* ribosomal RNA: spreading without a basic protein film. J Mol Biol 72:807–810

Portmann R, Koller T (1976) The divalent cation method for protein-free spreading of nucleic acid molecules. In: Ben-Shaul Y (ed) Sixth European Congr Electron Microscopy Jerusalem, vol 2. Tal, Israel, pp 546–548

Sogo JM, Rodeno P, Koller Th, Vinuela E, Salas M (1979) Comparison of the AT-rich regions and the Bacillus subtilis RNA polymerase binding sites in phage ϕ29 DNA. Nucleic Acids Res 7:107–120

Stettler UH, Weber H, Koller Th, Weissmann Ch (1979) Preparation and characterization of form V DNA, the duplex DNA resulting from association of complementary, circular single-stranded DNA. J Mol Biol 131:21–40

Vollenweider HJ, Sogo JM, Koller T (1975) A routine method for protein free spreading of double- and single-stranded nucleic acid molecules. Proc Natl Acad Sci USA 72:83–87

Williams RC (1977) Use of polylysine for adsorption of nucleic acids and enzymes to electron microscope specimen films. Proc Natl Acad Sci USA 74:2311–2315

Chapter 7 Transcription

7.1 Mapping of RNA Polymerase Binding Sites by Electron Microscopy

R. LURZ[1] and H. BURKARDT[2]

Contents

1. General Introduction 304
2. Objective 306
3. Procedure 306
4. Evaluation 307
5. References 308

1. General Introduction

DNA-dependent RNA polymerase catalyzes the synthesis of RNA copies (transcripts) of defined segments of DNA that usually correspond to genes or groups of genes (operons). Most types of transcripts are destined to be translated by ribosome into polypeptide gene products. In the first stage of the process of transcription, RNA polymerase recognizes and binds to a specific sequence on the DNA, a promoter, just upstream of the point at which transcription is initiated.

Many promoters have now been sequenced and shown to exhibit certain similarities (i.e., in the -10 and -35 regions and the general A-T richness upstream of the -35 region), but also considerable diversity; that is, there is no unique sequence that defines a promoter. One reason for this heterogeneity is that gene expression is usually regulated at the level of initiation of transcription and different promoter sequences and their flanking regions specify different affinities for RNA polymerase and regulatory proteins. A DNA sequence per se does not, therefore, define a promoter and in any given gene-sized DNA segment, there are usually several promoter-like sequences, only one of which may actually be used in vivo (for reviews, see Rosenberg and Court 1979, Bujard 1980).

If DNA is incubated with RNA polymerase in vitro, it forms stable initiation complexes at promoter sites. These binary complexes can be visualized by electron microscopy combined with protein-free DNA preparation methods. Since RNA polymerase

---

1 Max-Planck-Institut für Molekulare Genetik, Ihnestraße 63–73, D-1000 Berlin 33, Fed. Rep. of Germany
2 Institut für Mikrobiologie und Biochemie, Lehrstuhl für Mikrobiologie der Universität Erlangen/Nürnberg, Egerlandstr. 7, D-8520 Erlangen, Fed. Rep. of Germany

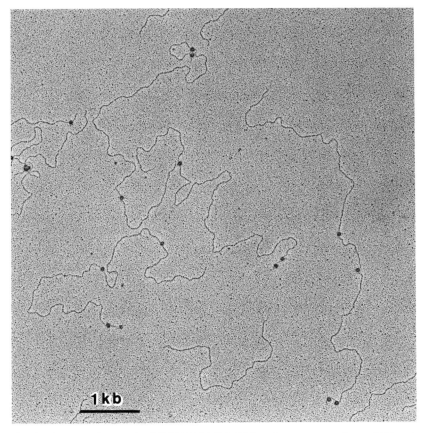

Fig. 1. Electron micrograph of RSF1010 DNA:RNA polymerase complexes. RNA polymerase was bound to ccc DNA and the binary complexes were cleaved with endonuclease *Bst*EII

has a molecular weight of about 5×10^5, the protein molecules are readily detected as dots on DNA strands (Fig. 1). This technique permits accurate localization (precision about ± 50 bp with a DNA fragment of about 10 kb) and determination of the relative strengths of RNA polymerase binding (RPB) sites on DNA, but gives no information on the orientation of transcription. Electron microscopy of RNA polymerase:DNA complexes is particularly suitable for demonstrating the existence of several RPB sites in one promoter region, as is the case for rRNA genes (Kiss et al. 1980).

As with other in vitro methods, results obtained need not necessarily reflect in vivo functions, e.g., if a promoter is positively regulated, it may not exhibit substantial affinity for RNA polymerase in vitro, in the absence of the specific positive regulator. However, in all studies made so far, excellent correlation between in vitro RPB sites and known active promoters has been observed.

A prerequisite for the localization of RPB sites by electron microscopy is the availability of highly purified holoenzyme. Another critical parameter of the method is the ratio of RNA polymerase to DNA in the binding mixture. If the ratio is low, the enzyme will bind only to strong binding sites and weak sites will not be detected; if too high, unspecific binding will occur (Kadesch et al. 1980). Under optimal conditions about 90% of the RNA polymerase molecules will bind to specific sites. In order to determine the optimal conditions for the binding of RNA polymerase to any given DNA, it is necessary to test several enzyme:DNA ratios, starting with 1:1 (w/w).

The mapping of RPB sites on circular genomes requires linearization of the DNA molecules at precise locations, usually by cleavage with restriction endonucleases. However, because RNA polymerase binds more strongly to negatively supercoiled circular (ccc) molecules than to linear or relaxed circular forms, it is preferable to carry out enzyme binding with supercoiled DNA and, after fixation of the enzyme: DNA complexes, to linearize the DNA prior to preparation for electron microscopy (Bagdasarian et al. 1981).

2. Objective

Localization of RNA polymerase binding sites on plasmid RSF1010.

3. Procedure

RNA polymerase holoenzyme and plasmid RSF1010 DNA are allowed to form complexes which are subsequently crosslinked by treatment with glutaraldehyde. Complexes are separated from unbound polymerase and glutaraldehyde by gel filtration on Sepharose 4B. Purified complexes are then cleaved with a restriction endonuclease and prepared for electron microscopy.

Mix in an Eppendorf tube 25 μl 2 × reaction buffer (60 mM Tris-Cl, pH 7.9, 100 mM KCl, 20 mM $MgCl_2$), 14 μl H_2O, 10 μl DNA (supercoiled RSF1010, 50 μg/ml) and 1 μl E. coli RNA polymerase holoenzyme (400 μg/ml) and incubate 10 min at 37°C.

Add 10 μl reaction buffer contaning 0.6% glutaraldehyde and incubate 15 min at 37°C.

Pass sample over a Sepharose 4B column (small Pasteur pipette, 2/3 filled and equilibrated with general restriction enzyme buffer: 10 mM Tris-Cl, pH 7.4, 10 mM $MgCl_2$, 6 mM KCl, 6 mM DTT), and collect the flow-through fractions (void volume determined previously with dextran blue).

Plasmid RSF1010 has unique cleavage sites for endonucleases BstII, PvuII, and EcoRI; divide solution containing the fixed complexes into three portions and add to each ca. 1 unit of one of these restriction enzymes.

Incubate solutions 30 min at 37°C, pass again over a Sepharose 4B column to remove substances in the enzyme solutions that interfere with DNA spreading (e.g., EcoRI preparations usually contain the detergent NP-40 which makes mica adsorption impossible), and prepare for electron microscopy using adsorption to a mica surface as described in Chap. 6, Sec. 8, c.

4. Evaluation

Photograph the complexes seen in the electron microscops at a magnification of about 15,000 ×. Some selection of DNA molecules is necessary; those with ⩽ 1 bound enzyme are not generally useful for analysis and probably represent molecules that were relaxed circles or linear molecules during the enzyme binding reaction and therefore exhibited low affinity for RNA polymerase. If most of the complexes show more than 5 bound enzymes, the ratio of enzyme:DNA was too high and much nonspecific binding will be evident. Ideally, the majority of DNA molecules should exhibit 2–5 bound RNA polymerases and molecules not falling into this category should be excluded from the analysis. DNA molecules with a loop and an enzyme at the crossover point must also be excluded because the position of the enzyme on such DNAs is ambiguous. For measuring see Chap. 6, Sec. 4.

It is possible to evaluate enzyme:DNA complexes by copying and measuring each photographed molecule (Chap. 6, Sec. 4) by hand and by orienting by eye. It is much easier and more objective, however, to make the analysis by computer. We currently use a program which was developed by G. Huber and P. Müller (Compulab, Erlangen) and provides mainly the following:

Molecules which differ in length from the mean length by more than a given fraction are excluded from the analysis. The molecules retained are then normalized to a unit length and oriented according to best fit of the polymerase loacations on a simultaneously produced map. The print-out provides (1) the positions of the bound enzymes on each DNA molecule used for the analysis; (2) the identity of the molecules that are excluded; (3) the number of peaks of RPB sites; (4) the positions of such peaks; (5) the number of RNA polymerases found in each peak; and (6) a summary histogram of the distribution of RNA polymerase binding sites along the molecule length (Fig. 2).

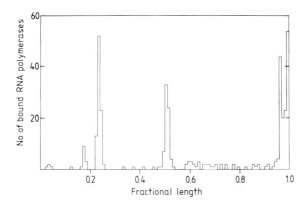

Fig. 2. Histogram of RNA polymerase binding sites on plasmid RSF1010. RPB sites were analyzed by computer. The *abscissa* represents the normalized molecule length. The *ordinate* shows the number of DNA molecules having an RNA polymerase bound at a given position

5. References

Bagdasarian M, Lurz R, Rückert B, Franklin FCH, Bagdasarian MM, Frey J, Timmis KN (1981) Specific-purpose plasmid cloning vectors. II. Broad host range, high copy number, RSF1010-derived vectors, and a host-vector system for gene cloning in *Pseudomonas*. Gene 16:237–247

Bujard H (1980) The interaction of *E. coli* RNA polymerase with promoters. TIBS 274–278

Kadesch TR, Williams RC, Chamberlin MJ (1980) Electron microscopic studies of the binding of *Escherichia coli* RNA polymerase to DNA. II. Formation of multiple promotor-like complexes at non-promotor sites. J Mol Biol 136:79–93

Kiss I, Boros I, Udvardy A, Venetianer P, Delius H (1980) RNA-polymerase binding at promoters of the rRNA genes of *Escherichia coli*. Biochim Biophys Acta 609:435–447

Rosenberg M, Court D (1979) Regulatory sequences involved in the promotion and termination of RNA transcription. Annu Rev Genet 13:319–353

7.2 Determination of the Startpoints and Orientation of in Vitro Transcripts by Electron Microscopy of R-Loops

R. EICHENLAUB and H. WEHLMANN [1]

Contents

1. General Introduction 309
2. Experiment 1: R-Loop Analysis of in Vitro Transcripts Made from the Replication
   Origin Region of Plasmid F 310
   a) Introduction 310
   b) Objective 310
   c) Procedure 310
   d) Results 311
3. Materials 313
4. References 313

1. General Introduction

Analysis of gene structure and expression generally requires the mapping of promoter sites, analysis of transcription products, and determination of the direction of transcription. Several procedures for the mapping of promoter sites on prokaryotic DNA molecules by electron microscopy are in current use:

1. Localization of RNA Polymerase Binding Sites. RNA polymerase and DNA are incubated together under conditions which promote the formation of specific transcription initiation complexes, after which they are visualized in the electron microscope (Bordier and Dubochet 1974, Koller et al. 1974, see also 7.1).

2. Analysis of Transcription Complexes. RNA polymerase and DNA are incubated under conditions which permit active transcription to occur. After subsequent addition of phage T4 gene 32 protein to extend the transcripts (see 6.3e), the DNA:RNA complexes are examined by electron microscopy (Delius et al. 1973, Stüber et al. 1978). Length measurements of the transcripts that are attached to the templates permit localization of the startpoints of transcription, determination of transcription lengths, and determination of transcription orientation.

---

[1] Universität Hamburg, Institut für Allgemeine Botanik, Arbeitsbereich Genetik, Ohnhorststraße 18, D-2000 Hamburg 52, Fed. Rep. of Germany

3. Analysis of R-Loops. DNA is transcribed in vitro. The transcripts are hybridized back to the template DNA under partially denaturing conditions that favor the formation of DNA:RNA hybrids. The resulting molecules consist of stretches of double stranded DNA interrupted by short segments (R-loops) comprised of a DNA:RNA hybrid and a single stranded DNA loop. The R-loops are stable and can be analyzed in the electron microscope (Thomas et al. 1976, Brack 1979, Wehlmann and Eichenlaub 1981).

2. Experiment 1: R-Loop Analysis of in Vitro Transcripts Made from the Replication Origin Region of Plasmid F

a) Introduction

Plasmids that are subjected to R-loop analysis are generally transcribed as supercoiled molecules after which they are linearized by cleavage with a restriction endonuclease and incubated with the transcripts under conditions that favor R-loop formation. However, for the identification of promoters on cloned DNA fragments, it is recommended that these be separated from the vector molecule prior to transcription by cleavage of the hybrid plasmid with an appropriate restriction endonuclease. This procedure prevents artifacts resulting from read-through transcription starting at promoters on the vector which proceeds into the cloned fragment or vice versa.

b) Objective

R-loop analysis of the E-5 fragment of plasmid F.

c) Procedure

In Vitro Transcription

To 5 µl of transcription buffer (4 ×), add 5 µl of rNTP mix, 10 µl (0.5–1 µg) of
  *Eco*RI-digested pML31 plasmid DNA, and 2.5–5 µg of RNA polymerase.
Incubate the reaction mixture for 10 min at 37°C.

Hybridization

Add 10 µl of the transcription mix to 40 µl of hybridization buffer and hybridize at
  56°C for at least 2 h before cooling the sample in ice.
Store at 4°C (no significant change in the structure of R-loops is observed over a
  4 week period).

Spreading

Add 2 µl of R-loop DNA to 50 µl of spreading buffer (final concentration of DNA in
  the spreading solution 0.5–1 µg/ml) plus about 0.1 µg of DNA that will serve as

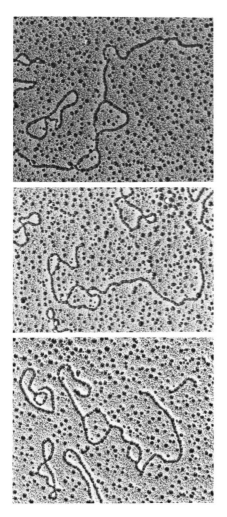

Fig. 1. R-loop molecules obtained using the cloned EcoRI fragment 5 (pML31) of the F-plasmid

an internal length marker (e.g., pBR322, 4362 bp) and spread on a hypophase of 20% formamide, 20 mM Tris-HCl, pH 8.5, 2 mM EDTA.

Pick up the DNA on Parlodion-coated copper grids, contrast, and shadow as described in Chap. 6.1.

Examine the grids in the electron microscope and photograph molecules containing R-loops (see Fig. 1 for typical R-loop molecules).

d) Results

Measure the lengths of the photographed molecules and loops and convert measurements to kb by comparison with the length of the internal length standard.

Very frequently, is is observed that the two branches of an R-loop differ in size, since the single stranded DNA part of the R-loop tends to collapse under the spreading

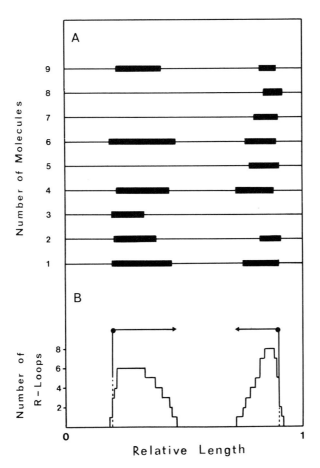

Fig. 2A, B. Evaluation of R-loop molecules. **A** Plot of R-loop molecules versus relative length; **B** cumulative histogram of R-loop molecules shown in **A** (startpoint and direction of transcription are indicated by *arrows*)

conditions. For length measurements, only the longer DNA:RNA hybrid part of a R-loop should be used. Plot the data obtained as shown in Fig. 2A. The orientation of R-loop molecules is not always unambiguous, but clear patterns generally emerge if a large number of molecules is analyzed. Moreover, linearization of the R-loop DNA with different restriction endonucleases and comparison of the R-loop patterns thereby obtained generally resolve all ambiguities. Analysis of R-loop molecules by means of a cumulative histogram as shown in Fig. 2B is also useful. The histogram represents the number of R-loops found for each position along the whole length of the DNA molecule. The steep incline on one side can be interpreted as the region in which transcripts are initiated, whereas the gradual decline on the other represents random termination of transcription. Thus, from the shape of the histogram, one can derive three important facts about each transcript, namely, its origin, orientation, and length.

3. Materials

Transcription buffer (4 ×): 80 mM Tris-HCl, pH 8.0, 40 mM $MgCl_2$, 600 mM KCl, 0.8 mM dithiothreitol.

Nucleotide-triphosphate mix (4 ×): 0.8 mM each of the four rNTP's in water

Hybridzation buffer: 87.5% (v/v) formamide (Fluka, puriss.), 125 mM PIPES (Aldrich), 24 mM Tris, 6.25 mM EDTA, 700 mM NaCl, pH adjusted to 7.8 with 5 N NaOH

Spreading buffer: 70% (v/v) formamide (Fluka, puriss.), 100 mM Tris-HCl, pH 8.5, 10 mM EDTA, 20–60 µg cytochrome c (Boehringer)

Hypophase buffer: 20% (v/v) formamide (Fluka, puriss.), 20 mM Tris-HCl, pH 8.5, 2 mM EDTA

4. References

Bordier C, Dubochet J (1974) Electron microscope localization of the binding sites of *Escherichia coli* RNA polymerase in the early promoter region of T7 DNA. Europ J Biochem 44: 617–624

Brack C (1979) Electron microscopic analysis of transcription: Mapping of initiation sites and direction of transcription. Proc Natl Acad Sci USA 76:3164–3168

Delius H, Westphal H, Axelrod N (1973) Length measurement of RNA synthesized in vitro by *Escherichia coli* RNA polymerase. J Mol Biol 74:677–687

Koller T, Sogo J, Bujard H (1974) An electron microscope method for studying nucleic acid-protein complexes. Visualization of RNA polymerase bound to the DNA of bacteriophage T7 and T3. Biopolymers 13:995–1009

Stüber D, Delius H, Bujard H (1978) Electron microscopic analysis of in vitro transcription complexes: Mapping of promoter of the coliphage T5 genome. Mol Gen Genet 166:141–149

Thomas M, White RL, Davis RW (1976) Hybridization of RNA to double-stranded DNA: Formation of R-loops. Proc Natl Acad Sci USA 76:3164–3168

Wehlmann H, Eichenlaub R (1981) Analysis of transcripts from plasmid mini-F by electron microscopy of R-loops. Plasmid 5:259–266

Chapter 8 DNA Replication

8.1 Localization of Origins of Plasmid DNA Replication

R. EICHENLAUB [1]

Contents

1. General Introduction 316
   a) Structure of Replicative Intermediates 316
   b) Isolation and Enrichment for Plasmid Replicative Intermediates 317
   c) Analysis of Replicative Intermediates 318
2. Experiment 1: Characterization of Plasmid Replicative Intermediates 318
   a) Objectives 318
   b) Strain 318
   c) Procedure 319
3. Materials 323
4. References 324

1. General Introduction

a) Structure of Replicative Intermediates

Plasmid DNA isolated from bacterial cells consists of three distinct conformations, namely, linear forms resulting from one or more double strand breaks in the original molecule, open circles containing one or more single strand breaks (OC), and covalently closed circles (CCC). Circular DNA forms may also be present as concatemers comprised of two or more interlocked monomeric OC forms, CCC forms, or mixtures of both. A supercoiled structure seems to be a prerequisite for DNA replication; supercoiling of template molecules is required for initiation of DNA synthesis as well as chain elongation in the case of plasmid ColE1 (Gellert et al. 1976). Electron microscopic examination of replicative intermediates from a variety of plasmids has shown that replication forks are bracketed on one side by a supercoiled region and on the other by a relaxed region, which represents the newly replicated part of the molecule (Fig. 1A). Introduction of a single strand break into one of the parental DNA strands of the replicative intermediate converts it into an open circle and enables its typical eye-structure (θ-type, see Fig. 1B; in contrast to the σ-type of a rolling circle) to be visualized.

---

[1] Universität Hamburg, Institut für Allgemeine Botanik, Arbeitsbereich Genetik, Ohnhorst-Straße 18, D-2000 Hamburg 52, Fed. Rep. of Germany

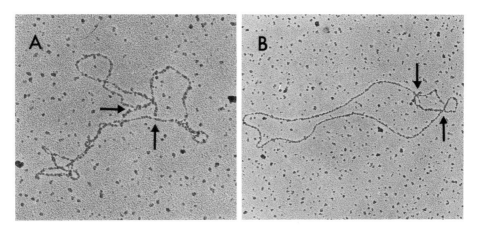

Fig. 1A, B. Replicative intermediates of plasmid pML31. *Arrows* indicate the replication forks. **A** Supercoiled intermediate; **B** relaxed intermediate

The physical properties of replicative intermediates allow their separation from nonreplicating molecules by velocity and equilibrium centrifugation in density gradients; intermediates have higher sedimentation coefficients that increase with degree of replication and have buoyant densities in CsCl gradients containing the intercalating dye ethidium bromide that are intermediate between those of OC and CCC DNA forms. In such gradients, molecules with small newly replicated regions band close to supercoiled DNA, while molecules with extensively replicated regions, band close to open circular plasmid DNA.

b) Isolation and Enrichment for Plasmid Replicative Intermediates

Plasmid DNA is ordinarily isolated by detergent (Brij 58 or Triton X-100) lysis of bacterial lysozyme-spheroplasts. Removal of the bulk of chromosomal DNA and cell debris occurs by a "clearing" centrifugation step followed by equilibrium centrifugation in dye-CsCl gradients. Plasmid DNA obtained in this way can be considered to originate from the cell cytoplasm. Replicating plasmid molecules, however, have been shown in some cases to be preferentially associated with the bacterial cell membrane (Falkow et al. 1971, Sherratt and Helinski 1973) and are therefore lost during the clearing spin. This can be avoided if membrane-DNA complexes are dissociated before the clearing step.

A lysis procedure described by Guerry et al. (1973), which was originally developed by Hirt (1967), uses cell lysis by sodium-dodecyl-sulfate (SDS). The procedure takes advantage of preferential precipitation of high molecular weight chromosomal DNA in the presence of SDS and high concentrations of NaCl. In this way, two crucial steps are accomplished: (1) the removal of the bulk of chromosomal DNA, and (2) the release of plasmid replicative intermediates from the bacterial cell membrane.

However, another problem remains. In a growing culture, only a small fraction of plasmid molecules are replicating at any given time. It is therefore necessary to increase the yields of replicative intermediates prior to their purification. A reduction in the rate of plasmid DNA synthesis is known to increase the proportion of replicating molecules and thymine deprivation of a thymine auxotroph and exposure of bacteria to hydroxyurea have both been successfully used for this purpose (Barnes and Rownd 1972, Perlman and Rownd 1976).

c) Analysis of Replicative Intermediates

The isolation of replicative intermediates by centrifugation techniques and their visualization in the electron microscope allows two aspects of plasmid replication to be analyzed, namely (1) the location of the origin of replication and (2) the mode of replication — unidirectional versus bidirectional replication.

A prerequisite for both types of analysis is the presence of readily identified reference points for the purpose of localization and orientation of replication forks on the molecules. One possibility is to partially denature replicative intermediates and to relate replication forks to readily denatured A-T-rich regions that have previously been mapped (denaturation map; Perlman and Rownd 1976, Schnös and Inman 1970). This procedure is, however, tedious and time-consuming and is not very precise. A second method takes advantage of the site-specific cleavage of DNA molecules by restriction endonucleases. The conversion of a circular replicating DNA molecule into a linear form with defined ends enables precise localization of the origin and determination of the direction of replication (Eichenlaub et al. 1977, 1979). Molecule orientation ambiguities are usually resolved by the use of two endonucleases that cleave the plasmid at different sites that are asymmetric to the origin.

2. Experiment 1: Characterization of Plasmid Replicative Intermediates

a) Objectives

Replicative intermediates of plasmid pML31 will be pulse-labeled, isolated by CsCl-ethidium bromide density gradient centrifugation, and subjected to sedimentation and electron microscopic analysis.

Note: Yields of plasmid DNA in general, and replicative intermediates in particular, from bacteria carrying pML31 are rather low; analysis of replicative intermediates of a high copy number plasmid, such as ColE1, is recommended for individuals with little experience in handling plasmids.

b) Strain

Escherichia coli MV17 *thy* Δ*trp*E5 (pML31).

c) Procedure

Day 1

Grow on overnight culture of *E. coli* MV17 (pML31) in M9s medium containing thymine (2 µg/ml).

Day 2

Inoculate 500 ml of M9s medium containing thymine in a 5 liter flask with 1–2 ml of the overnight culture (initial cell concentration: 10^7/ml).
Add 50 µCi of ^{14}C-thymine and incubate the culture at 37°C on a shaker with vigorous aeration until the cell density reaches 2×10^8/ml.
Spin the cells down at room temperature in a GS-3 rotor (Sorvall) at 5,000 rpm for 5 min.
Wash the cells by suspension in thymine-free M9s medium and recentrifuge and resuspend afterward in 200 ml of prewarmed (37°C) M9s medium *without* thymine.
Incubate the culture for 30 min at 37°C with aeration (thymine starvation step).
Cool the culture to 25°C.
Pulse label the culture for 30 s to 1 min by the addition of 1 mCi ^{3}H-thymidine and cold thymidine (final thymidine concentration 2 µg/ml).
Stop incorporation of label by addition of 5 M NaN_3 to a final concentration of 0.1 M and rapid freezing of the culture in an ethanol dry ice bath.
Thaw the culture and centrifuge in a GS-3 rotor (Sorvall) for 5 min at 5,000 rpm at 0°C.
Wash the cells with cold 10 mM NaCl and finally resuspend in 13 ml 25% sucrose in TES in a Sorvall SS-34 centrifuge tube.

Cell Lysis

Add 1.3 ml lysozyme (10 mg/ml in H_2O) and 2.6 ml 250 mM EDTA, pH 7.5, and keep the suspension in ice for 25 min.
Raise the temperature of the suspension to 20°C, add 2 ml 10% SDS in TES (final concentration 1% SDS) and leave at 20°C until lysis is complete and the cell lysate clears.
Add 4.4 ml 5 M NaCl (final concentration 1 M) and mix very gently.
Keep lysate in ice overnight.

Day 3

During the incubation of the lysate at 0°C, a white precipitate forms. Pellet this precipitate by centrifugation in an SS-34 rotor (Sorvall) for 30 min at 17,000 rpm and 0°C and dialyze the supernatant for 4 h against 1 l TES buffer (two changes) to remove the NaCl.

CsCl-Ethidium Bromide Grandient Centrifugation

Measure the volume of the dialyzed lysate (about 21 ml) and add water to a final volume of 30 ml.
Add 28.8 g of CsCl.

After CsCl is completely dissolved, add 2.4 ml ethidium bromide solution (5 mg/ml in water). This gives a total volume of 40 ml.

Transfer the solution to two 60 Ti centrifuge tubes (Beckman; 20 ml per tube) and fill up with paraffin oil. Centrifuge the tubes for 40 h at 40,000 rpm at 15°C.

Day 5

Puncture the centrifuge tubes at the bottom and collect 1 ml fractions.

Spot 50–100 μl of each fraction on Whatman 3 MM paper discs.

Dry filters under an infrared lamp, wash them twice with 10% TCA (trichloroacetic acid) and twice with ethanol.

Dry the filters and determine ^{3}H and ^{14}C radioactivity in a scintillation counter.

Plot ^{3}H and ^{14}C counts versus fraction number and also the ratio of ^{3}H/^{14}C.

A typical profile is shown in Fig. 2.

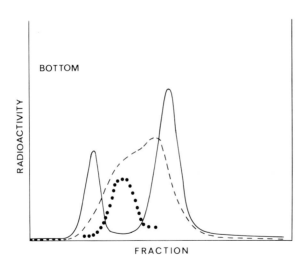

Fig. 2. Caesium chloride-ethidium bromide buoyant density centrifugation of plasmid replicative intermediates. ——— ^{14}C prelabel; - - - - - ^{3}H pulse label; ..... ^{3}H/^{14}C

Analysis of Replicative Intermediates by Sucrose Gradient Centrifugation

Pool the fractions enriched for the ^{3}H label in the region of the gradient with intermediate density between the supercoiled and open circular DNA peaks.

Extract the ethidium bromide with n-butanol saturated with TES buffer.

Dialyze 4 h against 1 l TES buffer (two changes) to remove the CsCl.

Day 6

Prepare 5–20% sucrose gradients (sucrose in TES with 0.5 M NaCl) in Beckman SW25 or SW27 centrifuge tubes and load each gradient with up to 5 ml of the dialyzed pooled fractions containing the replicative intermediates.

Centrifuge for 3 h at 25,000 rpm at 15°C.

Puncture the tubes at the bottom and collect 1 ml fractions.

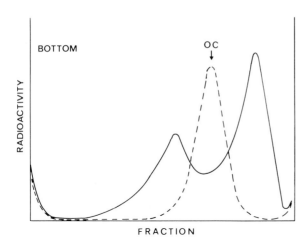

Fig. 3. Sedimentation analysis of plasmid replicative intermediates by sucrose gradient centrifugation. - - - - - ^{14}C prelabel; ——— ^{3}H pulse label

Spot 50–100 µl of each fraction onto Whatman 3 MM filter discs and wash and determine radioactivity on the filters as described above.

Plot ^{3}H and ^{14}C counts versus fraction number.

A profile of such a sucrose gradient is shown in Fig. 3.

Pool the fractions containing ^{3}H labeled DNA, which sedimented faster than OC plasmid DNA.

Dialyze against TES buffer at 4°C and concentrate DNA by ethanol precipitation.

Resuspend plasmid DNA in *Eco*RI restriction buffer and cleave with *Eco*RI restriction endonuclease at room temperature (37°C favors displacement of newly synthesized DNA chains by branch migration and causes loss of replicative intermediates).

Spread the replicative intermediates according to techniques as described in Chap. 6.1 and examine in the electron microscope. Include a sample of a standard DNA molecule (e.g., pBR322) as an internal contour length standard.

Day 7

Analyze electron micrographs of plasmid replicative intermediates of the type shown in Fig. 4 and draw scale representations of molecules on graph paper as shown in Fig. 5. Orient molecules according to "best-fit" with the shorter unreplicated arm on the left hand side.

Generally, a careful analysis of plasmid replicative intermediates requires 50–100 "good" molecules, i.e., molecules that have the correct unit length, few or no crossovers, and whose interpretation is completely unambiguous. However, 20–30 molecules usually suffice to identify the number and location of possible replication origins and the direction of replication from these origins. Analysis of replicative intermediates digested with another restriction endonuclease confirms these conclusions and clearly shows the orientation of the plasmid molecule vis-à-vis the origin(s). A second way to plot measurements derived from replicative intermediates is shown in Fig. 6. In this plot, the branch point positions of the molecules are plotted versus

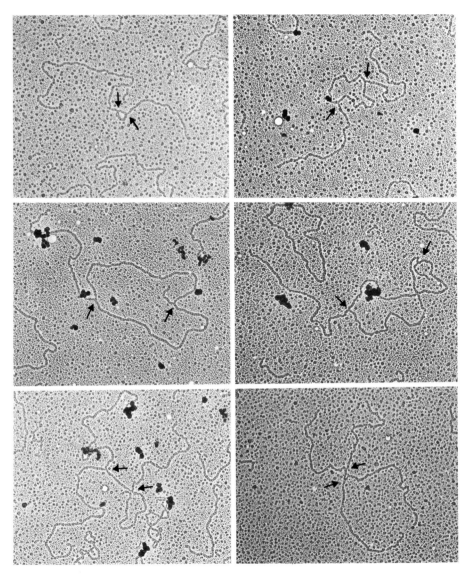

Fig. 4. Electron micrographs of plasmid replicative intermediates showing different extents of replication. Molecules have been linearized by digestion with restriction endonuclease *Eco*RI

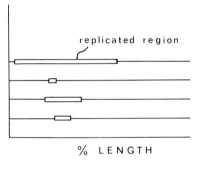

Fig. 5. Alignment of scale diagrams of replicative intermediates

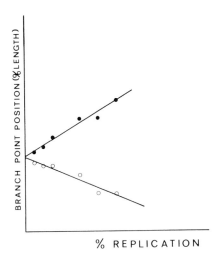

Fig. 6. Plot of branch point positions of replicative intermediates against percent replication. Regression lines indicate the position of the origin of replication (0% replication) at about 20% of the molecule length from the unique *Eco*RI site of the molecule and that the plasmid replicates bidirectionally

percent replication. Both types of plots should be subjected to linear regression analysis and the position of the origins(s) of replication located by extrapolation to 0% replication. Using a similar strategy for analysis of replicative intermediates, one can locate the termination point (ter) of replication. In those cases in which replication is unidirectional, the origin of replication is also by definition the termination site. In bidirectional replication systems, such as mini-F (Eichenlaub et al. 1977) and R6K (Lovett et al. 1975), replication in both directions may be symmetric with termination occurring at a point 50% of the molecule length from the origin, or asymmetric with termination at some other point on the molecule.

3. Materials

Medium

M9s medium consists of: 0.03 M Na_2HPO_4, 0.01 M NaCl, 0.02 M NH_4Cl, 0.025 M KH_2PO_4, 0.2% glucose, 0.4% casamino acids (Difco), 0.001 M $MgCl_2$, 0.5 µg thiamine/ml, pH adjusted to 7.5

Other Chemicals, Solutions, and Materials

^{3}H-thymidine
^{14}C-thymine
Thymine in H_2O, 1 mg/ml
25% sucrose in TES (100 mM Tris-HCl, pH 7.5, 50 mM NaCl, 5 mM EDTA)
Lysozyme in H_2O, 10 mg/ml (Sigma, Chemical Co.)
0.25 M EDTA, pH 7.5
10% SDS in TES
5 M NaCl in H_2O

CsCl
Ethidium bromide in H$_2$O (5 mg/ml)
n-Butanol saturated with TES
4 l TES buffer for dialysis
5% and 20% sucrose in TES with 0.5 M NaCl
Whatman 3 MM filter paper
Scintillation cocktail (1 l toluene + 5 g PPO)
10% trichloroacetic acid
Ethanol 96%
Beckman rotor 60 Ti with tubes and caps
Beckman rotor SW25 or SW27 with tubes
Fraction collecting device for 60Ti and SW25 or SW27 tubes
Restriction endonuclease *Eco*RI and *Eco*RI buffer

4. References

Barnes MH, Rownd R (1972) Effect of thymine limitation on chromosomal deoxyribonucleic acid synthesis in *Proteus mirabilis.* J Bacteriol 111:750–757
Eichenlaub R, Figurski D, Helinski DR (1977) Bidrectional replication from a unique origin in a mini-F plasmid. Proc Natl Acad Sci USA 74:1138–1141
Eichenlaub R, Sparks RB, Helinski DR (1979) Bidirectional replication of the mini-ColE1 plasmid pVH51. J Bacteriol 138:257–260
Falkow S, Tompkin LS, Silver RP, Guerry P, LeBlanc DJ (1971) The replication of R-factor DNA in *Escherichia coli* K12 following conjugation. Ann NY Acad Sci 182:153–171
Gellert M, O'Dea MH, Itoh T, Tomizawa J (1976) Novobiocin and coumermycin inhibit DNA supercoiling catalysed by DNA gyrase. Proc Natl Acad Sci USA 73:4474–4478
Guerry P, LeBlanc DJ, Fakow S (1973) General method for the isolation of plasmid deoxyribonucleic acid. J Bacteriol 116:1064–1066
Hirt B (1967) Selective extraction of polyoma DNA from infected mouse cell cultures. J Mol Biol 26:365–369
Lovett MA, Sparks RB, Helinski DR (1975) Bidirectional replication of plasmid R6K DNA in *Escherichia coli:* Correspondence between origin of replication and position of single-strand break in relaxed complex. Proc Natl Acad Sci USA 72:2905–2909
Perlman D, Rownd RH (1976) Two origins of replication in composite R-plasmid DNA. Nature 259:281–284
Schnös M, Inman RB (1970) Position of branch points in replicating DNA. J Mol Biol 51:61–73
Sherratt DJ, Helinski DR (1973) Replication of colicinogenic factor E1 in *Escherichia coli.* Properties of newly replicated supercoils. Europ J Biochem 37:95–99

8.2 Plasmid DNA Replication in Vitro

W.L. STAUDENBAUER [1]

Contents

1. General Introduction. 325
2. Experiment 1: Requirements of the Replication System and Effects of Inhibitors. . . . 326
      a) Introduction. 326
      b) Objectives 327
      c) Procedure 1: Preparation of Extract 327
      d) Procedure 2: Assay of Plasmid DNA Synthesis 328
      e) Procedure 3: Requirements of the Reaction. 329
      f) Procedure 4: Removal of Endogenous Nucleic Acids. 329
      g) Procedure 5: Effect of Inhibitors 330
      h) Results 330
3. Experiment 2: Characterization of the Reaction Product 331
      a) Introduction. 331
      b) Objectives 331
      c) Procedures and Results. 331
4. Materials and Methods 335
      a) Materials 335
      b) Preparation of ColE1 DNA 336
      c) Preparation of ^{32}P-Labeled ColE1 DNA 336
5. References. 337

1. General Introduction

Plasmids are autonomous replication units or replicons of bacteria. Because of their small size, low degree of genetic complexity, and ease of purification, they are attractive models for the study of genome function, in particular replication and its regulation. Plasmids differ from one another in their replication requirements. They may, for example, be grouped into three distinct classes according to their response to the inhibition of protein synthesis by treatment of host cells with chloramphenicol:

1. continuation of replication at an increased rate for many hours;
2. continuation of replicattion for a limited period of time at an ever decreasing rate;
3. immediate cessation of replication.

---

[1] Max-Planck-Institut für Molekulare Genetik, Abt. Schuster, Ihnestraße 63–73, D-1000 Berlin 33
Present address: Lehrstuhl für Mikrobiologie, Technische Universität München, Arcisstraße 21, D-8000 München 2, Fed. Rep. of Germany

Evidence has been presented that the replication of class 1 plasmids does not involve any plasmid-specific protein, whereas plasmids of class 2 and 3 require at least one plasmid-encoded protein for the initiation of DNA synthesis (Kolter and Helinski 1979). One may assume that the different kinetics of shut-off of replication caused by chloramphenicol treatment reflect differences in the stability of such proteins.

The elucidation of molecular mechanisms involved in plasmid DNA synthesis requires cell-free in vitro systems. It has been shown that concentrated extracts from plasmid-free *E. coli* cells can carry out replication of the class 1 plasmids ColE1, CloDF13, and RSF1030 (Sakakibara and Tomizawa 1974, Staudenbauer 1976). A similar system capable of replicating the class 2 plasmid R6K has been prepared from R6K-carrying bacteria ((Inuzuka and Helinski 1978). Until recently, a cell-free replication system for class 3 plasmids has not been developed, a failure that was attributed to the requirement for a cellular structure (membrane, nucleoid) which is absent from soluble extracts.

The class 3 plasmid R1 can now, however, be efficiently replicated in *E. coli* extracts supplemented with a plasmid DNA-directed protein synthesizing system (Diaz et al. 1981). Soluble replication systems are thus now available for all three groups of plasmids.

In this chapter, the preparation and characterization of a soluble in vitro system for the replication of the ColE1 plasmid will be described.

2. Experiment 1: Requirements of the Replication System and Effects of Inhibitors

a) Introduction

DNA synthesis may be of the replicative or repair type and it is crucial to exclude the involvement of repair synthesis when studying DNA replication in in vitro systems. DNA replication and DNA repair are carried out by different enzymes that can be distinguished by their distinct substrate and cofactor dependencies and sensitivities to inhibitors:

1. Replicative DNA synthesis is carried out mostly by DNA polymerase III (pol III), whereas repair synthesis is accomplished by DNA polymerase I (pol I). Pol III is strongly inhibited by arabinosylnucleoside triphosphates at concentrations that have practically no effect on pol I (Cozzarelli 1977).
2. Several enzymes functioning in DNA replication require rather high concentrations of ATP.
3. Initiation of plasmid replication involves a transcriptional event which, in the case of ColE1, is blocked by rifampicin and depends on all four ribonucleoside triphosphates.
4. Plasmid replication is remarkably sensitive to inhibitors of DNA gyrase (novobiocin, oxolinic acid), which have little or no effect on repair synthesis.

b) Objectives

1. To prepare a cell-free extract for ColE1 replication.
2. To show that the requirements of the system and its sensitivity to inhibitors correspond to replicative DNA synthesis.

c) Procedure 1: Preparation of Extract

Bacteria are grown in L-broth without antifoam.
Inoculate each liter of medium with 20 ml of a fresh overnight culture of *E. coli* C600.
Shake the cells at 37°C until they attain an OD_{600} = 1.0.
Check the pH of the culture and adjust to pH 8.0 with 5 N KOH.
Harvest the cells at room temperature by centrifuging at 7000 rpm for 10 min in a Sorvall G-3 rotor.
Resuspend the cells from 1 l of culture in 20 ml of buffer A and centrifuge at 10,000 rpm for 10 min in a Sorvall SS-34 rotor at room temperature.
Weigh the cells (from 1 l culture approx. 2 g of cells are obtained) and resuspend the cell pellet in an equal volume of buffer.
Measure the volume of the cell suspension and transfer it into a Beckman 50Ti polycarbonate tube.
Quick-freeze the tube in liquid nitrogen.
Thaw the cells in a 10°C water bath and add 1/50 volume of a freshly prepared lysozyme-EDTA solution (15 mg lyzozyme per ml 0.05 M EDTA, pH 8.0).
Close the tube with screw-cap and mix gently by inversion.
Incubate for 30 min in an ice bath.
Freeze the tube in liquid nitrogen and thaw again at 10°C. At this stage the lysate should become highly viscous.
Place tube in a chilled Beckman 50Ti rotor and centrifuge at 30,000 rpm for 30 min at 2°C.
Remove the supernatant with a Pasteur pipette, distribute it into Eppendorf tubes (about 0.4 ml per tube), and freeze immediately in liquid nitrogen. The protein content of the extract (as determined by the Lowry method) should be around 20 mg/ml.
An appropriate amount of extract is thawed in an ice bath immediately before use. It can be stored at $-70°C$ or in liquid nitrogen for months without loss of activity.

Extracts prepared by this method are capable of carrying out the replication of exogenous DNA of ColE1 or other class 1 plasmids. Since no plasmid-coded replication proteins are required, active extracts can be prepared from plasmid-free *E. coli* strains. Moreover, it has been observed that with certain strains, chloramphenicol treatment of the culture results in an increased replication capacity of the extract. In that case, 100 mg chloramphenicol (dissolved in 2 ml DMSO) is added to 1 l of culture at an OD_{600} = 1.0 and aeration at 37°C is continued for 2–4 h.

Extracts for in vitro replication of class 2 or class 3 plasmids are prepared in a similar fashion from plasmid-carrying *E. coli* strains without chloramphenicol treatment (Inuzuka and Helinski 1978, Diaz et al. 1981).

d) Procedure 2: Assay of Plasmid DNA Synthesis

Standard reaction mixtures (25 µl) contain final concentrations of 40 mM HEPES, pH 8.0, 100 mM KCl, 15 mM Mg-acetate, 2 mM ATP, 0.4 mM each of CTP, GTP, and UTP, 50 µM NAD, 25 µM each of dATP, dCTP, dGTP and ^{3}H-dTTP (specific activity approx. 500 cpm/pmol).

Prepare cocktail I (1.2 ml) from stock solutions (adjusted to pH 8.0, with Tris-base) as follows:

Stock solution	µl
0.5 M HEPES, pH 8.0	400
1 M Mg-acetate	150
0.1 M ATP	200
40 mM ATP	100
40 mM GTP	100
40 mM UTP	100
10 mM NAD	50
10 mM dATP	25
10 mM dCTP	25
10 mM dGTP	25
10 mM dTTP	25

Cocktail I can be stored frozen at $-20°C$. To assay DNA synthesis, cocktail II is prepared freshly by mixing the appropriate amount of cocktail I with 1/3 volume of template DNA (0.5 mg/ml) and 1/3 volume of ^{3}H-dTTP (1 mCi/ml).

Thaw an aliquot of the frozen extract in an ice bath.

Pipette into Eppendorf tubes 2.5, 5, 7.5, 10, 12.5, 15, 17.5, and 20 µl of extract in duplicate.

Adjust the volume with buffer A to 20 µl.

Add 5 µl of cocktail II to each tube, mix on vortex, spin few secs in Eppendorf centrifuge, and incuabate for 60 min an a 30°C water bath.

Terminate the rection by addition of 0.2 ml stop-mix and heat for 10 min at 80°C.

Add 0.5 ml 2 M trichloroacetic acid and put tubes for 10 min in an ice bath.

The reaction mixtures are then filtered through nitrocellulose filters.

Rinse the tubes twice with water and wash each filter with approx. 10 ml H_2O.

The filters are dried under a heat lamp and counted in a toluene-based scintillation fluid.

The amount of ^{3}H-dTMP incorporated is plotted against the protein concentration of the reaction mixture. Under optimal conditions, an incorporation of 100–200 pmol dTMP per 25 µl is obtained with extracts from *E. coli* C600. Normally, the incorporation plateaus at a protein concentration around 15 mg/ml (corresponding to about 15 µl extract per 25 µl reaction mixture). Occasionally an inhibition of DNA synthesis at higher protein concentrations is observed. High incorporation rates in diluted extracts are an indication of repair synthesis on damaged template DNA or broken chromosomal DNA.

e) Procedure 3: Requirements of the Reaction

Prepare a set of deficient replication cocktails (120 µl of cocktail I and 25 µl of the corresponding cocktail II are sufficient) by replacing the indicated component with an equal volume of water.

Tube	Reaction mixture
1	Complete (zero-time blank)
2	Complete
3	− Mg-acetate
4	− ATP
5	− CTP, GTP, UTP
6	− ATP, CTP, GTP, UTP
7	− NAD
8	− dATP, dCTP, dGTP
9	− ColE1 DNA

Pipette 15 µl extract and 5 µl buffer A into 2 × 9 Eppendorf tubes in an ice bath. Add 5 µl of the respective cocktail II in duplicate and mix on a vortex.

Stop-mix (0.2 ml) is added immediately to tube 1, which serves as a zero-time blank. The other tubes are incubated for 60 min at 30°C and assayed for acid-insoluble radioactivity as described above.

The zero-time blank (usually 300–500 cpm) is substracted and the results are expressed as percentage of the incorporation obtained with the complete reaction mixture.

The incorporation should depend completely on the presence of ribonucleoside triphosphates and Mg^{2+}. Omission of ATP alone should reduce the DNA synthesis by more than 90%. NAD has little or no stimulatory effect, but is added routinely as cofactor of *E. coli* DNA ligase. Residual synthesis without addition of all four deoxyribonucleoside triphosphates is due to the presence of low amounts of substrates in the crude extract. A significant amount of ATP-independent incorporation in the absence of exogenous plasmid DNA would indicate repair synthesis on chromosomal DNA fragments present in the extract.

f) Procedure 4: Removal of Endogenous Nucleic Acids

Chromosomal DNA (as well as any endogenous plasmid DNA) can be removed from the extract by the following procedure (Conrad and Campbell 1979):

Pour extract into a beaker with a stirring bar placed in an ice bath.

Add 1/10 volume of a 30% (w/w) solution of streptomycin sulfate in buffer A and stir for 30 min.

Remove the precipitate by centrifugation at 20,000 rpm in a Beckman 50Ti rotor for 10 min at 2°C.

Save supernatant and add solid ammonium sulfate (0.5 g/ml) over a period of 10 min while stirring slowly in an ice bath.

Continue stirring for 30 min and spin down the precipitate.
Dissolve the protein pellet in an amount of buffer B equal to half the volume of the original extract.
Dialyze against 200 ml buffer B for 3 h at 0°C with two changes of buffer. The dialyzed protein fraction is divided into aliquots and frozen in liquid nitrogen.

g) Procedure 5: Effect of Inhibitors

Prepare the following inhibitor solutions (10-fold concentrated) by diluting the stock solutions with the appropriate amount of buffer A:

Inhibitor	Concentration
Chloramphenicol	1.25 mg/ml
Rifampicin	0.25 mg/ml
Nalidixic acid	1.00 mg/ml
Oxolinic acid	0.50 mg/ml
Novobiocin	0.25 mg/ml
ara-CTP	2.5 mM

Pipette 15 µl extract into 2 × 8 Eppendorf tubes (0°C).
Add buffer A and inhibitor solutions as indicated below:

Tube	Reaction mixture	Inhibitor (µl)	Buffer A (µl)
1	Control (zero-time blank)	–	5.0
2	Control	–	5.0
3	+ Chloramphenicol	2.5	2.5
4	+ Rifampicin	2.5	2.5
5	+ Nalidixic acid	2.5	2.5
6	+ Oxolinic acid	2.5	2.5
7	+ Novobiocin	2.5	2.5
8	+ ara-CTP	2.5	2.5

Start the reaction by adding 5 µl cocktail II and incubate for 60 min at 30°C.
Stop the reaction and determine acid-insoluble radioactivity as before.
Subtract zero-time blank and express data as percentage of untreated control.

h) Results

Replicative synthesis of ColE1 DNA should be insensitive to chloramphenicol, but completely blocked by rifampicin, a specific inhibitor of RNA polymerase. Plasmid DNA replication is also highly sensitive to ara-CTP which acts as a preferential inhibitor of DNA polymerase III. Some residual incorporation (around 10% of the untreated control) is due to the synthesis of an early replicative intermediate carried out by DNA polymerase I (Staudenbauer 1978). Complete inhibition is observed with oxolinic acid (a structural analogue of nalidixic acid) and novobiocin, both of which prevent the supercoiling of plasmid DNA by DNA gyrase (Cozzarelli 1980). Nalidixic acid is a less potent inhibitor of DNA gyrase and reduces plasmid DNA synthesis by about 60%.

3. Experiment 2: Characterization of the Reaction Product

a) Introduction

After the reaction conditions of cell-free plasmid DNA synthesis have been optimized, it is necessary to characterize the DNA products. It has to be shown that the in vitro system carries out a complete round of DNA replication resulting in the formation of supercoiled monomeric plasmid DNA molecules. The closed-circular (CCC) and open-circular (OC) DNA forms can be identified by their sedimentation coefficients in high salts:

ColE1 DNA	Neutral (S)	Alkaline (S)
CCC Monomer (form I)	23.4	63.3
OC Monomer (form II)	17	19 (circle)
		17.3 (rod)

Direct evidence for a semiconservative replication process (as opposed to non-conservative repair synthesis) can be obtained by performing a "Meselson-Stahl" experiment, in which plasmid replication is carried out in the presence of a density labeled precursor. The daughter molecules formed after one round or replication will then consist of a density labeled progeny strand and an unlabeled parental strand. After a second round of replication, fully density labeled molecules are generated. The corresponding density shifts of the replicating DNA can be monitored by equilibrium centrifugation in CsCl gradients.

b) Objectives

1. To show that the plasmid DNA synthesized in vitro consists of fully replicated supercoils.
2. To analyze the density pattern of the DNA after incorporation of bromodeoxyuridine.

c) Procedure and Results

Velocity Sedimentation of Reaction Product

Prepare a 1 ml reaction mixture from the following ingredients:

600 μl extract
200 μl buffer A
120 μl cocktail I
 40 μl ColE1 DNA
 40 μl ^{3}H-dTTP

Incubate for 60 min in a 30°C water bath.
Stop the reaction by addition of 0.1 ml 0.5 M EDTA and cool the test tube on ice.

Separate the labeled DNA from nonincorporated ^{3}H-dTTP by gel filtration through a Sepharose 4B column (1.5 cm × 20 cm) in buffer C.

Collect 0.5 ml fractions, spot 25 µl aliquots on Whatman 3 MM discs and assay for radioactivity.

Pool the fast-moving peak and mix thoroughly with an equal volume of buffer-saturated phenol (reagent-grade phenol crystals stored at $-20°$C are satisfactory).

Centrifuge at 5000 rpm for 10 min to separate the phases.

The aqueous (upper) phase is removed and extracted 3 × with 2 vol diethyl ether.

Add 1/10 vol 3 M K-acetate and precipitate the nucleic acids with 2 vol ethanol overnight at $-20°$C.

The precipitate is collected by centrifugation in a polyallomer tube at 15,000 rpm for 20 min at $-10°$C in a Sorvall SS-34 rotor.

The pellet is washed once with cold 70% ethanol, dried in vacuum, and dissolved in 1 ml buffer C.

Prepare neutral CsCl solutions of densities 1.2 and 1.4 g/ml by adding 10 ml buffer C to 2.8 g CsCl and 6.1 g CsCl, respectively.

Linear gradients (4.2 ml) are formed in Beckman SW60 tubes by introducing 2.1 ml of each CsCl solution to the chambers of a gradient mixer.

Note that the denser solution is introduced in the front chamber, wheras the lighter solution is put in the rear chamber. The exit tubing must touch the upper wall of the centrifuge tube during the delivery of the gradient.

Prepare alkaline CsCl solutions by dissolving 2.8 g CsCl and 6.1 g CsCl with 10 ml 0.2 M NaOH.

Form linear gradients in Beckman SW60 tubes as described above. The tubes should be precoated with a solution of Ficoll – polyvinyl pyrrolidone – serum albumin (0.02% each) in order to minimize adhesion of the denatured DNA to the surface of the centrifuge tube.

Mix 0.2 ml of ^{3}H-labeled reaction product with ^{32}P-labeled ColE1 DNA (approx. 10,000 cpm) and layer the sample carefully on the top of the gradient.

Centrifuge at 45,000 rpm 1 h (alkaline CsCl).

Puncture tubes at the bottom with a 21-gauge needle and collect fractions directly on Whatman 3 MM discs. Two drops per disc gives about 40 fractions.

Dry discs under a heat lamp and measure radioactivity.

The sedimentation profile of the in vitro synthesized plasmid DNA usually shows three peaks in a neutral gradient (Fig. 1A). The predominant peak should cosediment with the CCC marker DNA. A distinct peak of faster sedimenting material (31 S) consists of CCC-CCC catenanae. They result from faulty termination of replication (Sakakibara et al. 1976). A minor portion of the label is found in OC DNA sedimenting at 17 S.

Upon denaturation in an alkaline gradient, the separation of CCC and OC DNA is greatly improved (Fig. 1B). One can also observe an additional peak at about 6 S. These DNA fragments are derived from early replicative intermediates which cannot be separated from supercoiled monomers in a neutral gradient.

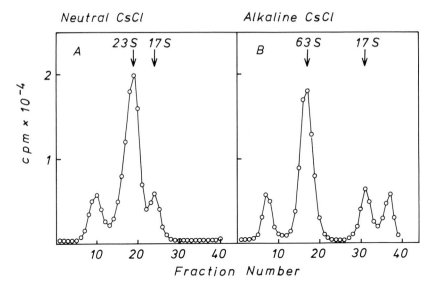

Fig. 1A, B. Velocity sedimentation of ColE1 DNA in neutral (**A**) and alkaline (**B**) CsCl gradients. Centrifugations were performed in a Beckman SW60 rotor at 45,000 rpm for 2.5 h (**A**) or 1 h (**B**) at 15°C. Fractions (2 drops) were collected on Whatman 3 MM discs and assayed for radioactivity. Sedimentation is *from right to left*

Density Labeling of Reaction Product

For density labeling, a modified reaction mixture is employed which contains instead of dTTP, bromodeoxyuridine triphosphate (BrdUTP) at a final concentration of 0.25 mM. This increased concentration is necessary to dilute out traces of endogenous dTTP present in the crude extract. ^{3}H-dCTP is used as a radioactive precursor replacing ^{3}H-dTTP.

Prepare a cocktail B from stock solutions (adjusted to pH 8.0, with Tris-base) as follows:

Stock solution	μl
0.5 M HEPES, pH 8.0	400
1 M Mg-acetate	150
0.1 M ATP	200
40 mM CTP	100
40 mM GTP	100
40 mM UTP	100
10 mM NAD	25
10 mM dATP	25
10 mM dCTP	25
10 mM dGTP	25
10 mM BrdUTP	250

Prepare 1 ml reaction mixture from the following ingredients:

600 μl extract
180 μl buffer A
140 μl cocktail I B
  40 μl ColE1 DNA
  40 μl ^{3}H-dCTP (1 mCi/ml)

Incubate the tube for 60 min at 30°C.

Stop the reaction with EDTA and pass the reaction mixture through a Sepharose 4B column as described above.

Recover radioactive DNA eluting with the void volume of the column.

Add ^{32}P-labeled ColE1 DNA (approx 20,000 cpm) as reference DNA (density 1.710 g/ml) to the pooled fractions and adjust the volume to 8.0 ml with buffer C.

Mix with 10.4 g CsCl and distribute the sample evenly into two Beckman 50Ti polypropylene screw-cap tubes.

Fill with paraffin oil and spin at 40,000 rpm for 36 h at 15°C.

Puncture the tubes at the bottom with a 21-gauge needle and collect 40–50 fractions (3 drops per fraction).

Spot 50 μl of each fraction on Whatman 3 MM discs and measure radioactivity.

Determine the density of the fractions from refractometer readings:

Density	Refractive index
1.700	1.3995
1.750	1.4041
1.800	1.4086

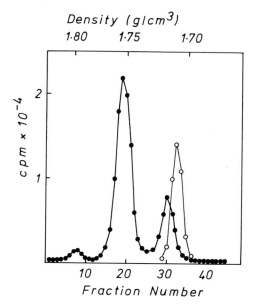

Fig. 2. Equilibrium centrifugation of bromodeoxyuridine-labeled ColE1 DNA. Centrifugation was performed in a Beckman 50Ti rotor at 40,000 rpm for 36 h at 15°C. Fractions (3 drops) were collected into Eppendorf tubes. Densities were determined from refractometer readings. 50 μl aliquots of each fraction were spotted on Whatman 3 MM discs and assayed for radioactivity. (●—●), ^{3}H-labeled reaction product; (○—○), ^{32}P-labeled reference DNA

Most of the bromodeoxyuridine-labeled DNA should band at the position of hybrid density (1.75 g/ml) and thus consist of a density labeled daughter strand and an unlabeled parental strand (Fig. 2). How can you show that this interpretation is correct? If the plasmid DNA has undergone more than one round of replication, part of the label is found at a density of about 1.80 g/ml as expected for DNA molecules, both strands of which have become labeled with bromodeoxyuridine. Radioactive label banding between the positions of hybrid DNA and reference DNA might either represent replicative intermediates or indicate repair synthesis. How can one distinguish between these possibilities?

4. Materials and Methods

a) Materials

Strains

E. coli C600
E. coli YS10 *thr leu min*A *str thi end* (ColE1)

Media
L-broth: 10 g tryptone, 5 g yeast extract, 5 g NaCl, 1 g glucose per liter
H-medium: 12 g Tris-base, 2 g KCl, 2 g NH_4Cl, 0.5 g $MgCl_2 \cdot 6H_2O$, 0.02 g Na_2SO_4, 7.5 ml conc HCl, 5 g glucose per liter

Buffers

Buffer A: 25 mM HEPES, pH 8.0, 125 mM KCl
Buffer B: 25 mM HEPES, pH 8.0, 125 mM KCl, 1 mM EDTA, 1 mM DTT, 10% (v/v) ethylene glycole
Buffer C: 10 mM Tris-HCl, pH 7.6, 1 mM EDTA, 10 mM NaCl

Stop-mix

0.5 N NaOH, 0.025 M Na-pyrophosphate, 1% (w/v) Na-dodecylsulfate, 0.5 mg calf thymus DNA per ml

Inhibitors

Stock solutions are prepared as indicated and stored at $-20°C$

Chloramphenicol	50 mg/ml DMSO
Rifampicin	10 mg/ml DMSO
Novobiocin	10 mg/ml DMSO
Nalidixic acid	10 mg/ml 0.1 N NaOH
Oxolinic acid	10 mg/ml 0.1 N NaOH

Oxolinic acid can be obtained under the trade name Nidantin from Gödecke AG, Freiburg, West Germany

b) Preparation of ColE1 DNA

Culture *E. coli* YS10 in 1 l L-broth at 37°C to an $OD_{600} = 1.0$.

Add chloramphenicol (100 μg per ml) and continue incubation for 12–16 h.

Harvest cells by centrifugation at 7.000 rpm for 10 min at 4°C in a Sorvall GS-3 rotor.

Resuspend pellets in 120 ml 50 mM Tris-HCl, pH 8.0, containing 25% (w/v) sucrose.

Add 40 ml of a freshly prepared lysozyme-EDTA solution (4 mg lysozyme per ml 0.15 M EDTA, pH 8.0) and incubate for 10 min in an ice bath.

Add 160 ml H_2O and 19 ml 10% (w/v) SDS and continue incubation for 20–30 min at 37°C until the lysate clears.

Add 85 ml 5 M NaCl and keep the lysate for 4 h in an ice bath.

Remove the precipitate by centrifugation at 10,000 rpm for 20 min at 4°C in a Sorvall GSA rotor and discard.

Add 1/4 vol of 50% (w/v) polyethylene glycol 6000 to the supernatant and store overnight at 4°C to allow precipitation of plasmid DNA.

Collect the precipitate by centrifugation at 10,000 rpm for 20 min at 4°C in a Sorvall GSA rotor and dissolve in 40 ml buffer C.

Add CsCl (1.0 g/ml) and stir at room temperature until CsCl is completely dissolved.

Add 1/5 vol ethidium-bromide solution (10 mg/ml) and remove any precipiate that forms by centrifugation at 15,000 rpm for 15 min at 15°C in a Sorvall SS-34 rotor.

Transfer the supernatant into two Beckman 60Ti polycarbonate tubes (25 ml per tube) and fill tubes up with paraffin oil.

Perform equilibrium centrifugation at 40,000 rpm for 40 h at 15°C.

Visualize the DNA bands using longwave UV light and remove the lower band representing supercoiled plasmid DNA from the top of the tube using a Pasteur pipette.

Extract ethidium-bromide 3 × with an equal volume of isopropanol that has been equilibrated with 2 M CsCl.

Remove the residual dye after dialysis against buffer C by passage through a 0.5 cm × 5 cm Dowex 50 column equilibrated with buffer C.

Adjust the eluant with buffer C to $OD_{260} = 10$ (0.5 mg DNA per ml) and store at 4°C.

c) Preparation of ^{32}P-Labeled ColE1 DNA

Culture *E. coli* YS10 in 100 ml H medium supplemented with 20 μg/ml vitamin B_1, 0.2% (w/v) casamino acids, and 1 mM Na_2HPO_4.

When the cells reach $OD_{600} = 1.0$, add chloramphenicol (100 μg/ml).

Shake culture for 90 min at 37°C before adding 1 mCi ^{32}P (inorganic phosphate).

Continue shaking for an additional 12–16 h.

Harvest cells by centrifugation at 7,000 rpm for 10 min in a Sorvall GSA rotor.

Wash cells twice with 100 ml buffer C.

Resuspend the pellet in a final volume of 3 ml buffer C and transfer the cell suspension to a Beckman 50Ti polycarbonate tube.

Add 60 μl of a freshly prepared lysozyme-EDTA solution (15 mg lysozyme per ml 0.05 M EDTA, pH 8.0) and incubate cells for 60 min at 0°C.

Add 0.8 ml 5 M NaCl and 0.2 ml 20% (w/v) Triton X-100 and incubate for 5 min in a 30°C water bath.

Centrifuge at 30,000 rpm for 30 min an a Beckman 50Ti rotor at 4°C.

Remove the supernatant from the clearing spin with a Pasteur pipette and apply to a Sepharose 4B column (2.5 cm × 20 cm).

Collect 1 ml fractions and pool the fast-running radioactivity peak.

After adjusting the volume of the pooled fractions to 9.5 ml with buffer C, add 9.2 g CsCl and 0.5 ml ethidium-bromide solution (10 mg/ml).

Distribute the sample evenly into two Beckman 50Ti polypropylene srew-cap tubes, fill with paraffin oil, and spin at 40,000 rpm for 36 h at 15°C.

Visualize the DNA bands with longwave UV and remove the lower band through the side of the tube with a 21-gauge hypodermic needle.

Remove ethidium-bromide as described above and store the ^{32}P-labeled DNA in buffer C at 4°C.

5. References

Conrad SE, Campbell JL (1979) Characterization of an improved in vitro DNA replication system for *E. coli* plasmids. Nucleic Acids Res 6:3289–3303

Cozzarelli NR (1977) The mechanism of action of inhibitors of DNA synthesis. Annu Rev Biochem 46:641–668

Cozzarelli NR (1980) DNA gyrase and the supercoiling of DNA. Science 207:953–960

Diaz R, Nordström K, Staudenbauer WL (1981) Plasmid R1 DNA replication dependent on protein synthesis in cell-free extracts of *E. coli*. Nature 289:326–328

Inuzuka M, Helinski DR (1978) Replication of antibiotic resistance plasmid R6K DNA in vitro. Biochemistry 17:2567–2573

Kolter R, Helinski DR (1979) Regulation of initiation of DNA replication. Annu Rev Genet 13: 355–391

Sakakibara Y, Tomizawa J (1974) Replication of colicin E1 plasmid in cell extracts. Proc Natl Acad Sci USA 71:802–806

Sakakibara Y, Suzuki K, Tomizawa J (1976) Formation of catenated molecules by replication of colicin E1 plasmid DNA in cell extracts. J Mol Biol 108:569–582

Staudenbauer WL (1976) Replication of small plasmids in extracts of *Escherichia coli*. Mol Gen Genet 145:273–280

Staudenbauer WL (1978) Structure and replication of the colicin plasmid. Curr Top Microbiol Immunol 83:93–156

Subject Index

Acinetobacter calcoaceticus 39
Agarose 27
– concentration in gel 27
– pore size in gel 27
Agarose gel electrophoresis 26 ff., 67 f., 71, 81 f., 94 f., 186 f., 194
–, characterization of plasmid DNA 26 ff.
–, determination of molecular weight of DNA 27 ff.
–, electrophoretic buffer of 27 f.
–, horizontal 27
–, migration of OC DNA, CCC DNA and linear DNA 28
–, migration velocity of DNA 27 f.
–, preparative 194 f.
–, vertical 27
–, voltage gradient 27
Agrobacterium 128, 152
Agrobacterium tumefaciens 54
Alcian blue 300
Alkaline extraction 16, 21 ff.
– of CCC plasmid DNA 21 ff.
Alkaline phosphatase 155, 161, 163, 166, 194
Aniline 262
Anthranilate-transferase 178
Antibiotic resistance 38 f., 47, 160
Antibiotic resistant plasmids → plasmids
Arabinosylnucleoside triphosphate 326
Autoradiograph
– of E. coli envA protein 242
– of gene products of pACYC184 and pACYC184 Tn5 mutants 232
– of traT region 97, 100
Autoradiography 84 ff., 245, 250, 266, 268

Bacillus subtilis 3, 39, 245
Bacterial pathogens 93
Bacteriophage lambda 113, 176 ff., 213, 236
– cos site of 190 f., 195 f.
– DNA 4, 33, 296

– –, concentration. 182 f.
– –, extraction 183
– –, in vitro packaging 191
– –, isolation 181 ff.
– –, restriction enzyme fragments 34 f.
–, gene cloning with 176 ff.
–, genes 176 f.
–, genetic map 176 f.
–, insertion vector 177 ff., 184 ff.
–, – L1 (NM607) 178 ff., 184 f.
–, – – L1-trpDEop (λi^{434} trpDE) 179 ff., 184 ff., 215, 219
–, – –, plaque morphology 187
–, – map 178
–, lac z gene 178
–, large scale production of lysate 182
–, morphogenesis of λ head 191 f.
–, –, self assembly of proteins 193 f.
– packaging mix 192 f.
– – packaging efficiency of 193 f.
–, plating efficiency on (P1 Cmts)::Tn1 lysogens 113
–, production of plate stock from a single plaque 181
– prophage 7, 111, 171
–, protein synthesis in minicells 212 ff.
–, purification of lysate 181, 183
–, replacement vector 177 f., 184 ff.
–, – L3 (NM596) 178, 184 ff., 215, 219
–, – – L3-trpDEop(λtryDE) 184 ff., 215, 219
–, – map 178
– repressor 176 f., 212 ff.
–, scaffolded prehead 191 f.
– self assembly in cell-free system 192
– subE gene 178
– vir 113
Bacteriophage NM596 → Bacteriophage λ replacement vector L3
Bacteriophage NM607 → Bacteriophage λ insertion vector L1

Bacteriophage P1 7, 111 ff.
—, circular permutation 111
—, Cmts 112 f., 115 ff.
—, — isolation 112, 115
Bacteriophage P1
— — Cmts map 116 f.
—, — ::Tn1 112 f., 115
— gene products encoded by res::Tn1 mutations 121 f.
—, headful mechanism 111
—, isolation of restriction deficient mutants (res⁻) by transposon mutagenesis 111 ff.
—, map 110 f.
—, — of res gene 122 f.
—, mapping of res::Tn1 DNA by EcoRI 112, 116 ff., 121
—, mediated transduction of markers in E. coli 3, 7 ff.
—, restriction/modification system 112
Bacteriophage PM2 290
Bacteriophage ΦX174 250, 269, 289 f.
—, replicative form (RF) 269
Bacteriophage T4
— DNA polymerase 255
— gene 32 protein (unwinding protein) 288 f., 309
— ligase 159
BAL31 exonuclease 74 ff., 141 ff.
— for mapping restriction endonuclease cleavage sites in circular genomes 74 ff.
— generation of deletion mutants in vitro 141
Benzyldimethyl-alkyl-ammonium chloride (BAC) 299 f.
Blunt ends → flush ends
Brij58 15, 18, 317
— lysis 19
Bromodeoxyuridine triphosphate 333
— labelling DNA 333
Bromophenol blue 27
Broth mating 41, 43

Carbon film 282 f.
CAT → chloramphenicol acetyltransferase
Cell capsules 93
Cell growth 15
Cell lysis 15, 18
Cell surface barriers 3
Cell surface components 93
Cell-free extract (system) 236 ff., 326 f.
— coupled transcription-translation in E. coli 236 ff.
— — identification of gene products 236 ff.
— — MG^{++} concentration 237, 240

— synthesis of proteins 237
Cesium chloride ethidium bromide (CsCl-EtBr) gradient centrifugation 16 ff., 46, 183, 194, 252, 317 f., 331
—, purification of DNA after 20
—, separation of DNA in 17
Cethyltrimethyl ammonium bromide 262
Chemical degradation approach → Maxam and Gilbert method
Chloramphenicol 4, 14 f., 330
— acetyltransferase (CAT) 208, 225 f.
— —, coding region on pACYC184 224 ff., 230 ff.
— — —, direction of transcription 225 f.
— — —, gene product analysis in minicells 230 ff.
— — —, mapping of Tn5 mutants by restriction enzyme analysis 226 ff.
— —, truncated proteins 225 f., 232 f.
— transacetylase 244
Chloroform 194
Chromosome mobilizing ability → R68.45
Citrobacter koseri 39
Cloning an E. coli Tn5 mutant into cosmids 194 ff.
Cloning with λ 176 ff.
Cloning with cosmid vectors 190 ff.
Cloning with plasmid vectors 160 ff.
Cohesive ends 33, 154 ff.
—, ligation of 154 ff.
—, —, influence of DNA concentration on 158
Cointegrate formation → R68.45
ColE1 64, 66, 143, 210, 290, 292, 295, 316, 326 f., 330 f., 333 f.
ColE2 64
Colony hybridization 92 ff., 200
— autoradiograph 97, 100
— for traT gene 93 ff.
Compatibility → Incompatibility
Competent cells 3 ff.
—, rapid preparation of 5 f.
Confluent lysis method 183
Conformation of DNA → DNA
Conjugation 2, 9 ff., 40 ff., 51, 80, 132, 143 f.
Conjugational transfer of chromosomal DNA in Pseudomonas aeruginosa 51
Conjugational transfer of plasmid DNA 3, 9 ff., 40 ff., 65, 205
—, frequency 10
—, importance in Gram⁻ and Gram⁺ bacteria 10
— of RP4 9 ff.
Construction of genetic maps 7

Subject Index

Cosmid vector 153, 161 f., 190 ff.
–, in vitro packaging 190 ff.
– pHC79 196
– – map 196
Coupled transcription-translation in vitro 205, 242
Cytochome C 284, 287, 299 f.
– spreading techniques 284 ff.
Cytosine 107

Deletion 141 f.
– in plasmids 143 ff.
– mutants 142 ff.
– –, efficiency of creation of 142
–, random sized 148
Density centrifugation → Cesium chloride ethidium bromide centrifugation
Density labelling of DNA 333
–, equilibrium centrifugation of 334 f.
Dephosphorylation 252 f.
Diethylpyrocarbonate (DEPC) 237, 262
Dimethylsulfate 261
DNA 250 (also see plasmid DNA)
– amplification 153
– cohesive ends 154 ff.
– conformation 15 f., 27, 65, 282, 294, 316
– –, covalently closed circle (ccc) 15 f., 65, 294, 316
– –, linear 15 f., 294, 316
– –, open circle (oc) 15 f., 65, 294, 316
– denaturation 85
– expression 153
– flush ends 154 ff.
– intermolecular joining 154 f., 161
– intramolecular joining 154 f., 161
– methylation 33
– mobility in agarose gel electrophoresis 27 f.
– molecular weight determination by agarose gel electrophoresis 27 ff.
– repair 326
– replication 315 ff.
– sequencing 74, 249 ff.
– –, chemical degradation approach (Maxam and Gilbert method) 250 ff., 261 ff., 275
– –, chemical sequencing vectors 276 ff.
– –, dideoxy sequencing on M13 template 268 ff.
– –, gel 251
– –, ΦX174 269
– –, primers for 269 f.
– –, –, preparing and running the gel 270 ff.

– –, –, reading the sequence 272
– –, radiolabelling in vitro 250
– –, rapid method 275 ff.
– –, shot gun 278
– –, subcloning strategy 276
DNase I 82
Dot blot hybridization 92
Droplet method 284 ff., 299
Duplex DNA
–, partial denaturation 296 ff.

Eckhardt lysis 29, 32 f., 39, 47, 81, 86, 200
Eclipse complex 3
Electron micrograph of
– DNA:RNA polymerase complex 305
– heteroduplex 295
– homoduplex 292
– partially denatured RP4 297
– pBR322 285, 301
– replicative intermediate 322
– R-loop 311
Electron microscopy 281 ff.
–, mapping RNA polymearse binding sites 304 ff.
–, nucleic acids 282 ff.
– –, cytochrome C spreading techniques 284 ff.
– –, support films for 282 f.
–, –, –. carbon film 282 f.
–, –, –, parlodion film 282 f.
–, R-loops 309 ff.
Elongation factor EF-Tu 214
End labelling of DNA 252, 255, 275
Enriched crude plasmid preparation (Hansen and Olsen) 29 ff.
Enterotoxin genes 93
Escherichia coli 3 ff., 29, 51, 93, 98, 127, 143, 153, 236, 318
– BHB290 191, 194
– BHB2688 191, 194
– C 46
– C600 6, 99, 107. 113, 116, 226, 329
–, cell-free extracts 236 ff.
– CSH51 54 ff.
– CSH52 54 ff.
– DS410 212 ff., 230
– envA gene 237, 240
– –, protein of 242
– K12 4, 6, 8 f., 11, 38, 40, 43, 45 ff., 126, 245
– lac gene 269
– minicells 186, 204, 212 ff., 224, 226, 230, 236
– –, expression 186
– – from DS410 212 ff., 230

– – –, proteins encoded by 214, 220
– –, gene product analysis of CAT gene 230 ff.
– –, generation by asymmetric cell division 212 f.
– –, isolation 216 f., 224, 230 f.
– –, labelling of plasmid containing 231
– –, lysis 231
– –, synthesis of bacteriophage and plasmid encoded proteins 212 ff.
– –, UV sensitivity of proteins produced by 217 ff.
Escherichia coli
– polymerase I 82, 255
– S605 226
–, transformation 3 ff.
–, – with pBR325 DNA 3 ff.
–, tryptophan operon 184
–, –, cloning into λ vector 184 ff., 178 f.
Ethanolic uranyl acetate 284
Ethidium bromide (EtBr) 16 f., 27 f., 300
Excluding recircularized vector 160
Exonuclease 74 ff., 141 ff.

Ficoll 27
Flush ends 33, 154 ff.
–, ligation of 154 ff.
–, –, influence of DNA concentration on 158
Formamide 100, 284, 287 f., 294
F-plasmid 9 f., 22, 51, 56, 86, 88, 143, 171 f., 277, 310 f.
–, replication origin region 310
–, R-loop molecules 311
F-primes 43

β-D-Galactopyranose 27
Isopropyl-β-D-thio-galactopyranoside (IPTG) 269
3,6-anhydro-L-galactose 27
β-Galactosidase 269, 277
5-Bromo-4-chloro-3-inolyl-β-galactoside 269
Gal operon 7
Gene cloning 152 f.
–, basic steps 153
– with bacteriophage lambda 176 ff.
– with cosmid vectors 190 ff.
– with plasmid vectors 160 ff.
Gene dosage effect 226
– for selection of multicopy plasmid Tn5 insertions from chromosomal Tn5 insertions 226
Gene expression 52
Gene library 153, 167, 190
–, construction 153, 160, 194 ff.
– in cosmids 194 f., 200

Gene transfer 2 ff.
Glutaraldehyde 306
Glycerol 27
Guanine 107
Gyrase 326, 330

Haemophilus influenzae 3
Heteroduplex analysis 282, 288, 291, 294 ff.
–, electron micrograph 295
– of bacteriophages 295 f.
– of plasmid DNA 294 ff.
Histogram of
– partial denaturation (A-T map) 298
– R-loops 312
– RNA polymerase binding sites (on RSF1010) 307
Homoduplex 74, 282, 291 ff.
– preparation 291 ff.
– –, electron micrograph 292
Homogenotization 135 f.
Homopolymer tailing 155, 161
Hybridization → colony hybridization
Hybridization → southern hybridization
Hydrazine 262
Hydroxylamine mutagenesis in vitro 106 ff.

Immunoprecipitation 242
Incompatibility (Inc) 47 ff., 80 ff.
– groups 47 ff., 99
– – FII 88
– – P 51 f., 126, 128 f.
– – P1 80 ff.
– – –, typing by southern hybridization 80 ff.
Inhibition of protein synthesis by chloramphenicol treatment 325 f.
Insertion 141, 152
– of Mu genome into RP4 129 ff.
Insertion elements (IS elements) 152, 291
– IS1 116 f., 119
– IS1a 162
– IS1b 162
– IS21 52 ff., 56 f.
Insertional inactivation 161
Intermolecular and intramolecular joining of DNA 154
Inversion 141
Inverted repeats 282, 291 ff.
In vitro packaging of recombinant λ-DNA in phage heads 6
In vitro transcripts 309 ff.
–, determination of orientation 309
–, determination of startpoints 309
Iridium 285
Isolation of minicells 224

Subject Index

Isolation of plasmid DNA → plasmid DNA
Isolation of plasmid relaxation complex
    → plasmid DNA

JRG7 126

Klebsiella nitrogen fixation gene clusters 226
Klebsiella pneumoniae 39

β-Lactamase 209, 242
Lactose 269
Lactose fermentation test 178, 184
Lambda → bacteriophage lambda
Leucine 54
−, gene for 54
−, −, mobilization by plasmid R68.45 54 ff.
Ligase 154, 186
Ligation 153 ff., 179, 186
−, electrophoretic pattern of 157 f.
− of cohesive-ended and flush-ended DNA
    fragments 154 ff.
−, T4 186
Lysopine dehydrogenase 225
Lysostaphin 39
Lysozyme 15, 39

M13 268 ff.
− mp2 269
− mp7 269
−, replicative form (RF) 269
Mapping by
− electron microscopy 304 ff.
− restriction enzyme analysis 116 ff.
Mapping of
− plasmid borne genes 127
− RNA polymerase binding sites 304 ff.,
    309
Marker exchange → homogenotization
Maxam and Gilbert sequencing 250 ff.,
    261 ff.
Maxicells 204 ff., 236
−, gene expression system 205 ff.
Meselson-Stahl experiment 331
Methylases 33
Microdensitometer 28
Microtiter plate 98
Minicells → E. coli minicells
Mini-cleared lysates 21 f.
−, preparation 22 ff.
Mini-F plasmid 323
Miniplasmid 166, 171 ff.
−, agarose gel electrophoresis 174
−, formation 171 ff.
Mobilization of
− E. coli plasmids 128 f.

− host chromosomal genes 51 ff.
− − by R68.45 51 ff.
− nonconjugative plasmids 43 ff.
Mobilizing plasmid → plasmid
Modification 4
Molecular weight → DNA
Molecular weight → restriction endonuclease
    fragments
Mu 126, 128 ff.
− insertion into RP4 129 ff.
Multicopy plasmid → plasmid
Mutagenesis 106 ff., 111 ff., 125 ff., 141 ff.
−, BAL31 exonuclease 141 ff.
−, hydroxylamine 106 f.
−, transposition of Tn1 111 ff.
−, transposition of Tn5 130 ff.

Nalidixic acid 330
Nick translation 70 f., 81 ff., 94, 96
nifH 128 f., 134
Nitrocellulose 85 f.
Nitrosoguanidine 106
Novobiocin 326, 330
Nucleases 3
Nucleic acid hybridization 81 f.
Nucleoside triphosphate (NTP) 250, 255
− labelling in vitro 250

Oligonucleotide primer 250
Origin of replication → replication origin
Origin of transfer replication (oriT) 10
Outer membrane proteins 93
Oxolinic acid 326, 330

Parlodion film 282 f.
Partial denaturation histogram (A-T map)
    297 f.
Partial denaturation of duplex DNA 296 ff.
Phosphoribosyl transferase 178
Piperidine 261 f., 264 ff.
Plaque morphology of λ 177, 179, 184,
    186 f.
Plasmid 14 ff., 26 ff., 38 ff., 47 ff., 106, 143,
    160 ff., 204, 325
− amplification 14 f.
−, antibiotic resistant 38 ff.
− classification 47, 325 f.
−, conjugative transfer of 40 ff., 65
−, −, frequency 43
− deletion mutants 143 ff.
− encoded proteins 204 ff.
− −, selective labelling with radioactive
        amino acids 204
− −, synthesis in maxicells 204 ff.
− −, synthesis in minicells 212 ff.

Plasmid
- enterotoxin 39
- in wild type strains 38 ff., 47 ff.
- mobilization 51 ff., 65, 75
- –, nonconjugative plasmids 43 f.
- multicopy 14, 21, 224 ff., 236
- –, coding regions 224
- model system for replication 14
- production of adhesion antigens 106
- production of toxins 106
- relaxation complex 64 ff.
- –, characterization 65
- –, isolation 65 f.
- –, sequencing 71
- –, relaxed 15
- replication 106 f., 171
- –, initiation 109
- surface exclusion 209
- –, transfer of 54
- –, frequency 54, 60
- transformation 3 ff., 45 ff.

Plasmid DNA 14 ff., 26 ff., 316 ff.
- characterization by agarose gel electrophoresis 26 ff.
- expression 204
- for heteroduplex studies 294
- for homoduplex studies 291
- isolation 14 ff.
- –, cell growth for 15
- –, cell lysis for 15
- –, ethidium bromide density centrifugation 10 ff.
- –, large scale 16 ff.
- –, rapid 21 ff.
- –, small scale 21 ff.
- –, purification 18 ff.
- replication 316 ff., 325 ff.
- –, bidirectional 318
- –, inhibition of 326 f.
- –, in vitro 325 ff.
- –, origins of 316 ff.
- –, termination point 323
- –, unidirectional 318
- replication forks 316
- replicative intermediates 316 ff.
- –, analysis 318 ff.
- –, CsCl-EtBr density centrifugation 320
- –, electron micrograph 322
- –, isolation and enrichment for 317 f.
- –, structure 316 f.
- –, sucrose density centrifugation 320
- –, supercoiled 316

p15A 210
pACYC184 56 ff., 81, 94, 205, 208, 224 ff.
- –, analysis of chloramphenicol acetyl transferase gene 224 ff.
- –, –, autoradiograph of gene products 232
- gene products of tetracycline resistance gene 233
- mobilization by R68.45 56 f.

pBR322 56, 155, 162 ff., 196, 276, 290, 301
- –, electron micrograph 285, 301
- R6-5 hybrid 166 ff.

pBR325 3 f., 35, 56, 128 f.
- mob 128 f., 133, 135
- mob-nif::Tn5 129
- mob::Tn5 134

pBR327 277
pC221 237, 242 ff.
- map 243
pCSV03 277 f.
pCW41 243 f.
- map 243
pCW46 243 f.
- map 243
pHSG415 75 ff.
- restriction enzyme map 76
pK1 116 209
pK1 118 209
pK1 120 209
pKB280 212 f., 215, 219
pKN80 161
pKN410 240, 242
pKT107 94, 97, 99, 205
- hybridization of traT 97
pKT260 (RSF1010::Tn3) 144, 147
- digestion by BAL31 147
pKT670 295
pKT710 292, 295
- –, electron micrograph 292
pKY2289 161
pLG510 240, 242
pMB1 210
pMB9 184
pML31 107, 317 f.
- –, replication defective derivatives 107 ff.
- –, replicative intermediates 317
pPK1 184, 186 f.
- cloning into λ 184 ff.
pPH1 86
- hybridization to IncP1 fragments 88
pSC101 65 ff., 210
- relaxation complex 72
- –, nick sites 69 ff.
- –, restriction enzyme map 72
pSM14 126
pSUP201 128, 133
pHSG415 75 ff.
- –, restriction enzyme map 76
pRK290 129
pUR222 276

- cloning and sequencing of DNA 276 f.
pWG 81 f.
- map 81
Platinum 285
Point mutation 141
Polar mutation 127
Polyacrylamide gel 186, 230, 250, 252, 259, 262, 268
- electrophoresis (PAGE) 209, 224, 258
-, thin sequencing gels 259, 264
Polycosmids 194, 200
Polylysine 300
Polynucleotide kinase 252 f.
Promotor 237, 304
Proteases 65
Protein analysis by SDS-PAGE 245
Proteinase K 69 f.
Protein monolayer technique 284
Protein synthesis by minicells 109
Protein synthesis in cell-free extracts 109
Protein-nucleic acid complexes 282, 299 ff.
Pseudomonas aeruginosa 39, 51, 54
Pseudomonas pseudocaligenes 39
Pseudomonas putida 143

R 65
R1 326
R6-5 94, 97, 99, 161 ff., 171 f., 215, 292, 295
- cloning fragments 162 ff.
- complement resistance protein 205 ff.
- -, identification by maxicell system 205 ff.
- genetic map 94, 204
- hybridization to traT 97
- miniplasmid 172
- -, agarose gel electrophoresis of 174
- pBR322 hybrids 166 ff.
- restriction endonuclease map 162, 206
- tra genes 205, 208 f.
- -, proteins 208 f.
R6K 65, 97, 323
- hybridization to traT 97
R64 43
R68 51 ff., 127
- restriction endonuclease map 52 f.
R68.45 10, 51 ff.
- chromosome mobilizing ability (Cma) 51, 54
- cointegrate formation with vector plasmid 56
- frequency of plasmid transfer in E. coli 54 ff.
- mobilization of
- -, bacterial chromosomes 52
- -, E. coli leu gene 54 ff.

- -, E. coli vector plasmids 52
- pACYC184 cointegrate formation 56 ff.
- restriction endonuclease map 52 f.
R100 126
R124 97
- hybridization to traT 97
R483 126
R751 81 f., 86, 88
- histogram of RNA polymerase binding sites 307
- hybridization of IncP1 fragment 88 ff.
Radioactive labelled DNA 82, 258
- isolation 82 ff., 258 ff.
Radioactive labelling of DNA 82, 252
- by end labelling 252 ff.
- by nick translation 70 f., 81 ff.
- in vivo 82
Rapid isolation of plasmid DNA → plasmid DNA
Recombination 2, 8, 152 f.
-, reciprocal 9
Relaxation complex 64 ff.
-, characterization 65 ff.
-, induction 66, 68
-, initiation of plasmid replication 65
-, nick sites 65, 69 ff.
-, plasmid conjugational transfer 65
-, rapid isolation 67 ff.
Replication → DNA replication
Replication → plasmid DNA replication
Replication origin 153, 171 f.
-, cloning 171 ff.
Replicative form (RF) of ΦX174 and M13 269
Restriction endonuclease 26, 33 ff.
- class I 33
- class II 33
- cleavage sites 74
- -, map of 74
- -, mapping by BAL31 74 ff.
- -, mapping by electron microscopy 74
- -, mapping by exonucleolytic degradation 75
- -, mapping by partial digestion 75
Restriction fragments 33
-, molecular weight determination 33 ff.
-, -, λ 34 f.
-, -, pBR325 35
Restriction-modification system 33, 106
Reverse transcriptase 255
R-factor 10
Rhizobium 128 f., 135
- leguminosarum 54, 135
- meliloti 32, 54, 128 f., 134 f., 225
- - component II of nitrogenase 225
- - gene library 127

Rhizobium meliloti
– – nifH1 134
– transposon mutagenesis 135
Rhodopseudomonas sphaeroides 54
Ribosomal proteins 214
Rifampicin 326
RK2 51, 65, 127, 171
R-loop 282, 288, 309 ff.
– evaluation 312
RNA polymerase 214, 304 ff., 309
– binding sites (RPB sites) 304 ff.
RNA polymerase
– binding sites (RPB sites)
– –, histogram 307
– –, localization 309
– –, mapping by electron microscopy 304 ff.
– holoenzyme 306
RNA sequencing 262
Rolling circle 36
RP1 51, 127
RP4 7 ff., 43, 51, 81, 86, 88, 111 ff., 115 f., 126 ff., 143, 297
–, conjugational transfer 9 ff.
–, electron micrograph 297
–, genetic map 127
–, hybridization to IncP1 fragment 88 ff.
– ::Mu 135
–, partial denaturation histogram (A-T map) 298
–, promoting host chromosomal gene transfer 127
–, restriction endonuclease map 89
– ::Tn5 128, 135
–, transduction of 7 ff.
R-plasmid 10
RSF1010 65, 143 ff., 171, 305 ff.
– mob genes 144, 147
– mobilization 143 f.
– –, restriction endonuclease map 144
– restriction endonuclease map 143
RSF1030 326

S302.10 56 ff.
Salmonella 127
Salmonella typhimurium 7
Sarcosyl 15, 18
– lysis 18 f., 194, 197
SDS-lysis 29 f., 56, 60 f.
SDS-PAGE 245
Self-transmissible resistance elements → R-plasmids
Semiconservative replication 331
Sepharose 2B 67
Sequence homology 152

Sequence reading 266 f., 272 f.
Sequencing → DNA
Sequencing → RNA
Sequencing gel 251 f., 259, 262, 264, 267, 275
– preparation 265 f.
Serratia marcescens 39
Size fractionation 194
Sodium-dodecyl-sulfate (SDS) 18, 65 ff., 317
Southern blotting 85 ff., 94 ff.
Southern hybridization 80, 85 ff., 97
– for typing incompatibility groups 80, 85 ff.
Spectinomycin 15
Spheroplast 4, 15
Spreading method 284, 286 ff.
Staggered nicks 33
Staphylococcus aureus 39, 243 ff.
Sticky ends 33
Streptococcus pneumoniae 3
Sucrose 15, 27
– gradient 28, 65, 67, 320
– – centrifugation 194, 230, 252
Suicide plasmid 135
Sulfonamide 44
Suppressible mutations 106 ff.
– isolation 106 ff.
Surface mating 42 ff.

T4 DNA polymerase → bacteriophage T4
T4 ligase → bacteriophage T4
T-DNA of octopine tumours 225
Thymus terminal transferase 255
Three factor crosses by transduction 7
Touch blot hybridization 92
Transcription 303 ff.
–, complex 282
–, direction 309, 312
–, termination 312
Transduction 2 ff., 80
– frequency 9
–, general 7, 111
– of RP4 8 f., 132
–, plasmid manipulation 7
–, specialized 7, 111
– three factor crosses 7
Transfection 2 ff., 179 ff., 184 ff.
– by λ-DNA 4, 6
– frequency 6, 181
– of E. coli cells 4, 6
Transformation 2 ff., 21 f., 45 ff., 80, 106, 142, 147, 190
–, Ca^{++} concentration 4
– frequency 46

Subject Index

- heat shock 4
- in gram⁻ and gram⁺ bacteria 3
- of chromosomal DNA 4
- of nonmobilizable plasmids 56
- of plasmid DNA 3 ff., 45 ff.
- rapid procedure 5 f.

Transposition 111 ff.
Transposon (Tn) 39, 111, 125 f., 152, 291
- in bacteria other than E. coli 127 f.
- , polar mutations caused by 127
- Tn1 111 ff., 126
- - in P1 111 ff.
- - mutagenesis 112 ff.
- - restriction enzyme map 119
- Tn3 143 f., 147 f., 205, 208 f.
- Tn5 126, 128, 132 ff., 196, 200, 224 ff.
- -, insertion of 230
- -, insertion into mobilizable E. coli vectors 133 ff.
- -, mapping in CAT gene by restriction enzyme analysis 228 f.
- -, restriction enzyme map 227
- -, transposition into RP4::Mu 132 f.
- Tn7 126
- Tn9 112, 116 f., 126
- -, restriction enzyme map 117, 119
- -, tandem 116 f.
- Tn10 126, 162
- Tn601 (Tn903) 162, 292, 295

Transposon mutagenesis 111 ff., 125 ff., 144, 167, 225 f.
-, random 128 f., 136
- of Rhizobium 135 ff.
-, site specific 128 f.

TraT gene 93 ff.
-, autoradiograph of 100
-, protein of 93
-, related sequences 98

Tri parental cross 43
Tris-acetate 27
Tris-borate 27
Triton X100 15, 18, 65, 317
-, lysis by 20
Tryptophan operon → E. coli

Universal primer 269

Vector → bacteriophage
Vector → cosmid
Vector → plasmid
Velocity sedimentation in CsCl gradient 333

Whole cell lysates 29

Molecular Genetics of the Bacteria Plant Interaction

Editor: **A. Pühler**
1983. 154 figures. XIV, 394 pages
(Proceedings of Life Sciences)
ISBN 3-540-12798-4

Contents: General Introduction. – The **Rhizobium**-Plant Interaction. – The **Agrobacterium**-Plant Interaction. – Plant Pathogenic Bacteria and Related Aspects. – Subject Index.

Molecular Genetics of the Bacteria Plant Interaction contains the most recent research results in the field of symbiotic nitrogen fixation, plant tumorgenesis and plant pathogenic bacteria. It summarizes the genetic techniques and tools which are used to analyze gram-negative bacteria such as **Rhizobium, Agrobacterium,** and plant pathogenic **Pseudomonas** species. In addition, the involvement of plant genes in nodule-formation as well as the genetic basis of plant cells transformed by **Agrobacterium** are discussed. The articles are written by leading study groups the world over who define their research problems, discuss solutions and speculate on further developments in this rapidly developing field.

Springer-Verlag
Berlin
Heidelberg
New York
Tokyo

E. A. Birge

Bacterial and Bacteriophage Genetics

An Introduction
Corrected 2nd printing. 1983. 111 figures.
XVI, 359 pages. (Springer Series in Microbiology)
ISBN 3-540-90504-9

Contents: Unique Features of Prokaryotes and Their Genetics. – The Laws of Probability and Their Application to Prokaryote Cultures. – Mutations and Mutagenesis. – T4 Bacteriophage as a Model Genetic System. – The Genetics of Other Intemperate Bacteriophages. – Genetics of Temperate Bacteriophages. – Transduction. – Transformation. – Conjugation. – The F Plasmid. – Plasmids Other Than F. – Regulation. – Repair and Recombination of DNA Molecules. – Gene Splicing, the Production of Artificial DNA Constructs. – Future Developments. – Index.

From the reviews:
"The book succeeds in what it sets out to do. ... There is no doubt it will be a useful adjunct in teaching introductory courses in molecular genetics." *Nature*

"... clearly covers the subject that an introductory genetics testbook should offer beginning genetics students." *ASM News*

"... the author presents his thoughts in a very clear, understandable, but at the same time, most precise way, and this is why this book is recommended first of all for students and teachers, but also for all those who are familiar with the topic. All might benefit from it." *Theoretical and Applied Genetics*

Springer-Verlag
Berlin
Heidelberg
New York
Tokyo